CAMBRIDGE LIBRARY COLLECTION

Books of enduring scholarly value

Astronomy

From ancient times, humans have tried to understand the workings of the world around them. The roots of modern physical science go back to the very earliest mechanical devices such as levers and rollers, the mixing of paints and dyes, and the importance of the heavenly bodies in early religious observance and navigation. The physical sciences as we know them today began to emerge as independent academic subjects during the early modern period, in the work of Newton and other 'natural philosophers', and numerous sub-disciplines developed during the centuries that followed. This part of the Cambridge Library Collection is devoted to landmark publications in this area which will be of interest to historians of science concerned with individual scientists, particular discoveries, and advances in scientific method, or with the establishment and development of scientific institutions around the world.

Essays from the Edinburgh and Quarterly Reviews

First published in 1857, this work comprises assorted noteworthy writings by the mathematician and astronomer Sir John Herschel (1792–1871), reflecting his diverse scientific and literary interests. It includes a piece on terrestrial magnetism, a review of William Whewell's writings on the history and philosophy of science, and several addresses to the Royal Astronomical Society. Of particular interest is Herschel's commentary on Adolphe Quetelet's work on probability, which advocated applying statistics and probability calculus to social and political questions. Herschel's article not only influenced the growth of social science in Britain, but also played an important role in James Clerk Maxwell's development of a statistical treatment of heat phenomena. Also included in this collection are Herschel's translations of poems by Schiller (accompanied by the original German) as well as examples of his own verse. In an intriguing appendix, Herschel outlines a method for compiling vocabularies of indigenous peoples.

Cambridge University Press has long been a pioneer in the reissuing of out-of-print titles from its own backlist, producing digital reprints of books that are still sought after by scholars and students but could not be reprinted economically using traditional technology. The Cambridge Library Collection extends this activity to a wider range of books which are still of importance to researchers and professionals, either for the source material they contain, or as landmarks in the history of their academic discipline.

Drawing from the world-renowned collections in the Cambridge University Library and other partner libraries, and guided by the advice of experts in each subject area, Cambridge University Press is using state-of-the-art scanning machines in its own Printing House to capture the content of each book selected for inclusion. The files are processed to give a consistently clear, crisp image, and the books finished to the high quality standard for which the Press is recognised around the world. The latest print-on-demand technology ensures that the books will remain available indefinitely, and that orders for single or multiple copies can quickly be supplied.

The Cambridge Library Collection brings back to life books of enduring scholarly value (including out-of-copyright works originally issued by other publishers) across a wide range of disciplines in the humanities and social sciences and in science and technology.

Essays from the Edinburgh and Quarterly Reviews

With Addresses and Other Pieces

John Herschel

CAMBRIDGE
UNIVERSITY PRESS

University Printing House, Cambridge, CB2 8BS, United Kingdom

Cambridge University Press is part of the University of Cambridge.
It furthers the University's mission by disseminating knowledge in the pursuit of education, learning and research at the highest international levels of excellence.

www.cambridge.org
Information on this title: www.cambridge.org/9781108069656

This edition first published 1857
This digitally printed version 2014

ISBN 978-1-108-06965-6 Paperback

ESSAYS

FROM THE

EDINBURGH AND QUARTERLY REVIEWS,

WITH

ADDRESSES AND OTHER PIECES.

BY

SIR JOHN F. W. HERSCHEL, BART., K.H.

M.A., D.C.L., F.R.S. L. & E., HON. M.R.I.A., F.R.A.S., F.G.S., M.C.U.P.S.;

MEMBER OF THE INSTITUTE OF FRANCE;

CORRESPONDENT OR HONORARY MEMBER OF THE IMPERIAL, ROYAL, AND NATIONAL ACADEMIES OF SCIENCES OF

BERLIN, BRUSSELS, COPENHAGEN, GÖTTINGEN, HAARLEM, MASSACHUSETTS (U.S.), MODENA, MOSCOW (NAT. CUR.), NAPLES, PETERSBURG, STOCKHOLM, TURIN, VIENNA, AND WASHINGTON (U.S.);

THE ITALIAN AND HELVETIC SOCIETIES;

THE ACADEMIES, INSTITUTES, ETC. OF ALBANY (U.S.), BOLOGNA, CATANIA, DIJON, LAUSANNE, NANTES, PADUA, PALERMO, ROME, VENICE, UTRECHT, AND WILNA;

THE PHILOMATHIC SOCIETY OF PARIS; ASIATIC SOCIETY OF BENGAL; SOUTH AFRICAN LITERARY AND PHILOS. SOCIETY; LITERARY AND HISTORICAL SOC. OF QUEBEC; HISTORICAL SOCIETY OF NEW YORK; ROYAL MEDICO-CHIRURGICAL SOCIETY; SOC. OF ARTS, MANUFACTURES, AND COMMERCE, AND INST. OF CIVIL ENGINEERS, LOND.; BATAVIAN SOC. OF EXP. PHIL. IN ROTTERDAM; GEOGRAPHICAL SOC. OF BERLIN; ASTRONOM. AND METEOR. SOC. OF BRITISH GUIANA; ETC. ETC. ETC.

LONDON:

LONGMAN, BROWN, GREEN, LONGMANS, & ROBERTS.

1857.

LONDON:
Printed by SPOTTISWOODE and Co.
New-street Square.

CONTENTS.

ERRATA.

Page 402., lines 6 and 7 from bottom, *for* "two principles" *read* "a principle." Page 677., line 1, *for* "dii nction" *read* "distinction."

ESSAYS,

&c.

AN ADDRESS

TO THE SUBSCRIBERS TO THE WINDSOR AND ETON PUBLIC LIBRARY AND READING ROOM, DELIVERED AT THE FIRST GENERAL MEETING OF THE SUBSCRIBERS, HELD AT THE CHRISTOPHER INN, ETON, ON TUESDAY, JAN. 29. 1833, BY SIR J. F. W. HERSCHEL, K. H., PRESIDENT OF THE INSTITUTION.

GENTLEMEN,—I ought, perhaps, to apologize to you for addressing you on this occasion from a written paper. I know that to do so is not altogether in consonance with the habits of our countrymen when assembled on public occasions, and I should certainly not claim such an indulgence on this, if I had no better reason to assign than a mere want of the readiness and fluency of a practised speaker. But I consider this an important occasion; and as I have thought long, and with deep interest, on the advantages of a public and national description, which may be expected from institutions of this nature, as well as the evils to which they may become obnoxious if not conducted on proper principles, and with a view to the general result, I am very desirous that what I have to say should not lose the force I wish it to carry, by coming before you mixed up with my own imperfections as a speaker;—and I should be very sorry that the real interest of those topics on which I mean to beg your attention, should be frittered away in unmeaning or hyperbolical expressions, which, in the excitement of the moment, I might have the bad taste to think very eloquent, but which would really have no other effect than to distract attention from the plain common sense of the matter.

I will tell you, Gentlemen, in the first place, why I think this occasion so important. We are assembled here in performance of our part of a process which is going on at present more slowly than might be wished, but which *is* at length fairly entered upon, and must advance with more rapidity as example sanctions and persuasion urges it; and will, I trust, in a very few years, be in active progress in every town, village, and parish in the kingdom—a process on which it is no exaggeration to say, that the future destinies of this empire will very mainly depend—because on it depends, by a natural and indissoluble link, our capacity as a nation for a high degree of civil liberty. The process I mean is the active endeavour, on the part of every one who can lend a hand to it, to improve the standard of moral and intellectual culture in the mass of the people.

I am not going, Gentlemen, to read you a political lecture—far less to meddle with the topic of party politics, which my soul abhors—but I think it must be clear to every one, that in giving, as has been recently done, to the popular part of our constitution, a more extended and intimate contact with the people at large, a step has been taken, which when tried by the event—whether it shall have proved a step in advance towards a higher and purer form of civil liberty, or in a retrograde sense towards licence and its necessary consequence, arbitrary power—will take its character in the alternative from the degree in which that element shall be found to prevail—(*that*, the most important of all political elements)—which I have called the *capacity* of a nation for liberty, and this capacity in all ages and nations I consider to be directly measured by the extent to which moral and intellectual culture are diffused among all ranks and conditions of men. And why?—because these,—which in their ultimate meaning reduce themselves to *benevolence* and *wisdom*, acquired, as far as they can be acquired, by a free access to the best sources of instruction,—these, I say, are the only principles of self-government which *can* replace effectually, by their intimate presence in the bosom of each individual, a lightened coercion of the governing power from without; and the only ones which can afford any rational assurance that a system of legislation, founded openly and avowedly on public opinion, shall turn out a prudent, or even a safe one. Indeed, I might go farther, and assume it as a principle which, were it necessary, could be supported by many instances, ancient and modern—that the capacity for liberty, thus defined and measured, must for ever, and of necessity, as human society is constituted, *command* sooner or later that degree of freedom which is commensurate with it—that *more*, attempted to be prematurely forced upon it, is sure to degenerate into licence and call back the chain; while *less* cannot possibly be permanently withheld by any combination of the governing powers. Regarding then, as every reasonable man must do, a high and enlarged degree of rational liberty as the first of temporal blessings, it cannot be a matter of small interest to witness the establishment among us of institutions which have either for their avowed object, or for their direct, though perhaps not immediately intended—perhaps not in all cases perceived—tendency, to foster and encourage the only means by which it can be permanently and beneficially secured.

I shall therefore, I hope, be excused, if I take advantage of the honour you have done me by placing me in this chair, to offer a few observations on the more immediate objects which it is desirable we should aim at, so that in pursuance of individual and local advantages, we may not lose sight of the general end, but rather endeavour to accommodate our future proceedings to the furtherance of that end, even though it should involve the surrender of some slight superfluity on our own parts, some resignation of what may be considered mere literary luxuries as a sacrifice to public utility.

I may take it for granted, I think, when I look upon the circle around me, that in forming this institution we have all of us a higher end in view than the mere amusement of the passing hour. We are desirous to have at our disposal a fund of instruction, by drawing from which, as from a fountain, we may enlarge our knowledge, improve our taste, correct our judgment, and confirm our principles; and I will not pay my hearers so ill a compliment, or rather I will not lay on them so unmerited a reproach, as not to assume that this is the principal immediate object in view with us all. Now it must be borne in mind, and I cannot impress it too strongly on your attention, that this principal object must always, and especially in the beginning, while our funds are limited, be in a certain degree at variance with a subordinate, but still very agreeable, and by no means useless part of our system — I mean the Periodical department — that of the Journals, and Magazines, and floating literature of the day. And I say this without any intention of depreciating either the entertainment or excitement of that sort of reading, or any real utility which may justly be ascribed to it — or its especial utility to us, in its power of inducing some to give us their support who otherwise might not feel disposed to do so;

but simply to caution you thus early in our existence against any future tendency to give an undue extension to this department, so as to divert any large part of the bulk of our resources from their higher and far more useful destination, the increase of our library by the annual addition of sterling and standard works, such as form the main body of our literature and science—such as have outlived ephemeral applause, and risen above cotemporary neglect—and will continue to represent to all ages the intellectual greatness of the country which produced them.

It is therefore with great satisfaction that I have heard it resolved this evening, to throw open the library to a class of subscribers, at a lower rate than that which confers the privilege of access to the newspapers and periodical works. This is entirely as it should be. Such reading is a luxury and an indulgence, and should be paid for accordingly; the other is a necessary, and should be afforded as cheaply and extensively as possible.

I augur everything from the approbation the proposal has met with, but I should be sorry, I confess, that we should stop short at that point. My own impression is, that we should make a still farther step, and provide a considerable stock of books for a class of subscribers who should subscribe *nothing but the reading of them*—books of which we should supply the perusal gratis to all who choose to apply for them, leaving perhaps some very trifling deposit, to ensure their return. I do not mean, of course, that our most expensive works, or valuable books of reference, should be so lent out, but, on the contrary, that cheap editions, or second-hand copies, should be expressly set apart for that use. The choice of the works to be admitted into this department, too, would call for some discrimination. And this brings me to a part of my subject on which I must beg your earnest attention.

There is a want too much lost sight of in our estimate of the privations of the humbler classes, though it is one of the most incessantly craving of all our wants, and is actually the impelling power which, in the vast majority of cases, urges men into vice and crime. It is the want of amusement. It is in vain to declaim against it.—Equally with any other principle of our nature, it calls for its natural indulgence, and cannot be permanently debarred from it, without souring the temper, and spoiling the character. Like the indulgence of all other appetites, it only requires to be kept within due bounds, and turned upon innocent or beneficial objects, to become a spring of happiness; but gratified to a certain moderate extent it must be, in the case of every man, if we desire him to be either a useful, active, or contented member of society. Now I would ask, what provision do we find for the cheap, and innocent, and daily amusements of the mass of the labouring population of this country? What sort of resources have they to call up the cheerfulness of their spirits, and chase away the cloud from their brow after the fatigue of a day's hard work, or the stupifying monotony of some sedentary occupation? Why, really very little—I hardly like to assume the appearance of a wish to rip up grievances by saying *how* little. The pleasant field-walk and the village-green are becoming rarer and rarer every year. Music and dancing (the more's the pity) have become so closely associated with ideas of riot and debauchery, among the less cultivated classes, that a taste for them for their own sakes can hardly be said to exist, and before they can be recommended as innocent or safe amusements, a very great change of ideas must take place. The beer-shop and the public-house, it is true, are always open, and always full, but it is not by *those* institutions that the cause of moral and intellectual culture is advanced. The truth is, that under the pressure of a continually condensing population, the habits of the city have crept into the village—the demands of agriculture have become sterner and more imperious, and while hardly a foot of ground is left uncultivated, and unappropriated, there is positively not space left for many of the cheerful amusements of rural life. Now, since this appears to be unavoidable, and as it is physically impossible that the amusements of a condensed population should continue to be those of a scattered one, it behoves us strongly to consider of some substitutes. But perhaps it may appear to some almost preposterous to enter on the question. Why, the very name of a labourer has something about it with which amusement seems out of character. Labour is work, amusement is play; and though it has passed into a proverb, that one without the other will make a dull boy, we seem to have altogether lost sight of a thing equally obvious, that a community of "dull boys" in this sense, is only another word for a society of ignorant, headlong, and ferocious men.

I hold it, therefore, to be a matter of very great consequence, independent of the kindness of the thing, that those who are at their ease in this world, should look about and be at some pains to furnish available means of harmless gratification to the industrious and well-disposed classes, who are worse provided for than themselves in every respect; but who, on that very account, are prepared to prize more highly every accession of true enjoyment, and who really want it more. To do so, is to hold out a bonus for the withdrawal of a man from mischief in his idle hours—it is to break that strong tie which binds many a one to evil associates and brutal habits—the want of something better to amuse him, by actually making his abstinence become its own reward.

Now, of all the amusements which can possibly be imagined for a hard-working man, after his daily toil, or in its intervals, there is nothing like reading an entertaining book, supposing him to have a taste for it, and supposing him to have the book to read. It calls for no bodily exertion, of which he has had enough, or too much. It relieves his home of its dulness and sameness, which, in nine cases out of ten, is what drives him out to the ale-house, to his own ruin and his family's. It transports him into a livelier, and gayer, and more diversified and interesting scene, and while he enjoys himself there, he may forget the evils of the present moment, fully as much as if he were ever so drunk, with the great advantage of finding himself the next day with his money in his pocket, or at least laid out in real necessaries and comforts for himself and his family, — and without a headache. Nay, it accompanies him to his next day's work, and, if the book he has been reading be anything above the very idlest and lightest, gives him something to think of, besides the mere mechanical drudgery of his every day occupation—something he can enjoy while absent, and look forward with pleasure to return to.

But supposing him to have been fortunate in the choice of his book, and to have alighted upon one really good, and of a good class. What a source of domestic enjoyment is laid open! What a bond of family union! He may read it aloud, or make his wife read it, or his eldest boy or girl, or pass it round from hand to hand. All have the benefit of it, all contribute to the gratification of the rest, and a feeling of common interest and pleasure is excited. Nothing unites people like companionship in intellectual enjoyment. It does more, it gives them mutual respect, and to each among them self-respect—that corner-stone of all virtue. It furnishes to each the master-key by which he may avail himself of his privilege as an intellectual being, to

> "Enter the sacred temple of his breast,
> And gaze and wander there a ravished guest;
> Wander through all the glories of his mind,
> Gaze upon all the treasures he shall find."

And while thus leading him to look within his own bosom for the ultimate sources of his happiness, warns him at the same time to be cautious how he defiles and desecrates that inward and most glorious of temples.

I recollect an anecdote told me by a late highly-respected inhabitant of Windsor, as a fact which he could personally testify, having occurred in a village where he resided several years, and where he actually was at the time it took place. The blacksmith of the village had got hold of Richardson's novel of "Pamela, or Virtue Rewarded," and used to read it aloud in the long summer evenings, seated on his anvil, and never failed to have a large and attentive audience. It is a pretty long-winded book; but their patience was fully a match for the author's prolixity, and they fairly listened to it all. At length, when the happy turn of fortune arrived, which brings the hero and heroine together, and sets them living long and happily, according to the most approved rules, the congregation were so delighted as to raise a great shout; and, procuring the church keys, actually set the parish bells ringing. Now let any one say whether it is easy to estimate the amount of good done in this simple case. Not to speak of the number of hours agreeably and innocently spent — not to speak of the good fellowship and harmony promoted — here was a whole rustic population fairly won over to the side of good—charmed—and night after night, spell-bound within that magic circle which genius can trace so effectually, and compelled to bow before that image of virtue and purity which (though at a great expense of words) no one knew better how to body forth with a thousand life-like touches than the author of that work.

If I were to pray for a taste which should stand me in stead under every variety of circumstances, and be a source of happiness and cheerfulness to me through life, and a shield against its ills, however things might go amiss, and the world frown upon me, it would be a taste for reading. I speak of it of course only as a worldly advantage, and not in the slightest degree as superseding or derogating from the higher office, and surer and stronger panoply of religious principles, but as a taste, an instrument, and a mode of pleasurable gratification. Give a man this taste, and the means of gratifying it, and you can hardly fail of making a happy man, unless, indeed, you put into his hands a most perverse selection of books. You place him in contact with the best society in every period of history — with the wisest, the wittiest, with the tenderest, the bravest, and the purest characters who have adorned humanity. You make him a denizen of all nations—a cotemporary of all ages. The world has been created for him. It is hardly possible but the character should take a higher and better tone from the constant habit of associating in thought with a class of thinkers, to say the least of it, above the average of humanity. It is morally impossible but that the manners should take a tinge of good breeding and civilization, from having constantly before one's eyes the way in which the best bred and the best informed men have talked and conducted themselves in their intercourse with each other. There is a gentle, but perfectly irresistible coercion in a habit of reading well directed over the whole tenor of a man's character and conduct, which is not the less effectual because it works insensibly, and because it is really the last thing he dreams of. It cannot, in short, be better summed up, than in the words of the Latin poet—

"Emollit mores, nec *sinit* esse feros."

It civilizes the conduct of men—and *suffers* them not to remain barbarous.

The reason why I have dwelt so strongly upon the point of amusement, is this—that it is really the *only* handle, at least the only innocent one, by which we can gain a fair grasp of the attention of those who have grown up in a want of instruction, and in a carelessness of their own improvement. Those who cater for the passions, especially the base or malignant ones, find an easy access to the ignorant and idle of every rank and station; but it is not so with sound knowledge or rational instruction. The very act of sitting down to read a book is an effort, it is a kind of venture,—at all events, it involves a certain expenditure of time which we think might be otherwise pleasantly employed,—and if this is not instantly, and in the very act, repaid with positive pleasure, we may rest assured it will not be often repeated; and, what is worse, every failure tends to originate and confirm a distaste. If then we would generate a taste for reading, we must, as our only chance of success, begin by pleasing. And what is more, this must be not only the ostensible, but the *real* object of the works we offer. The listlessness and want of sympathy with which most of the works written expressly for circulation among the labouring classes, are read by them, if read at all, arises mainly from this, that the story told, or the lively or friendly style assumed, is *manifestly* and *palpably* only a cloak for the instruction intended to be conveyed—a sort of gilding of what they cannot well help fancying must be a pill, when they see so much and such obvious pains taken to wrap it up.

But try it on the other tack. Furnish them liberally with books not written expressly for them as a class, but published for their betters (as the phrase is), and those the best of their kind. You will soon find that they have the same feelings to be interested by the varieties of fortune and incident —the same discernment to perceive the shades of character—the same relish for striking contrasts of good and evil in moral conduct, and the same irresistible propensity to take the good side—the same perception of the sublime and beautiful in nature and art, when distinctly placed before them by the touches of a master—and, what is most of all to the present purpose, the same desire having once been pleased, to be pleased again. In short, you will find that in the higher and better class of works of fiction and imagination duly circulated, you possess all you require to strike your grappling-iron into their souls, and chain them, willing followers, to the car of advancing civilization.

When I speak of works of imagination and fiction, I would not have it supposed that I would turn loose among the class of readers to whom I am more especially referring, a whole circulating library of novels. The novel, in its best form, I regard as one of the most powerful engines of civilization ever invented; but not the foolish romances which used to be the terror of our maiden aunts—not the insolent productions which the press has lately teemed with under the title of fashionable novels—nor the desperate attempts to novelize history which the herd of Scott's imitators have put forth, which have left no epoch since the creation untenanted by modern antiques—and no character in history unfalsified—but the novel as it has been put forth by Cervantes and Richardson, by Goldsmith, by Edgeworth, and Scott. In the writings of these, and such as these, we have a stock of works in the highest degree enticing and interesting, and of the utmost purity and morality—full of admirable lessons of conduct, and calculated in every respect to create and cherish that invaluable habit of resorting to books for pleasure. Those who have once experienced the enjoyment of such works, will not easily learn to abstain from reading, and will not willingly descend to an inferior grade of intellectual privilege; they have become prepared for reading of a higher order, and may be expected to relish the finest strains of poetry, and to draw with advantage from the purest wells of history and philosophy. Nor let it be thought ridiculous or overstrained to associate the idea of poetry, history, or philosophy with the homely garb and penurious fare of the peasant. How many a rough hind, on Highland hills, is as familiar with the "Paradise Lost," or the works of his great national historians, as with his own sheep-hook. Under what circumstances of penury and privation is not a high degree of literary cultivation maintained in Iceland itself—

> "In climes beyond the solar road,
> Where savage forms o'er ice-built mountains roam,
> The muse has broke the twilight gloom,
> To cheer the shivering native's dull abode."

And what is there in the character or circumstances of an Englishman that should place him, as a matter of necessity, and for ever, on a lower level of intellectual culture than his brother Highlander, or the natives of the most inhospitable country inhabited by man? At least, there is always this advantage in aiming at the highest results—that the failure is never total, and that though the end accomplished may fall far short of that proposed, it cannot but reach far in advance of the point from which we start. There never was any great and permanent good accomplished, but by hoping for and aiming at something still greater and better.*

I have taken up a good deal of your time on this subject, and could still enlarge upon it; but I will content myself with one or two observations in the way of caution, in the event of our adopting this or any similar project, of placing a certain portion of our library at the disposal of gratuitous readers. In the first place, then, it appears to me quite an indispensable feature of such a plan, that no work, in any department of reading, should be

* A taste for reading once created, there can be little difficulty in directing it to its proper objects. On this point I refer with pleasure to some excellent observations in a little work entitled, "Hints and Cautions on the Pursuit of General Knowledge; being the substance of Lectures delivered to Mechanics' Institutions at Southampton and Salisbury, by John Bullar." (Longman & Co., London, 1833) Pp. 23. et seq. But the first step necessary to be taken is to set seriously about arousing the dormant appetite by applying the stimulant; to awaken the torpid intellectual being from its state of inaction to a sense of its existence and of its wants. The after-task, to gratify them, and while gratifying to enlarge and improve them, will prove easy in comparison.

allowed a place in the portion so set apart, which is not of acknowledged and admitted excellence: nothing ephemeral—nothing trashy—nothing, in short, which shall have the slightest tendency to lower the high standard of thought and feeling which should be held up. The educated and cultivated reader may bear a great deal, and throw off what is unworthy of the rest. The illiterate and ignorant is placed in danger by anything short of the very best.

The other caution which I would hold out is, that an extreme scrupulousness should be exercised, with reference to the admission of works on Politics and Legislation into such a department. Indeed, I should strongly advocate their exclusion from it altogether. This is not from any jealousy of the discussion of political subjects by all classes of Englishmen, which, in the present age, would certainly be a very superfluous feeling; but simply for this reason, that the true and useful object of such an institution is not to establish a school of politics, nor to propagate opinions (which every one who puts a political book into the hands of another must inevitably do), but to lay a broad foundation, by generally enlarging the information and cultivating the mental powers, to enable every man, however humble his station, to form his own opinion on this and a great many other subjects of deep import, (since opinions he must and will have) with a generally better chance of forming a right one than he has at present. We shall be taking on ourselves a deep responsibility, and one for which I may conscientiously for my own part say, I am not prepared, by any step which may tend to interfere, one way or the other, with the free formation of public opinion on such subjects—nor indeed can I conceive a more probable cause of disagreement among ourselves, which is of all things the most to be deprecated, than the discussions which might arise on this point—the only way to keep clear of which, is to exclude such works altogether.

On the other hand, I see not the slightest objection to the admission of a large class of works of this nature into that department of our library destined for the use of pecuniary subscribers—always reserving a strong objection against works of a violent party character. Indeed, I can hardly imagine a more useful addition to it than an assemblage of the best works on Political Economy, as a science, and a subject of rational enquiry entirely distinct from politics—a subject, it is true, on which much dispute subsists—but on which, among all its complication and difficulty, a dawn of light has begun to appear, and on which it is of the highest importance that every one calling himself an educated man, should possess some knowledge, and some habit of exercising a logical discrimination, were it only to enable him to detect the fallacies which are continually brought forward.

I might now, Gentlemen, proceed to dilate on the advantages generally to the more educated and better informed, of those accessions to their education and information, which is included in the very notion of a large access to a well-chosen library; but time is short, and I am sure they are already appreciated. I shall therefore, now, cease to trespass longer on your patience, and finish what I have to say, with the sincerest wishes for the progress of the institution, and its increase in everything which can add to the gratification of its members, and the general improvement of the neighbourhood in which it has arisen.

MECHANISM OF THE HEAVENS.

1. *Mechanism of the Heavens.* By Mrs. Somerville. London. 8vo. 1832.
2. *Mécanique Céleste.* By the Marquis De la Place, &c. Translated, with a Commentary, by Nathaniel Bowditch, LL.D., &c. Volume I. Boston. 1839.

(From the Quarterly Review, 1833. No. 99.)

The close of the last century witnessed the successful termination of that great work, commenced by Newton, and prosecuted by a long succession of illustrious mathematicians, by which the movements of the planetary system were reduced under the expression of dynamical laws, and their past and future positions, with respect to their common centre and to each other, rendered matter of strict calculation. A wonderful result, which will for ever form a principal epoch in the history of mankind, was at length arrived at in the announcement of the fact, that a brief and simple sentence, intelligible to a child of ten years of age, accompanied with a few determinate numbers, capable of being written down on half a sheet of paper, comprehends within its meaning the history of all the complicated movements of our globe, and the mighty system to which it belongs—the mazy and mystic dance of the planets and their satellites

—"cycle on epicycle, orb on orb"—from the earliest ages of which we have any record, nay, beyond all limits of human tradition,—even to the remotest period to which speculation can carry us forward into futurity. By the announcement of this law and the establishment of these data, an indefinite succession of events is thus combined into one great fact, and may be considered as a single feature in creation, independent of the lapse of time, and registered only in the unprogressive annals of eternity.

In the course of the investigations which have terminated in this result, another fact, of a no less high and general order, has come to light, of which Newton could have formed no anticipation, —that, namely, of the stability of our system, and the periodic nature and restricted limits of its fluctuations, which preclude the possibility of such deviations from a mean or average state as may lead to the subversion of any essential feature of that happily balanced order which we observe at present to subsist in it. This noble theorem forms a beautiful and animated comment on the cold and abstract announcement of the general law of gravitation. A thousand systems might have been formed of which the motions would, for a time have been regular and orderly enough, but which would either have ended in a collision of parts subversive of the original conditions, or would pass through a succession of phases or states, endless in variety, among which some would be found no less incompatible with life than such collisions themselves—whether from extreme remoteness or proximity of the source of light and heat, or from violent and sudden alternations of its influence—or in which, at all events, that beautiful and regular succession of seasons—that "grateful vicissitude" we admire and enjoy, and those orderly and established returns of phenomena which afford at once the opportunity and the inducement to trace their laws, would have been wanting; while in their place might have reigned a succession of changes reducible to no apparent rule; variety without progressive improvement; years of unequal length and seasons of capricious temperature; planets and moons of portentous size and aspect, glaring and disappearing at uncertain intervals, and every part of the system wearing the appearance of anarchy, though, in fact, obeying, to the letter, the same general law of gravitation, which must yet have for ever remained unknown to its inhabitants.

Among infinite systems equally possible, such, we have no reason to doubt, might exist,—but our own is not, nor can it ever, in its own natural progress, pass into such a one. In the choice of its arbitrary constants, (to use the language of geometers,) in the establishment of the relations of magnitude, speed, and distance of its parts, such a case is expressly provided against. In the circulation of its members all in one direction—in the moderate amount of the eccentricities and inclinations of all the planetary orbits, and the extremely small ones of those of its more important bodies, but more especially in the mode in which the general system is broken up into several subordinate ones, and in the individual attachment and allegiance of each member to its immediate superior, we must look to the safeguards of this glorious arrangement.

This last-mentioned condition may require some illustration. Had the Earth and Mars, for instance, formed a binary combination separated by an interval no greater than the moon's actual distance from the earth, there is no doubt that such a double planet might have continued to circulate round the sun nearly as the earth and moon do at present. But with such a combination the moon could not have coexisted, without a complete breach of the law of regular periodicity. Its path would be alternately commanded by one and the other of its great equipollent centres, whichever, for the moment, occupied the most advantageous position; and should its primitive velocity be so adjusted that it could neither throw itself to a sufficient distance from both to escape from the influential attraction of either, and become a separate planet, nor attach itself so closely to one of them as to be carried about it as a mere appendage, it must continue to wind for ever an intricate and sinuous course around and between them, in which occasional collision with one or other would, by no impossible or improbable contingency, afford a tragic epoch in the history of so ill-adjusted a system.

It is, moreover, well worthy of remark, that the mode in which the stability of our system is accomplished is by no nice mathematical adjustment of proportions,—no equilibrated system of counterpoises satisfying an exact equation, and which the slightest deviation in any of the data from its strict geometrical proportion would annul. Such adjustments, it is true, are not incompatible with the law of gravitation, even in a system composed of several bodies. Geometers have demonstrated, for example, that three or even more bodies, exactly adjusted in their weights and distances, and in the velocities and directions of their motions at any one instant, might continue for ever to describe conic sections about each other, and about their common centre of gravity. But without supposing any such adjustment of the weights and distances of the members of a system subjected to the law of gravitation, and taking them as they are actually in our own, there is yet another supposition in which the absence of secular perturbation might have been ensured,—that, namely, in which the planetary motions should be performed all in one plane, and all in perfect circles about the sun,—realising, in fact, the old Aristotelian notion of celestial movements, all which he considered to be of necessity exactly circular. We do not remember to have seen any mention made of the possibility of this case.

It follows, however, immediately, from the general proposition demonstrated by Lagrange and Laplace, which establishes an invariable relation among the eccentricities of any number of perturbed orbits; viz., that the sum of the squares of all the eccentricities, each multiplied by an invariable coefficient, is itself invariable, and subject to no change by the mutual action of the parts of the system. For it is evident, that had the orbits been all originally circular, or if at any one instant of time each of the eccentricities were, by some external agency, destroyed, so as to render these orbits at once all circles, after which the system should be abandoned to its own reactions, the sum in question would also vanish at that instant, and therefore at every subsequent instant, which would be impossible, (since none of the coefficients are negative,) unless each several eccentricity were to remain for ever evanescent *per se*, or each several orbit a perfect circle.*

If we depart from the law of gravitation, and inquire whether, under other conceivable laws of central force, a system might not exist essentially and mathematically free from the possibility of perturbation, and in which every movement should be performed in undeviating orbits and unalterable periods, we have not far to search. Newton has himself demonstrated, in his "Principia," or at least,

* Perturbations would still take place, but could not accumulate; but would, at the contrary, annul each other at each successive conjunction. (*H.* 1857.)

it follows almost immediately from the 89th proposition of his first book and its corollary, that this wonderful property belongs to a law of attractive force in the *direct* proportion of the distance; and, however extravagant such a supposition may appear, if we consent to entertain it as a mere mathematical speculation, it is impossible not to be struck with the simplicity and harmony which would obtain in the motions of a system so constituted. Whatever might be the number, magnitudes, figures, or distances of the bodies composing an universe under the dominion of such a law—in whatever planes they might move, and in whatever directions their motions might be performed—each several body would describe about the common centre of gravity of the whole, a perfect ellipse; and all of them, great and small, near and remote, would execute their revolutions in one common period, so that, at the end of every such period, or *annus magnus*, of the system, all its parts would be exactly re-established in their original positions, whence they would set out afresh, to run the same unvarying round for ever.

We may please ourselves with such speculations, and enjoy the beauty and harmony of their results, in the very same spirit with which we rejoice in the contemplation of an elegant geometrical truth, or a property of numbers, without presumptuously encroaching on the province of creative wisdom, which alone can judge of what is really in harmonious relation with its own designs. The stability of our actual system, however, rests on a basis far more refined, and far more curiously elaborate. It depends, as we have before observed, on no nice adjustments of quantity, speed, and distance. The masses of the planets, and the constants of their motions, might all be changed from what they are, (within certain limits,) yet the same tendency to self-destruction in the *deviations* of the system from a mean state, would still subsist. The actual forms of their orbits are not ellipses, but spirals of excessive intricacy, which never return into themselves; yet this intricacy has its laws, which distinguish it from confusion, and its limits, which preserve it from degenerating into anarchy. It is in this conservation of the principle of order in the midst of perplexity—in this ultimate compensation, brought about by the continued action of causes, which appear at first sight pregnant only with subversion and decay—that we trace the Master-workman with whom the darkness is even as the light.

This momentous result has been brought to light slowly, and, as it were, piecemeal. The individual propositions of which it consists have presented themselves singly, and at considerable intervals of time, like the buried relics of some of those gigantic animals which geologists speak of, each, as it emerged, becoming a fresh object of wonder and admiration, proportioned to the labour of its extraction, as well as to its intrinsic importance; and these feelings have at length been carried to their climax by finding the disjointed members fit together, and unite into a regular and compact fabric.

It is to our continental neighbours, but more especially to the geometers of France, that we owe the disclosure of this magnificent truth: Britain took little share in the enquiry. As if content with the glory of originating it, and dazzled and spell-bound by the first great achievement of Newton, his countrymen, with few and small exceptions, up to a comparatively late period, stood aloof from the great work of pursuing, into its remote details, the general principle established by him. We are far from being disposed to attribute this remarkable supineness to the prevalence of any of the meaner or more malignant feelings of national pride, prejudice, or jealousy. Some irritation and distaste for the continental improvements might be, and no doubt were, engendered, and, to a certain extent, continued, by the controversies which excited so lively a sensation among the cotemporaries of Newton; but, on the other hand, it could not have been, at first, reasonably presumed, (what proved afterwards to have been really the case,) that the applicability of Newton's mode of investigation should terminate almost at the very point where he himself desisted from applying it—still less that algebraic processes, which were regarded by him as mere auxiliaries to geometrical construction and demonstration, should be destined to acquire such strength and consistency as to supersede all others, and leave them on record only as scientific curiosities. It is rather to the barrier thrown by our insular situation in the way of frequent personal communication between our mathematicians and those abroad, to the want of a widely diffused knowledge of the continental languages, and to the consequent indifference in the reading part of the public as to the direction which thought was taking, in the loftier regions of its range, in other lands than our own, that we are inclined to refer what cannot but appear an extraordinary defect of sympathy in so exciting a course of discovery. Much, too, must be attributed to that easy complacency with which human nature is too apt to regard progress already made as all that can be made,—which dwells with admiring and grateful satisfaction on achievements performed and laurels won, while it neglects to body forth the possibilities of a yet richer and more glorious future,—suffers a short breathing time to become prolonged into a state of languor and indifference; and consigns to other and fresher aspirants, the toil and the reward of penetrating farther into those thorny and entangled thickets of unexplored research which bound our actual horizon and, by the force of habit and repose, come at length to hedge in our thoughts and wishes.

Whatever might be the causes, however, it will hardly be denied by any one versed in this kind of reading, that the last twenty years of the eighteenth century were not more remarkable for the triumphs of both the pure and applied mathematics abroad, than for their decline, and, indeed, all but total extinction at home. From the publication of Waring's profound, but cumbrous and obscure, "Meditationes Algebraicæ," and Landen's researches on the motions of solids, and his remarkable discovery of the rectification of the hyperbola by two ellipses, we may search our libraries in vain for investigations of the slightest moment in the higher analysis, or, indeed, for any evidence of its abstruser parts being so much as known to our mathematical writers. While the academical collections of Turin, Paris, Berlin, and Petersburg, were teeming with the richest treasures of the analytic art, poured forth with unexampled profusion, our own presented the melancholy contrast of entire silence on all the great questions which were then agitating the mathematical world,—a blank, in short, which the respectable names of Vince and Hellins only served to render more conspicuous.

It was with the commencement of the present century that a sense of our deficiencies, and of the astonishing and disreputable distance to which we had fallen behind the general progress of mathematical knowledge in all its branches, began to make itself felt; but to remedy the evil was more difficult than to discover its existence. Great bodies move slowly. It requires time, where national tastes and habits are concerned, to turn the current of thought out of its smooth-worn track into untried and, at first, abrupter channels; and, besides, the means were wanting. A total deficiency of all elementary books in our own language in which the modern improvements could be studied, precluded beginners from obtaining any glimpse beyond the narrow circle in which their teachers had revolved. The student is guided in his early choice of books by sanction and by usage. He may not, without hazard, venture to chalk out for himself a course of reading unusual and remote; and, rejecting the writers of his own country, choose foreigners for his instructors. To come to such a resolution presupposes a discrimination and a preference which is incompatible with entire unacquaintance with his subjects. It was only, therefore, when, although well instructed and perfect in the usual routine, he found himself arrested at the very first page of any of the elaborate works of the foreign geometers which chance might throw into his hands, that he could acquire the painful but necessary conviction of having all to begin afresh, much even to unlearn; to forget habits, to change notations, to abandon points of view which had grown familiar, and, in short, put himself once more to school.

The late Professor Woodhouse seems to have been among the first of our countrymen who experienced this inward conviction with its natural concomitant, the desire to propagate forward to other minds the rising impulse of his own. His papers on the independence of the analytical and geometrical modes of investigation, and on the evidence of imaginary symbols, as well as his treatise on the principles of analytical calculation, contributed largely to produce this effect; and in his Trigonometry, in which, for the first time, this important part of geometry was placed before the English reader in a purely analytical form, and with all that peculiar grace and flexibility which belongs to it in that form, he conferred a most essential benefit on the elementary mathematics of his country. We owe also to him a treatise on the Calculus of Variations, not indeed very luminous, nor very extensive, but which had one pre-eminent merit, that of appearing just at the right moment, when the want of any work explanatory of what is merely technical in that calculus was becoming urgent.

An increasing interest in mathematical subjects was now also manifested by the occasional appearance of papers of a higher class in our learned Transactions, (such as that of Dr. Brinkley, now Bishop of Cloyne, on the exponential developments of Lagrange, — a memoir of curious and elaborate merit, and, though somewhat later in point of time, the curious investigations of Mr. Babbage on the theory of functional equations,) as well as of distinct works on subjects of pure analysis. The most remarkable of these is the "Essay on the various Orders of Logarithmic Transcendants," by the late W. Spence of Greenock, the first formal essay in our language on any distinct and considerable branch of the integral calculus, which had appeared since the publication of Hellins's papers on the "Rectification of the Conic Sections." A premature death carried off, in Spence, one who might have become the ornament of his country in this department of knowledge. His posthumous essays, which were not, however, collected and published till 1819, prove him to have been both a learned and inventive analyst. He appears to have studied entirely without assistance, and to have formed his taste and strengthened his powers by a diligent perusal of the continental models. In consequence, he was enabled to attack questions which none of his countrymen had entered upon, such as the general integration of equations of finite differences, and others of that difficult and elevated class.

Among our Scottish countrymen, indeed, the torch of abstract science had never burnt so feebly nor decayed so far as in these southern abodes; nor was a high priest of the sublimer muse ever wanting in those ancient shrines, where Gregory and Napier had paid homage to her power. The late Professor Robison, though his taste for the older geometry led him to undervalue both the evidence and the power of the modern analysis, was yet a mathematician of no inconsiderable note. The remarkable papers of Professor Playfair on Porisms show how deeply the mind of that sound mathematician and elegant writer was imbued with the spirit of the analytical methods, and a sense of their superior power — a power, however, which he was content to admire and applaud, rather than ready to wield. It may indeed be questioned whether, by any researches of his own, however successful, he could have given a stronger impulse to the public mind in this direction than what his admirable review of the "Mécanique Céleste" communicated.

To this school also we owe the only British geometer who, at this period, seems to have possessed not only a complete familiarity with the resources of the higher analysis, but also the habit of using them with skill and success in inquiries of moment in the system of the world — we mean Professor Ivory. The appearance of his "Memoirs on the Attraction of Spheroids," which are deservedly considered masterpieces of their kind, and which at once placed their author in the high rank among the geometers of Europe which he has ever since maintained, was almost simultaneous with that of Spence's work,—a coincidence which might seem to warrant the most sanguine hopes of the speedy re-establishment of our mathematical glories. But the national taste and acquirements had sunk so low, that the stimulus of these examples was yet for a while unfelt. The "Essay on Logarithmic Transcendants" attracted little immediate notice, and the Memoirs of Ivory, though received abroad with the respect and admiration they so justly merited, met with slender applause and no imitation at home. Their effect was, to seat their author on a solitary eminence, equally above the sympathy and the comprehension of the world around him. Since that period, however, a change has been slowly but steadily taking place in mathematical education. Students at our universities, fettered by no prejudices, entangled by no habits, and excited by the ardour and emulation of youth, had heard of the existence of masses of knowledge, from which they were debarred by the mere accident of position. There required no more. The *prestige* which magnifies what is unknown, and the attraction inherent in what is forbidden, coincided in their impulse. The books were procured and read, and produced their natural effects. The brows of many a Cambridge moderator were elevated, half in ire, half in admiration, at the unusual answers which began to appear in examination papers. Even moderators are not made of impenetrable stuff: their souls were touched, though fenced with sevenfold Jacquier, and tough bull-hide of Vince and Wood. They were carried away with the stream, in short, or replaced by successors full of their newly-acquired powers. The modern analysis was adopted in its largest extent, and at this moment we believe that there exists not throughout Europe a centre from which a richer and purer light of mathematical instruction emanates through a community, than one, at least, of our universities.

One of the immediate consequences of the increased demand for a knowledge of the continental analysis, and the manner in which it is made subservient to physical inquiry, was a rapid and abundant supply of elementary works. Lacroix's lesser treatise (we wish it had been the greater) has been translated, with note and comment, from the French, and Meier Hirsch's admirable work on the "Theory of Algebraic Equations," from the German; and, in addition to these transplanted authorities, (the former of which may be regarded as having greatly contributed, by its numerous examples, to the final *domestication* of the peculiar notation of the differential calculus among us,) a host of indigenous ones on almost every branch of the pure and applied mathematics have emanated chiefly, but by no means entirely, from the press of the Cambridge University, which has thus signalised itself in a manner equally useful to the country and honourable to its directors. Many of these works bear, it is true, strong and singular marks of the *transition state* of the science in which they were produced, but, on the whole, they contain a copious body of instruction; and although we have still nothing approaching in extent and excellence to the elementary works of Euler, or to the superb digest of analytical knowledge contained in the great work of Lacroix, to which we have before alluded, yet, at least, our students can no longer complain of being left wholly without a guide, or without preparation for a profounder course of reading, should they feel disposed to enter upon it.

Another consequence, no less natural and obvious, of this altered state of feeling and instruction, has been the gradual formation of what, at length, begins to merit the appellation of a British School of Geometry. We are far indeed from hoping soon to outstrip those who have so much the start of us, but the race is at least less hopeless than heretofore. The interval between the competitors has begun sensibly to diminish, and we need, at least, no longer fear being disgracefully distanced. We no longer perceive the same shyness, on the part of our mathematical champions, in entering on the great and vexed questions of the lunar and planetary perturbations, the theory of the tides, and others relating to the system of the world; nor the same indifference on that of the bystanders whether they are successful or not. The eminent geometer whom we have before named is no longer the only one among us who adventures himself fairly and boldly within this magic circle. On the contrary, we have recently witnessed the publication, by one of our countrymen, of several profound memoirs on the most intricate and important parts of the terrestrial and planetary theory; on that of another, the novel, and, since Newton's time, the *unique* fact, of a new planetary inequality, not only detected, as so many have been, by British observation, but successfully referred to its origin, and subjected to exact calculation by British analysis, and that by no trifling effort or command of its resources.

We are very sure that in speaking so decidedly as we have felt compelled to do, of the long-subsisting superiority of foreign mathematics to our own, we run no hazard of wounding any feeling we would wish to spare. Had our prospects, indeed, remained in the same deplorable state into which, but a very few years ago, they seemed to have settled, we should perhaps have preferred silence to the discouraging task of attempting to arouse an apathy so profound; — but a better era is evidently advancing. The auguries are favourable. We hail them with delight, and we feel at the same time assured that our Airys, our Lubbocks, our Hamiltons, and our Challises, the hope of our reviving geometry, will bear us out in the view we have taken, and acknowledge with gratitude and pleasure the sources whence they have drawn those principles they are now using so emulously and so well.

Meanwhile the anomalous state of our mathematical literature which we have above described, explains, very naturally, what must have struck most mathematical readers as a remarkable feature in it, — we mean, the scanty supply of English works illustrative of the celestial mechanism, whether in the nature of express commentary and avowed illustration of the immortal work of Laplace, or in the form of independent treatises, calculated to bring the whole subject before the reader in a more compendious and explanatory manner than was compatible with Laplace's object, with the greatness and sweeping generality of his outline, or the close and laboured filling-in of his detail. The "Elementary Illustrations of the Celestial Mechanics" of Laplace, by the late celebrated Dr. Young, will hardly, we apprehend, be regarded by any reader as supplying satisfactorily the one of these desiderata; and although the "Physical Astronomy" of Professor Woodhouse approaches much nearer to what is requisite for the other, yet it by no means satisfies all, or nearly all, the conditions which such a work should accomplish. The detail of processes and developments into which it enters, though ample for elucidating the principles of the methods employed, is yet hardly sufficient to give a complete and effective grasp of the subject matter, while the combination of historical detail with theoretical elucidation, which it keeps in sight, tends to embarrass the reader, by constantly shifting his point of view, and calling off his attention to inquire how mistakes have heretofore been committed and rectified; a most instructive thing in itself, no doubt — but calculated rather to render such a work a useful companion in a course of original reading, than to enable it to supply the place of many books, and offer, in a moderate compass, a compendium of what is known.

The works whose titles head the present article supply to the English reader, so far as they extend, both these desiderata, and supply them in a manner that leaves little to wish for. They are both, moreover, otherwise extremely remarkable in respect of the quarters from which they emanate. A lady, our own countrywoman, is the authoress of one; and to an American, by birth and residence, and to the American press, we stand indebted for the other. If anything were wanting to put our geometers effectually on their mettle, it would we think be found in such a coincidence.

Mrs. Somerville is already advantageously known to the philosophical world by her experiments on the magnetizing influence of the violet rays of the solar spectrum; a delicate and difficult subject of physical inquiry, which the rarity of opportunities for its prosecution, arising from the nature of our climate, will allow no one to study in this country except at a manifest disadvantage. It is not surprising, therefore, that the feeble, although unequivocal indication of magnetism, which she undoubtedly obtained, should have been regarded by many as insufficient to decide the question at issue. To us their evidence appears entitled to considerable weight; but it is more to our immediate purpose to notice here, the simple and rational manner in which those experiments were conducted — the absence of needless complication and refinement in their plan, and of unnecessary or costly apparatus in their execution — and the perfect freedom from all pretension or affected embarrassment in their statement. The same simplicity of character and conduct, the same entire absence of anything like female vanity or affectation, pervades the whole of the present work. In the pursuit of her object, and in the natural and commendable wish to embody her acquired knowledge in a useful and instructive form for others, she seems entirely to have lost sight of herself; and, although in perfect consciousness of the possession of powers fully adequate to meet every exigency of her arduous undertaking, it yet never appears to have suggested itself to her mind, that the acquisition of such knowledge, or the possession of such powers, by a person of her sex, is in itself anything extraordinary or remarkable. We find accordingly, beyond the name in the title-page, nothing throughout the work introduced to remind us of its coming from a female hand. Even the tempting opportunity of deprecating criticism, which a preface affords, is neglected; nor does anything apologetic, in the tone of her admirably-written preliminary discourse, betray a latent consciousness of superiority to the less-gifted of her sex, or a claim either on the admiration or forbearance of ours, beyond what the fair merits of the work itself may justly entitle it to. There is not only good taste, but excellent good sense in this. Whether admiration be due, or allowances needed, we accord both the one and the other, with perfect readiness, when left to the workings of our own good feeling. On the other hand, whenever we see such things as the poems of a minor, or the learning of a lady, introduced by an appeal, direct or indirect, to our good nature, we enter on our task of perusal with no very pleasant impression that this amiable weakness of our disposition is about to be largely taxed — an expectation in which, sooth to say, we are rarely disappointed.

In the present instance, however, we are neither called on for allowances, nor do we find any to make; on the contrary, we know not the geometer in this country who might not reasonably congratulate himself on the execution of such a work. The volume is dedicated to Lord Brougham, and appears to have been originally undertaken, at his instance, for publication by the Society for the Diffusion of Useful Knowledge; but the views of the author extending with its progress, it outgrew its first destination, and assumed an independent form. The nature of these views — the scope and object of the work — will perhaps be best understood from Mrs. Somerville's own words: —

"A complete acquaintance with physical astronomy can only be attained by those who are well versed in the highest branches of mathematical and mechanical science; such alone can appreciate the extreme beauty of the results, and the means by which these results are obtained. Nevertheless, a sufficient skill in analysis to follow the general outline — to see the mutual dependence of the several parts of the system — and to comprehend by what means some of the most extraordinary conclusions have been arrived at — is within the reach of many who shrink from the task, appalled by difficulties which perhaps are not more formidable than those incident to the study of the elements of every branch of knowledge; and possibly overrating them, by not making a sufficient distinction between the degree of mathematical acquirement necessary for making discoveries and that which is requisite for understanding what others have done. That the study of mathematics, and their application to astronomy, are full of interest, will be allowed by all who have devoted their time and attention to these pursuits; and they only can estimate the delight of arriving at truth, whether it be in the discovery of a world or of a new property of numbers." — p. 7.

Let us now see how far the conduct of Mrs. Somerville's work corresponds with these views. In so doing, it is obvious that we are not to look for original discovery, the ambition of which is disclaimed, and which indeed would be misplaced in a work of the kind — nor even for absolute novelty in the methods of arriving at known results. The subject has been, in fact, so copiously handled, and by such a host of the most profound and accomplished mathematicians, that such novelty is now no longer to be expected, nor indeed desired in any fresh exposition of it. It is sufficient if all the results which it imports to know are clearly and perspicuously derived from their principles — the artifices of calculation on which their deduction rests, distinctly explained, and the processes actually pursued to such an extent as to give the reader a *thorough practical insight* into the developments of the subject. This, we think, is fully accomplished in the work before us, for all those parts of the general subject which it professes to embrace, that is to say, the general exposition of the mechanical principles employed — the planetary and lunar theories, and those of Jupiter's satellites, with the incidental points arising naturally out of them. The development of the theory of the tides, and precession of the equinoxes, the attraction of spheroids, and the figure of the earth, appear to be reserved for a second volume. A certain degree of inconvenience is incurred by this in the investigation of those irregularities in the motions of the moon and satellites depending on the oblate form of their planets, which compels an anticipation of results not previously demonstrated; but this inconvenience is one more easily perceived than avoided.

In Mrs. Somerville's preliminary dissertation, a general view is taken of the consequences of the law of gravitation, so far as they have hitherto been traced, whether as relates to the elliptic motions and mutual perturbations of the planets and their satellites, and the slow variations in the forms of their orbits thereby produced, or to the figures assumed by each of them individually, in consequence of the combination of their rotations on their axes with the attractions of their particles on each other and that of neighbouring bodies, together with the nutations, precessions, and librations of their axes themselves, arising from external actions, or, lastly, to the equilibrium and oscillations of the waters and atmospheres which cover their surfaces, comprehending the theory of the tides, and the great geological question of the general stability of the ocean. These, and the important points which are essentially dependent on such investigations — their application to those greater operations of geography to which the term geodesy is usually applied — to the determination of standards of weight and measure — to the fixation of chronological epochs — and a multitude of other interesting inquiries, are treated with a condensation, but at the same time a precision and clearness, which render this preliminary dissertation a model of its kind, and a most valuable acquisition to our literature. We have indeed no hesitation in saying, that we consider it by far the best condensed view of the Newtonian philosophy which has yet appeared. We do not, of course, mean to include the "Système du Monde" of Laplace himself, which embraces a far wider range, both of illustration and detail, and of which Mrs. Somerville's preface may in some sort be regarded as an abstract, but an abstract so vivid and judicious as to have all the merit of originality, and such as could have been produced only by one accustomed to large and general views, as well as perfectly familar with the particulars of the subject.

As specimens of Mrs. Somerville's style of writing, we shall extract a few sentences almost from the commencement of this discourse: —

"Science, regarded as the pursuit of truth, which can only be attained by patient and unprejudiced investigation, wherein nothing is too great to be attempted, nothing so minute as to be justly disregarded, must ever afford occupation of consummate interest and subject of elevated meditation. The contemplation of the works of creation elevates the mind to the admiration of whatever is great and noble, accomplishing the object of all study which, in the elegant language of Sir J. Mackintosh, is 'to inspire the love of truth, of wisdom, of beauty, especially of goodness, the highest beauty,' and of that supreme and eternal

Mind which contains all truth and wisdom, all beauty and goodness. By the love, or delightful contemplation, of these transcendent aims, for their own sake only, the mind of man is raised from low and perishable objects, and prepared for those high destinies which are appointed for all those who are capable of them."

We rejoice at this testimony to the intrinsic worth of scientific pursuits, and the pure and ennobling recompense they carry with them, from such a quarter. The female bosom is true to its impulses, and unwarped in their manifestation by motives which, in the sterner sex, are continually giving a bias to their estimates and conduct. The love of glory, the desire of practical utility, nay, even meaner and more selfish motives, may lead a man to toil in the pursuit of science, and adopt, without deeply feeling, the language of a disinterested worshipper at that sacred shrine—but we can conceive no motive, save immediate enjoyment of the kind so well described in the passage just quoted, which can induce a woman, especially an elegant and accomplished one, to undergo the severe and arduous mental exertion indispensable to the acquisition of a really profound knowledge of the higher analysis and its abstruser applications.

What follows is no less pleasing in another point of view:—

"The heavens afford the most sublime subject of study which can be derived from science: the magnitude and splendour of the objects, the inconceivable rapidity with which they move, and the enormous distances between them, impress the mind with some notion of the energy that maintains them in their motions with a durability to which we can see no limit. Equally conspicuous is the goodness of the great First Cause in having endowed man with faculties by which he can not only appreciate the magnificence of his works, but trace with precision the operation of his laws; use the globe he inhabits as a base wherewith to measure the magnitude and distance of the sun and planets, and make the diameter of the earth's orbit the first step of a scale by which he may ascend to the starry firmament. Such pursuits, while they ennoble the mind, at the same time inculcate humility, by showing that there is a barrier which no energy, mental or physical, will ever enable us to pass; that however profoundly we may penetrate the depths of space, there still remain innumerable systems, compared with which, those which seem so mighty to us must dwindle into insignificance, or even become altogether invisible."

We shall extract only one other passage from this discourse, as an example of the manner in which our fair authoress treats the less familiar topics, to which this part of her work is devoted. It is that in which the stability of the equilibrium of the seas and the permanence of the axis of the earth's rotation are considered.

"It appears from the marine shells found on the tops of the highest mountains, and in almost every part of the globe, that immense continents have been elevated above the ocean, which [ocean] must have engulphed others. Such a catastrophe would be occasioned by a variation in the position of the axis of rotation on the surface of the earth; for the seas tending to the new equator would leave some portions of the globe, and overwhelm others. But theory proves that neither nutation, precession, nor any of the disturbing forces which affect the system, have the smallest influence on the axis of rotation which maintains a permanent position on the surface, if the earth be not disturbed in its rotation by some foreign cause, as the collision of a comet, which may have happened in the immensity of time. Then, indeed, the equilibrium could only have been restored by the rushing of the seas to the new equator, which they would continue to do till their surface was everywhere perpendicular to the direction of gravity. But it is probable that such an accumulation of the waters would not be sufficient to restore equilibrium, if the derangement had been great; for the mean density of the sea is only about a fifth part of that of the earth, and the mean depth, even of the Pacific Ocean, is not more than four miles, whereas the equatorial radius of the earth exceeds the polar radius by twenty-five or thirty miles: consequently the influence of the sea on the direction of gravity is very small; and as it thus appears that a great change in the position of the axis is incompatible with the law of equilibrium, the geological phenomena must be ascribed to an internal cause. Thus, amidst the mighty revolutions which have swept innumerable races of organized beings from the earth — which have elevated plains, and buried mountains in the ocean—the rotation of the earth, and the position of the axis on its surface, have undergone but slight variations."

We will only pause to remark here, that an argument, which appears to us much more conclusive against the fact of any disturbance having, in remote antiquity, taken place in the axis of the earth's rotation, is to be found in the amount of the lunar irregularities which depend on the earth's spheroidal figure. However insufficient the mere transfer of the mass of the ocean from the old to the new equator might be to ensure the permanence of the new axis, the enormous abrasion of the solid matter of such immensely-protuberant continents, as would, on that supposition, be left, by the violent and constant fluctuation of an unequilibrated ocean, would, (according to an ingenious remark of Professor Playfair,) no doubt, in the lapse of some ages, remodel the surface to the spheroidal form ; but the lunar theory teaches us that the *internal strata*, as well as the *external outline*, of our globe, are elliptical, their centres being coincident and their axes identical with that of the surface,—a state of things incompatible with a subsequent accommodation of the surface to a new and different state of rotation from that which determined the original distribution of the component matter.

Mrs. Somerville's work is divided into four books, of which the first is devoted to the establishment of those general relations which prevail in the equilibrium or motion of bodies, or systems of bodies, whether solid or fluid, which are necessary to serve as a groundwork for the subsequent investigations;—the second, to the planetary theory, the elliptic motions and mutual perturbations of the bodies of our system, and the secular changes which take place in their orbits. The third book is given to the lunar theory; and the fourth to that of Jupiter's satellites, which is now for the first time introduced in any regular and extensive form to the English reader. From some confusion in the arrangement, or at least the numbering of the chapters of this book, it would seem to have been the original intention of the authoress to have thrown these two divisions of her subject into one, probably under the general head of the theory of Satellites. The actual arrangement is, on every account, infinitely preferable.

In the treatment of the statical and dynamical principles developed in the first part, the processes of the first book of the "Mécanique Céleste" are pretty closely but by no means servilely adhered to. Laplace's demonstration, for instance, of the fundamental principle of the composition of forces is suppressed, and its place supplied by one more elementary; and again, in the investigation of the equation of continuity of a fluid, the excessive difficulty and complication of the analysis by which he arrives at this result is evaded, and the whole subject in consequence greatly simplified by adopting a different and easier method of estimating the volume of an elementary molecule of the fluid in its displaced position. The whole of this portion of the work is also copiously illustrated by diagrams, which, however readily dispensed with by those whom long habit has rendered familiar with analytical mechanics, are yet extremely useful in assisting the conception of less experienced readers. We could wish that a little more assistance of this kind had been afforded, and altogether a little more explanatory illustration bestowed on that chapter which treats of the rotatory motion of a solid mass. The subject needs it. There is a difficulty of conception in the notion of an axis of rotation shifting its position within a solid from instant to instant, as well as that of pressures exerted by the revolving matter, on such an imaginary and fugitive line, which is very embarrassing to one not accustomed to such speculations, though easily removed by dilating a little on the subject, and placing it in different and familiar points of view. We have always considered this part of analytical mechanics as among the most beautiful and exquisite of its applications. It is usually, however, regarded by beginners as more abstruse than its real difficulties authorize. This arises partly from the obscurity of conception we have alluded to, but partly, also, from a more technical cause,—the frequent changes of coordinates which its analytical treatment involves. This is a difficulty of the same kind as transposition, in a musical performance, from one key to another; and as a musician can never expect to become a ready performer till practice has made such difficulties vanish, so the mathematical student can never feel at complete ease in the higher applications, till all such mere technical evolutions cease to be complained of as difficulties, or even felt as inconveniences.

We could have wished, too, that instead of entering, in this part of the work, on the theory of the tides, which is by far the most complicated and infinitely the least satisfactory part of the general subject, that of the attractions of spheroids had been traced, at least so far as to demonstrate the theorems which are afterwards taken for granted in the development of those terms of the mercurial and lunar theory, and that of Jupiter's satellites, which depend on the oblate figure of the primary. As it is only a single term in the development of the series expressing the deviation of the law of gravity in the spheroid from that in the sphere which is wanted, this might have been very easily done, and at the same time the reader prepared to enter more fully into this interesting part of the subject, in a more advanced state of his knowledge.

In the second book the planetary theory is given with a fulness commensurate with its importance. Its first chapters are of course devoted to the theory of elliptic motion, which is concisely, but very perspicuously stated. The equations used are the beautiful integrals of the general differential equations first obtained, if we remember rightly, by Lagrange, and used by him with such wonderful effect for ascertaining the variations of the elements. They are the same which Laplace derives in the 18th article of his second book, by a process which we should be inclined to tax with excessive and useless generality, were it not quite necessary to show that this important part of the theory had been probed to the quick, and every resource which analysis could furnish exhausted on it. Mrs. Somerville, however, very properly derives them by the ordinary processes of direct integration. The usual properties of elliptic motion, with the series for the developments of the anomalies and radius vector afterwards required, are there demonstrated, and a few pages added on the determination of the elements.

We should have been glad to have found in this part of the work some outline of the powerful and elegant researches of Gauss on the determination of the orbits of the celestial bodies, and especially some more practical method of determining those of comets than Laplace's. The subject of the motion of comets is, however, summarily dismissed; and even the beautiful theorem of Lambert, which expresses the time of describing a parabolic arc in terms of the radii vectores of its extremities and its chord, is omitted.

The fine idea of Lagrange, by which the perturbations of a planet are expressed by means of a variable ellipse, and all its inequalities referred to changes in the elliptic elements which are supposed to be in a state of continual fluctuation, has introduced a degree of simplicity and symmetry into the analytical treatment of the planetary theory such as could hardly have been hoped for, and divested it of all that was repulsive, and much that was merely laborious in its investigation. It is in this view of the subject alone, that a neat conception can be formed of the distinction between variations truly secular, and those inequalities of long periods which were originally confounded with secular changes. The former class are those which are independent of the mutual configurations of the planets one among the other, and in their theory no other quantities enter than the elements themselves and the time; all those variables on which depend the situations of the planets in their orbits, such as their longitudes, latitudes, and distances from the sun, being excluded. The reactions contemplated in this part of the theory are not so much those of planet on planet, as of orbit on orbit. Nothing can be more exquisite in analysis, nothing more refined in conception, than this investigation, on which depend all those grand propositions respecting the stability of the system to which we have already alluded. In the conduct of this part of her subject, Mrs. Somerville has chiefly adhered to the analysis of Lagrange, as stated by Laplace in the supplement to the third volume of the "Mécanique Céleste," only in that important and difficult part of it which concerns the invariability of the axes as affected by the squares and products of the disturbing forces, availing herself of the subsequent elaborate investigations of Poisson.

The periodical part of the perturbations of the elements is next investigated; not, however, with a view to the ultimate derivation of formulæ for the practical computations of the longitudes and latitudes of the disturbed planets, which, though practicable, is not so easy in this view of the subject as in that of Laplace, which depends on the principle of successive approximations from the differential equations of the troubled orbit; and, so to speak, consists in a continual gathering up of the loose and ravelled ends of the skein which appear in the form of unperiodic terms out of their proper place. The chief advantage of Lagrange's view of the subject, when applied to the periodical terms, consists in the clear insight which it gives us into the nature of those equations of long period, such as, for instance, the secular equations, as they were formerly called, of Jupiter and Saturn, and the secular acceleration of the moon, which appear to alter the mean motion, and therefore to affect the axes of their orbits. They, in fact, do so; but such alterations are all periodical, and no way interfere with the general truth of their ultimate and average invariability. It ought to be remarked, however, that in the case of highly eccentric orbits, such as those of comets, which may approach very near the greater bodies of our system, deviations from the mean motion and fluctuations of the periodic time may go to such an extent, and the compensation may be put off so long, that, although theoretically true, the proposition of the permanence of the axis may cease to have any useful or practical meaning. This is remarkably exemplified in the comet of Halley, whose periodic return is affected by inequalities of a great many months, nay, even whole years.

In the actual development of the perturbation of a planet in longitude there is a term introduced, at the very first step, proportional to the time.

This is, in fact, the representative of that part of the planetary action which, like the mean effect of the ablatitious force in Newton's lunar theory, tends to diminish or increase the average intensity of gravitation to the central body, and thereby alter the mean motion and period from what they would be had the disturbing planet no existence. The nature of this term, which appears very obscure as it is disposed of in the "Mécanique Céleste," is placed by Mrs. Somerville in a much clearer light. —(p. 299.)

The developments of the perturbations in longitude, latitude, and distance, though tedious, intricate, and laborious, offer no points of real difficulty, except—first, in respect of the terms proportional to powers of the time introduced by integration, for the treatment of which we are referred to Laplace's memoir, in which this difficulty was first obviated; secondly, in respect of terms which, from the near commensurability of the mean motions, acquire small divisors by integration. These are, of all which occur in the planetary theory, the most troublesome. In the case of Jupiter and Saturn they give rise to the "great equation" of those planets, to which Mrs. Somerville has devoted a masterly chapter, where it is treated with much clearness, and in a very compact and well-digested form. On the whole, we consider the development of the planetary theory, as we have it thus brought before us, to be extremely well performed, and, in fact, a most useful and valuable summary of the subject.

The lunar theory differs in many essential points from the planetary. This is owing to the rapid motion of the apsides and nodes of its orbit, in consequence of which it is impossible to treat it, as we do those of the planets, as an ellipse, subject to small and slow variations: this necessitates a totally different analytical treatment of the problem. That which has been universally followed since its first employment by D'Alembert, consists in expressing, not as for the planets, the longitude, &c., in functions of the time, but *vice versâ*, making the moon's longitude itself the independent variable, and expressing the time and the other co-ordinates in terms of this. The reversion of the first series, and substitution of the result in the others, will then enable us to express all the co-ordinates in functions of the time.

Nothing, however, can be well imagined more formidable than the actual execution of these operations; at the same time that, when the delicacies of the management of the co-efficients depending on the motions of the apsides and nodes are once understood, the whole is little more than a mechanical process, demanding only unwearied patience for its accomplishment. In the treatment, therefore, of this part of the subject, an author, whose object is merely to exhibit a clear view of processes, and a summary of results, is limited to a narrow path, affording little scope for the exercise of any faculty but judgment in deciding where to stop. Mrs. Somerville seems to have considered it her duty here to err on the safe side; so that the equations of her lunar theory are, in fact, little else than a transcript, *mutatis mutandis*, of those of Laplace, and co-extensive with his formulæ. She has, however, had recourse to the gigantic work of Damoiseau for the expression of the longitude in terms of the time, the deduction of which, by the actual reversion of Laplace's series, would have been a work of infinitely too much labour, and which every one but those who make it their especial object to surpass all who have gone before them in this most intricate inquiry, must be content to receive on his authority.

The last division of the work is devoted to the theory of Jupiter's satellites—a curious and elegant system, in which the near approach to commensurability in the mean motions of the three interior satellites gives rise to peculiarities of a very remarkable nature both in the analysis and its results. In this system also the great ellipticity of the central body causes a material deviation in its attraction from the law of gravity, the effect being to introduce a term in the expression of the *perturbative function*, varying inversely as the cube of the distance. As we have before observed, the investigation of this term is not given, and we must, moreover, take this opportunity to notice that, by an inaccuracy of wording, which is repeated wherever the same point is referred to in other parts of the work, this term is always spoken of as expressing "the attraction of the excess of matter at the equator of the central body," whereas, in fact, it expresses no *attractive force* at all, but an artificial quantity, being the significant perturbative term in the development of that useful function in the theory of the attraction of spheroids, which expresses the sum of the molecules of the attracting body, divided each by its respective distance from the point attracted, and which is constantly employed by Laplace in this theory, in preference to the direct expression of the attraction itself, for the convenience and symmetry of analysis. We are the more particular in noticing this point, as the most considerable fault we have to find with the work before us consists in an habitual laxity of language, evidently originating in so complete a familiarity with the *quantities* concerned, as to induce a disregard of the *words* by which they are designated, but which, to any one less intimately conversant with the actual analytical operations than its author, must have infallibly become a source of serious errors, and which, at all events, renders it necessary for the reader to be constantly on his guard. It would not be difficult to support this charge (which is rather a grave one) by citations, but we should be extremely unwilling to leave, at the conclusion of our article, any impression less agreeable than that of the unfeigned delight, and we may add, astonishment, with which the perusal of the work has filled us.

We must not, however, stop without saying something of Mr. Bowditch's performance; though what we do say must be short. The idea of undertaking a translation of the whole "Mécanique Céleste," accompanied throughout with a copious running commentary, is one which savours, at first sight, of the *gigantesque*, and is certainly one which, from what we have hitherto had reason to conceive of the popularity and diffusion of mathematical knowledge on the opposite shores of the Atlantic, we should never have expected to have found originated — or, at least, carried into execution, in that quarter. The first volume only has as yet reached us; and when we consider the great difficulty of printing works of this nature, to say nothing of the heavy and probably unremunerated expense, we are not surprised at the delay of the second. Meanwhile the part actually completed (which contains the first two books of Laplace's work) is, with few and slight exceptions, just what we could have wished to see—an exact and careful translation into very good English — exceedingly well printed, and accompanied with notes appended to each page, which leave no step in the text of moment unsupplied, and hardly any material difficulty either of conception or reasoning unelucidated. To the student of "Celestial Mechanism" such a work must be invaluable, and we sincerely hope that the success of this volume, which seems thrown out to try the feeling of the public, both American and British, will be such as to induce the speedy appearance of the sequel. Should this unfortunately not be the case, we shall deeply lament that the liberal offer of the American Academy of Arts and Sciences, to print the whole at their expense, was not accepted. Be that as it may, it is impossible to regard the appearance of such a work, even in its present incomplete state, as otherwise than highly creditable to American science, and as the harbinger of future achievements in the loftiest fields of intellectual prowess. Here, at least, is an arena on which we may contend with an emulation unembittered by rivalry. — "Whatever," says Delambre, "be the state of political relations, the sciences ought to form, among those who cultivate them, a republic essentially at peace within itself," — a sentiment applicable, doubtless, to all, but pre-eminently so to that calm, dispassionate pursuit of truth which forms the very essence of the abstract sciences.

TERRESTRIAL MAGNETISM.

1. *Allgemeine Theorie des Erdmagnetismus.* GAUSS.
2. *Resultäte aus den Beobachtungen des Magnetischen Vereins.* 1838. Herausgegeben von C. F. GAUSS und W. WEBER. Leipzig. Im Verlage der Weidmannschen Buchhandlung.
3. *Intensitas Vis Magneticæ terrestris ad Mensuram certam revocata.* GAUSS. Göttingen, Sumptibus Dieterichianis. 1833.
4. *Lettre du Baron von Humboldt à son Altesse R. le Duc de Sussex.* Berlin. 1836.
5. *Royal Society Report of the Committee of Physics, including Meteorology, on the objects of Scientific Inquiry in those Sciences.*

(FROM THE QUARTERLY REVIEW, No. 131. JUNE, 1840.)

AMONG the great branches of science which the present generation has either seen to arise as of new creation, or to spring forward by a sudden and general impulse into a fresh and more luxuriant state of development, there is none more eminently practical in its bearings and applications than that of terrestrial Magnetism. It might naturally have been expected that the directness and importance of these applications would have secured to it, at all times, a more than ordinary share of attention, and at all events have preserved it from that state of torpor into which, during the latter years of the eighteenth century, it had begun to lapse; especially since the general subject of magnetism continued from time to time to receive large and valuable accessions both in the line of theory and experiment. But terrestrial magnetism is a science of *observation*, in contradistinction to one of *experiment*, and this character, along with some remarkable peculiarities which it possesses as such, sufficiently explain a neglect that might otherwise appear singular, and even in some degree blameworthy. No single observer, whatever be his zeal and industry—no series of observations, however long continued and exact, made at a single place, can add much to our knowledge of the highly intricate laws and relations which prevail in it. For this purpose the assemblage and comparison of observations, made in every region of the globe and extending over long periods of time, are requisite. In order to master so large a subject multitude must be brought to contend with mass, combination and concert to predominate over extent and diffusion, and systematic registry and reduction to fix and realize the fugitive phenomena of the passing moment, and place them before the eye of reason in that orderly and methodical arrangement which brings spontaneously into notice both their correspondences and their differences.

For similar reasons that the progress of all

sciences which are properly and purely sciences of observation, such as astronomy, meteorology, &c., has necessarily been hitherto more slow, and interrupted by longer intervals of dormancy, than those in which appeal can be made to experiment. An experiment, if it lead to any new view or striking conclusion, may be instantly followed up, while the mind is excited and alert, by others adapted to its verification or extension; while, for corroborative observations or interesting conjunctures, we must wait—a condition especially adapted to blunt the keenness of inquiry and obscure the connexions of thought. An experiment misstated or misinterpreted, may be repeated, rectified, and studied with better attention and success. An observation omitted, leaves a blank which never can be filled;—inaccurately or erroneously stated, it poisons the stream of knowledge at its source, and exercises an influence the more baleful, as it tends, in proportion to its apparent importance, to warp our theories, and thereby prevent, or at least retard, the detection of its faultiness.

Nor does the progress of such sciences suffer less from our ignorance of what is and what is not of primary importance in the natural development of phenomena—of what ought to be diligently recorded, and what may be allowed to pass without notice. Hence it happens that great masses of knowledge are daily perishing before our eyes without the possibility of recovery, because, in fact, our eyes are not open to them, and we have nothing to awaken our attention to their transient display. It is on this account that a theory is of so much more consequence, and forms in fact so much larger a part of our knowledge in these sciences of observation than in those conducted by the way of experiment. In the latter, facts are realities; they stand of themselves, may be reproduced, touched, and handled, and admit us, as it were, by appeal to our senses, into the most direct and intimate knowledge which we can attain of their efficient causes. To such substantial forms, theories sit loosely, as an airy investiture, easily accommodating themselves to the changes of attitude and general growth of the body they adorn and symmetrize; while, to the incoherent particles of historical statement which make up the records of a science of observation, theories are as a framework which binds together what would otherwise have no unity. They give to a collection of fleeting impressions the power of presenting itself to our intellect as an existing whole. In these, then, it is perhaps not using too strong an expression to assert that the theory *is* the science. In it alone we must look for indications that we are on the safe track towards the detection of efficient causes,—from it only we can receive hints to guide us in our choice both of things *to be* observed, and of the best and most available mode of making and recording our observations,—and to it we must look for our only means of reproducing the past, and recovering the lost history of bygone time. It is when they first become capable of performing this office, that theories begin to assume their places as corner-stones in the temple of science, a building always altering, always enlarging, and combining in every age, in its several departments, every form of architecture from the rudest to the most refined which that age admits.

In erecting the pinnacles of this temple, the intellect of man seems quite as incapable of proceeding without a scaffolding or circumstructure foreign to their design, and destined only for temporary duration, as in the rearing of his material edifices. A philosophical theory does not shoot up like the tall and spiry pine in graceful and unencumbered natural growth, but, like a column built by men, ascends amid extraneous apparatus and shapeless masses of materials; nor is that column in its fair and harmonious proportions more different in its aspect when erect and complete from what it was when so surrounded and overborne, than such a theory, presented to us in its simplicity, from the tentative, transient, and empirical conceptions which have helped to its construction.

In the science of physical astronomy, the scaffolding has been long stripped away, and its theory stands august and stately, with that air of nature which marks it as the intellectual shadowing forth of a sublime reality. In that of terrestrial magnetism, a science which is not without its analogous features, we are yet busied in building and pulling down, casting and recasting our design, piecing together our scaffolding, and securing our foundations for a far greater and more massive edifice than was at first contemplated. But already some portions have begun to assume a symmetry, and to convey to the experienced eye glimpses, if not of the plan and dimensions, at least of the general style and character of the future whole; glimpses, however, not obtained by viewing it from the lower ground of its first foundation, but by ascending to a higher level, and surveying it from the "coign of vantage" afforded by the more advanced and rapid progress of its nearest related experimental science, electro-dynamics, or from the commanding heights of physical astronomy, to which, as already remarked, it stands in no remote connexion of analogy. To the former of these it owes its essential character and the direction of its leading lines, since it is there we are to look for the *vera causa*, of the Newtonian philosophy. From the other it has already begun to borrow largely, in point of style and manner, in the adoption of its mode of treating the complicated problems which occur in the estimation of its resultant forces on the most general suppositions as to the distribution of the magnetic power through the substance and over the surface of the globe.

Regarded as a branch of that great assemblage of facts and theories which relate to the physical constitution of this our planet—the forces which bind together its mass, and animate it with activity —the structure of its surface,—its adaptation for life,—and the history of its past changes—

the nature, movements, and infinitely varied affections of the air and ocean, and all which our continental neighbours understand by their term *physique du globe* (a phrase of which our "terrestrial physics" is rather a faint and inexpressive reflection), the science of terrestrial magnetism occupies a large and highly interesting place. Its relations lie amongst those mysterious powers which seem to constitute the chief arcana of inanimate nature, and its phenomena form a singular exception to the character of stability and permanence which prevails in every other department of the general subject. The configuration of our globe — the distribution of temperature in its interior — the tides and currents of the ocean — the general course of winds and the affections of climate — whatever slow changes may be induced in them by those revolutions which geology traces — yet remain for thousands of years appreciably constant. The monsoon, which favours or opposes the progress of the steamer along the Red Sea, is the same which wafted to and fro the ships of Solomon. Eternal snows occupy the same regions, and whiten the same mountains — and springs well forth at the same elevated temperature, from the same sources, now as in the earliest recorded history. But the magnetic state of our globe is one of swift and ceaseless change. A few years suffice to alter materially, and the lapse of half a century or a century to obliterate and completely remodel the form and situation of those lines on its surface, which geometers have supposed to be drawn, in order to give a general and graphical view of the direction and intensity of the magnetic forces at any given epoch.

It is this feature which constitutes, in fact, the great and peculiar difficulty of the subject. Were the magnetic forces at every point of the earth's surface invariable, like the force of gravity, or nearly so, we should long ago have been in possession — and that without extraordinary effort — of complete, or nearly complete, magnetic charts. The report of every seaman and traveller would have added something permanent to our accumulating stock of knowledge, and truth would have emerged, even from inaccurate determinations, by the conflict and mutual destruction of opposite errors. As it is, the case is widely different. The changes are so rapid that it becomes necessary to assume epochs which ought not to be more than ten years apart, to which every observation should be reduced. But to do this, it is requisite to know the rate of change for each locality; information we are so far from possessing that there are great regions of the globe over which we do not even know in what direction the change is taking place.

For want of this information, nothing can be more disheartening than the mass of confusion and apparent error which, under the title of magnetic observations, comes to be *discussed* whenever some laborious and self-devoted inquirer girds himself to the task of comparison and reduction. The instruments with which all the earlier, and many modern, magnetic observations have been made, were of rude construction, or otherwise incapable of yielding much accuracy. The effect of unknown change has thus in innumerable cases become entangled with presumed instrumental error, so as to render it very difficult to decide whether or not to retain, and how, if retained, to employ, the observations so made. Hitherto, however, when it has been possible to apply a correction for lapse of time, the result has been, generally speaking, favourable to the exactness of even very early magnetic determinations, at least on land; so that such early records, like the ancient eclipses in astronomy, become, as time flows on, of great importance and value, which will not fail to be felt hereafter, when theory shall find itself strong enough to leap the interval, and declare the magnetic state of the globe a century or two back. But all earlier observations at sea, or rather all up to a comparatively recent period, are vitiated by another source of error, arising from the iron of the ship, and that in a manner the more hopelessly irrecoverable, because the error so induced is not constant, but varies, not only with every change of geographical situation, but with every alteration in the position with respect to the points of the compass in which the ship is lying at the moment of observation. Fortunately for magnetic science, this vexatious source of error, first detected by Captain Flinders, has been greatly alleviated, and in ordinary cases nearly destroyed, by Mr. Barlow's ingenious adaptation of a *compensating iron*, purposely placed near the compass, so as to counteract, by an equal and opposite attraction, that of all the rest of the iron in the vessel. And even in what might at first appear the desperate case of a vessel *built entirely of iron*, the recent elaborate and admirably conducted inquiries of Mr. Airy have furnished the means of reducing to a mere trifle, or annihilating altogether, the complicated errors arising from two distinct sources of magnetism, the one transient, induced in the soft iron of the vessel by the earth's influence; the other permanent, originating in the rolled and hardened plates, and other masses deviating from the condition of pure soft iron employed in its construction.

In neglect or in spite of these difficulties, the exigencies of navigation have necessitated the construction, from time to time, of charts expressive of the variation of the compass, or the angle by which the needle *declines* from the true meridian at every point of the earth's surface (whence the term *declination*, now used instead of *variation*). The first chart of this sort, based upon the idea of employing for their construction a series of curves drawn through the points of equal declination (in itself a scientific invention of no mean order), is due to Halley. It was constructed by him with infinite labour and research, by the collection of all such observations as that age had furnished.

This chart, and the very remarkable papers by which its communication to the Royal Society was preceded (in 1683 and 1692), to say nothing of his own personal labours and devotion in his memorable voyages of magnetical discovery to St. Helena, must ever form a leading epoch in the science of terrestrial magnetism, and justly entitle him to be regarded as the father and founder of that science, considered as a body of knowledge, bound together by laws and relations. To him we owe the first appretiation of the real complexity of the subject, and the first attempt at a rational *coup-d'œil* of the whole in the announcement of a theory, which, though rude and unabstract in the form of its statement, and rendered thereby liable to obvious and fatal objections *in limine*, has at least the merit of affording a handle for exact reasoning and distinct comparison with facts; joined to that of giving a not unplausible account (the postulates being granted) of several important features of the phenomena. Especially it is designed to account for the existence of not two only, but four points, or rather regions, of apparent convergence of the magnetic needle, two in each hemisphere, and for the changes going on in every part of the globe, in the direction assumed by it with respect to the meridian, both which, the latter as an undeniable physical fact, the former as an unavoidable conclusion from the course of the variation lines in his chart, are broadly declared by him in these papers. It is wonderful indeed, and a striking proof of the penetration and sagacity of this extraordinary man; that, with his means of information, he should have been able to draw such conclusions, and to take so large and comprehensive a view of the subject as he appears to have done. The following passage in his paper of October 19. 1692, will be considered as having especial interest at the present time, when the spirit of inquiry is excited on the subject to a degree never before known, and when an expedition of magnetical exploration and discovery, forming part of by far the most extensive combined scientific operation the world has ever witnessed, has recently left our shores.

"The nice determination of this and several other particulars in the magnetic system is reserved for a remote posterity. All that we can hope to do is to leave behind us observations that may be confided in, and to prove hypotheses which after ages may examine, amend, or refute. Only here I must take notice to recommend to all masters of ships, and all other lovers of natural truths, that they use their utmost diligence to make, or procure to be made, observations of these variations in all parts of the world, as well in the south as the north latitude (after the laudable manner of our East India commanders), *and that they please to communicate them to the Royal Society in order to leave as complete a history as may be to those that are hereafter to compare all together, and to complete and perfect this abstruse theory.*"

We may refer with complacency to such a passage from the pen of our illustrious countryman, himself a seaman*, at the moment that his brother officers of a later age, Ross and Crozier, on their adventurous voyage, and imbued with his own spirit, are engaged in realizing his anticipations, "making observations of these variations in all parts of the world," and "communicating them to the Royal Society,"† and in conjunction with the directors of our magnetic observatories, maintaining and perpetuating our national claim to the furtherance and perfecting of this magnificent department of physical inquiry.

The theory, or rather hypothesis, of Halley, to which reference is made above, and which regards this our globe as a great piece of clockwork, sphere within sphere, by which the poles of an internal magnet are carried round in a cycle of determinate but unknown period, may be regarded, in respect of the secular variations of the magnetic phenomena, in the light of a specimen of that sort of scaffolding to which we have figuratively alluded. With such additional epicycles as the progress of magnetical discovery might necessitate from time to time, it might serve to represent several of the leading phenomena much in the same way as the Ptolemaic orbs served to convey something more than a vague and general idea of the celestial movements. But even as the rude and cumbrous celestial mechanism of Hipparchus and his succes-

* Halley held a captain's commission in the navy.

† Their observations up to the end of 1839 are already received.

sors has tapered into the lofty and florid "mécanique céleste" of modern times, so the pursuit of those slow and intricate changes in the magnetic elements of each particular terrestrial locality, which presented themselves to Halley under the aspect of mechanical revolutions, begins to assume, in the eyes of modern theorists, under the influence of more general views as to the origin and distribution of the magnetic forces, the semblance of those ever varying and never overstepping, those inherently equipoised and self-bridled oscillations which, so far as we can see, afford the best expression of the planetary movements.

The variation chart of Halley had been hardly forty years completed when, by the effect of these secular changes, it had already become obsolete, and, to satisfy the wants of navigation, it became necessary to reconstruct it. This was performed by Messrs. Mountain and Dodson about the middle of the last century, and their labours are highly deserving of notice by reason of their having attempted to execute this task systematically for several equidistant epochs, viz., for 1711, 1722, 1733, 1744, by the aid of observations drawn from official and other records which were furnished them in great abundance by the Commissioners of the Navy, and the East India, African, and Hudson's Bay Companies. Thus they expected to be enabled, by comparing the charts so obtained, to form a *predicted* chart for 1755; a bold and praiseworthy attempt, which, however, was baffled by the discordances offered by the observations before them, discordances owing doubtless to the causes above enumerated. They appear therefore to have given up this course in despair, and to have formed their final chart for 1756 in a way little calculated to inspire confidence, viz., by mixing together observations of different dates, and by the exercise of a pretty arbitrary discretion in accepting some and rejecting others.

In this unsatisfactory state, the subject of the magnetic variation appears to have remained until 1811, when, on the occasion of a prize proposed by the Royal Danish Academy, M. Hansteen, whose attention had for many years been turned to the magnetic phenomena, undertook its re-examination, with a view to determine how far it might be possible to reconcile the observations accumulated up to that time with the supposition of two magnetic poles revolving round the pole of the world in indefinite periods, an opinion which had been defended by Euler, Churchman, and others— or whether, as Halley had asserted, four such poles were necessary—or, lastly, whether any such suppositions as to the revolutions of polar points be competent at all to represent the phenomena. His work, "Ueber den Magnetismus der Erde," published in 1819, is in every way most remarkable. With indefatigable labour he has traced back the history of the subject, and filled up the interval from Halley's time, and even from an earlier epoch (1600), with charts constructed for that epoch, and a great many intermediate ones, up to 1800, so as to present before us in one view, as far as it can now be done, the succession of states or phases through which this element has been passing during the last two centuries. The result, apart from all theoretical considerations and ideas of poles, axes, &c., is most curious and instructive. The whole system of variation lines, with their intricate convolutions, loops, ovals, intersections and asymptotic branches, are seen to be sweeping westward—not, however, as it were bodily, but each in its progress undergoing most singular modifications of form and flexure, and gliding by gradations, which it now becomes possible to trace, but which without such restorations would baffle every attempt of the imagination, through all varieties of conjugate oval, cusp, and node in which the geometry of curves luxuriates. It would be interesting, but far beyond our limits, to show how beautifully this sort of moving magnetic panorama explains, or rather how easily it enables us to conceive, the puzzling facts presented by the history of the variation at particular spots:—by what a felicity of accident, for example, the whole mass of West India property has been saved from the bottomless pit of endless litigation by the invariability of the magnetic declination in Jamaica and the surrounding archipelago during the whole of the last century, all surveys of property there having been conducted solely by the compass (Robertson, *Phil.*

Trans. 1806)—by what a curious *absorption* of a conjugate oval and transition to another system it has happened that the needle has passed, within the period of recorded observation in London and Paris, from 11° east of the true meridian to 24° west, having attained the former direction by a gradual movement eastward—there remaining awhile stationary—thence receding with a westward movement to the direction last indicated, where it again became stationary about 1806 or 1807, and is now again on the move towards the east, by which, curious changes taking place immediately under their eyes, the secular variation of the magnetic elements has been forced on the attention of the philosophical world. We might specify a multitude of interesting cases of the same nature.

M. Hansteen declares himself in favour of four poles and no more, thus adopting so far the Halleian hypothesis. But he is obliged to complicate it with additional cycles, by declaring each pole to have a separate and independent movement and period—a modification which goes a great way towards divesting them of any attribute of physical reality. But, on the other hand, Mr. Barlow, who, so recently as 1833, has published a variation chart, perhaps the most elaborate which has yet been produced, declares quite as strongly against them. "I can see (says he, speaking of the variation lines in the Pacific ocean) no possible position of four poles which can lead to such a configuration." And, again, in discussing their course in the Indian seas, he considers it "equally inconsistent with the notion that all these phenomena are due to the action of four *or more* magnetic poles." For this hypothesis he accordingly substitutes one more general, "That there is no determinate pole to which all needles point, but that each place has its own particular pole and polar revolution, governed probably by some one general but unknown cause." On this we have only to remark that it amounts to giving up altogether the hypothesis of "poles," and "magnetic axes," since there is no conceivable law of change in the magnetic lines to which a proposition so general will not apply. It declares, in effect, that the true law of these changes is still to seek; a position in which we fully agree. It is clear that the possibility or impossibility of representing the magnetic action of the globe on *every* point of its surface by that of two or more *fixed* points within it, must depend on the geometrical resultant of the sum of its molecular attractions and repulsions, passing or not passing through an invariable attractive and another invariable repulsive point, or being equivalent to several others so passing, a condition, in the abstract, generally incapable of fulfilment, and in the highest degree improbable in any particular case. In effect, we may conceive the magnetic force of the earth, on a boreal molecule at its surface, as being the difference of two forces, whereof the austral, or attractive, is the total attraction of a solid of unknown form and density, but approaching to a sphere whose particles attract with a force identical in law with gravity ; and the boreal, or repulsive, is the total repulsion of a solid exactly similar and equal, whose molecules repel with equal forces, but of which each particle is removed from the corresponding particle of the attractive solid by an infinitesimal quantity, according to an unknown law of displacement. From this view of the matter (which strikes us as new and as offering some advantages), it follows, without any calculation, that the total magnetic action of the earth on a needle at a *given* place is equivalent to that of *one* infinitely small magnet of infinite power placed at a point not very remote from the centre. But it by no means follows, except in the single case of an *equal* and *parallel* separation of the opposite magnetisms in each molecule of an homogeneous sphere, but quite the reverse, that one and the same such magnet, or any finite combination of such, should possess this property for every point in the surface. We cannot help concluding, therefore, that it is lost labour to make further attempts to reconcile the phenomena with any hypotheses of this nature.

In considering the distribution of the earth's magnetic action over its surface, the variation-lines have hitherto received by far the greater, and, theoretically speaking, an undue share of attention, by reason of their nautical importance. They have the disadvantage (as a graphical representation of phenomena) of offering nothing distinct to the imagination except their own unaccountable flexures — and rather tend to complicate than to aid conception of the play of forces in which they originate. It has been proposed to substitute for them a system of lines perpendicular at every point to the direction of the needle. This would be a great improvement were it practicable to construct such lines from direct observation, which it unfortunately is not, by reason of a difficulty purely mathematical — our inability to integrate differential equations, whose variable co-efficients are only given by observation.

It is otherwise with what are called the isoclinal and isodynamic lines. Their course, graphically projected, speaks not only to the eye, but immediately to the mind. It is only, however, within a comparatively short period that charts of their course have been constructed. The work of Mr. Hansteen exhibits the specimens of such charts, or fragments of them, for 1600, 1700, and 1780, which, so far as they can be depended on (and he considers them entitled to considerable confidence), confirm the general westward tendency of the magnetic system, though in a manner less striking than in the case of the variation or *isogonal* lines, by reason of their gentler flexures and more general parallelism to the equator of the globe.

The direction taken by the magnetic needle is determined by the two elements, its horizontal position, or declination from the meridian, and the dip or inclination. Complete charts of the dip and declination, therefore, did such exist, would afford complete knowledge of this direction over the globe. But another important element remains, viz., the intensity of the total magnetic force, or of the power with which, when withdrawn from its position of equilibrium, it tends to revert to it. The discovery that this power is not equal in all parts of the globe, as a matter of observed fact (for theoretically it may be said to have been always understood), is of comparatively recent date. Major Sabine, to whom we are indebted for a report on this subject, "Seventh Report of the British Association," remarks, that this important fact "remained, at the commencement of this century, unattested by a single published observation," while such has been the diligence with which they have been since accumulated, that the charts with which that report is accompanied, representing the course of the isodynamic lines (lines of equal intensity) over both hemispheres, rest on no less than 753 distinct determinations at 670 stations, collected, arranged, and discussed, with a care, precision, and luminous order which it is difficult to estimate too highly. We consider this report, indeed, as one of the most finished things of the kind that has ever been produced, and as having accomplished in the completest manner the objects proposed by that association in calling for such reports, by so comprehending in one view the results of our knowledge and the amount of our ignorance, as to afford the greatest possible stimulus to further inquiry. It is, indeed, impossible to inspect these charts without perceiving that a new branch of magnetic science has been created, and here for the first time embodied. The observations on which they are grounded are, for the most part, those of Humboldt in his travels and voyages in Equinoctial America — of Hansteen and Due in their magnetic journey through Siberia, in which they traversed the whole north of Europe and Asia, and carried their researches to the polar circle; and of Erman, who, with the same object, encircled the globe by a mixed land and sea voyage, setting out from Petersburg, embarking in Kamtschatka, and returning by the Cape. Major Sabine's personal contributions to the same stock, also, are both numerous and important, the scenes of his labours having the unique interest of having been chosen in the most inaccessible, the most desolate, and the most unhealthy regions upon earth — such as Spitzbergen, Melville Island, St. Thomas's, &c. The general result is, that the isodynamic lines appear to be arranged on the globe in forms which strongly remind us of the lemniscate curves exhibited by crystals exposed to polarized light, when referred to a sphere traversed in all directions through its centre by the polarized rays — somewhat wanting in symmetry, it is true, but, especially as regards the two northern systems of isodynamic ovals, very definitely marked out; while in the south, unequivocal traces, shadowing out the existence of two similar ovals, point to a distribution of magnetism in that hemisphere analogous to what obtains in the northern. Observations are yet wanting to determine whether this system of lines be in a similar state of secular progress westward over the globe, with those of the dip and variation (though that such is the case can hardly be doubted), and whether and what changes of form and mutual relation they undergo in its course.

The direction taken by a needle freely suspended, and the force by which it tends to settle in that position being known on every accessible point of the earth's surface to a certain degree of approximation, the next step, in the inductive process of discovery, is to embody this knowledge in a law mathematically stated, and either derived from some rational theory of magnetic action, or at least shown to be not inconsistent with such a theory. In the remarkable work which we have selected as part of the subject-matter of these pages, "Allgemeine Theorie," &c.,* M. Gauss has succeeded in obtaining such a formula by a mixed process of theoretical investigation in general, and empirical adaptation in particular, which represents, in a most striking and unexpected manner indeed, the whole mass of these complicated phenomena, so far as they have been yet developed. Setting out with the most general suppositions as to the distribution of magnetism over the surface and

* This work will be found extremely well translated in "Taylor's Scientific Memoirs."

through the substance of the earth, and assuming only that the magnetic force follows the same law of decrease with that of gravity, he applies the Laplacian method of representing the attraction of a spherical or spheroidal solid to the expression of the resultant magnetic force considered as resolved into three components, one perpendicular to the horizon at any point producing the dip, the other two in the horizontal plane. The whole investigation, after the examples of Laplace and Poisson, is made to turn upon the properties and development of that peculiar function which represents the sum of the active molecules, whether attractive or repulsive, each divided by its distance from the point attracted or repelled,—a function which much wants a name *, and for which we would venture to propose that of the "*integral proximity*" of the attracting mass. The differential co-efficients of this function express the resolved components of the total magnetic action; and the art of the analyst is shown in the elegant and masterly manner in which he succeeds in obtaining laws and relations susceptible of practical verification, without compromising the generality of this auxiliary function, and involving himself in the difficulties

* It has been since termed the *potential* function, a term extremely objectionable, as conveying an idea quite foreign to the general signification of the function, which is a purely geometrical one, or at least one to which the conception of mechanical force or power is altogether extraneous. (*Author*, 1856.)

which would attend its expression in terms of any presumed law of distribution of magnetic power, such as, for instance, its concentration in poles, axes, &c. Some of these relations are propositions of considerable interest — as, for example, M. Gauss demonstrates that whatever be the law of magnetic distribution — if there be any series of stations forming a polygon of inconsiderable dimensions compared with the area of the globe, the dip, horizontal direction, and intensity at each of these stations, must satisfy a certain very simple equation of condition, by which, if all but one of them be given, that one may be calculated — and taking the case of a triangle formed by Paris, Göttingen, and Milan, he finds the condition to be, in fact, exactly satisfied by the actual elements furnished by observation for those stations. Another of these propositions may be instanced as still more general and remarkable, viz., that the knowledge of the value of that particular component of the horizontal magnetic force only which acts in the direction of the meridian, supposing that knowledge complete, and to extend to every point of the earth's surface, would enable us to assign the nature of the function expressing the *integral proximity*, and thence to deduce every other particular of terrestrial magnetism.

The development of this function, and thence of the three magnetic components depending on that function in terms of the latitude and longitude of the point acted on, without any compromise of its generality, is performed by the aid of those co-efficients introduced by Laplace in the analysis of the attraction of spheroids and the figure of the earth, which have been found to facilitate in so high a degree these difficult investigations. The *form* of these developments as functions of the sines and cosines of arcs, arranged in successive orders by their powers and products, is thus generally assigned, but the special values of the co-efficients remain to be discovered; and this can only be done in two ways, viz., *à priori*, by a knowledge of the actual law of the distribution of magnetism in the earth, and the performance of the requisite integrations, or *à posteriori*, by comparing the developments of each component force with actual observation, and thus, by the usual aids which analysis, assisted by the theory of probabilities, supplies in such cases, eliciting the numerical values of those co-efficients which suit the observations best. This method is familiar to geometers by the extensive application which has been made of it in the lunar theory, in which the forms of the equations, or, as they are termed, their arguments, being assigned by theory, the comparison of their series (with unknown co-efficients) with an extensive series of observations, has been resorted to as a means of determining the values of those co-efficients, otherwise too complicated to be directly investigated. Such is the process followed in this case by M. Gauss, assisted, however, and stripped of the worst part of its otherwise almost insuperable labour and difficulty, by a choice of data in the highest degree ingenious and artificial—which is renderéd possible by the possession of the charts above alluded to—and to which, as a fine example of the kind of power placed in the hands of geometers by the method of graphical representation in general, we are desirous to draw especial attention. It consists in comparing the series expressing the elements in question, not with their values as actually assigned by observation at real stations, but with values *graphically interpolated by the aid of the charts*, to correspond to a set of imaginary stations so distributed over the globe as to afford the greatest possible facility to the calculations, and to break up the mass of unknown quantities, which, in the general case, would be hopelessly entangled one with another, into groups of easy management. Thus, in the case before us, M. Gauss distributes his stations over seven parallels of latitude, so as to divide each parallel into twelve equal parts. It has been usual to consider such charts and graphical representations as mere helps to the imagination, or as rough registers, giving by inspection approximate values for ready practical use; but this we consider to be quite an underestimate of their importance. We regard such projections, when carefully executed, not only in this, but in every other science in a similar stage of progress, as necessary instruments and adjuncts to the highest applications of theory—as the only means we possess, or ever can possess, of purifying great masses of observational data from the effects of local influence and personal or casual error. They furnish, in short, and will, henceforward, as this their important office becomes better understood, every day more and more furnish that intermediate step between observation and theory which has long been wanting to the perfection of both. They enable the theorist in particular, to choose his ground above all individual place and circumstance, and to select his data, not where casualty or convenience shall have led the observer to collect them, but in pure accordance with the requirements of his geometry, and the simplification of his calculus. In consonance with this view of the subject, we anticipate the time when no computist will ever take the trouble to compare formulæ with single observations in their crude state, *for the purpose of determining elements*, such comparison being reserved for finally testing the validity of theories.

The charts used by M. Gauss for this purpose were, that of the dip published by Horner (*Physicalisches Wörterbuch*, b. 6.), and those of the variation and intensity, by Barlow and Sabine, already mentioned. We may be proud as Englishmen to have furnished two out of the three digested masses of data for this vast undertaking, especially as it is to the appearance of the last of these charts that M. Gauss expressly ascribes his having been induced to enter upon the formidable calculations it involves.

The success of this remarkable attempt we consider as signally encouraging. M. Gauss has himself compared his resulting formula with actual observation, at ninety-one of the best stations in every variety of latitude and longitude, and in all the particulars of dip, variation, and intensity. In one instance only does the error in the dip exceed 4°; in only two does that of the variation amount to 5°; while the intensity is represented throughout within an extremely minute fraction of the whole, with exception of two stations, Port Famine and Santa Cruz, where there is no doubt some error of observation.

This comparison becomes more interesting, and assumes almost the character of ocular evidence, when, as is done in the report made by the Committee of the Royal Society now before us, charts constructed from the formulæ alone are placed side by side with those derived from observation. This comparison with his own variation chart, constructed from observations made between 1827 and 1830, has been made by M. Erman, and accompanies a most interesting letter from him, appended to the report in question, and a similar comparison with Major Sabine's chart of the total intensity is also annexed; and the resemblance in both cases between the type and the antitype is so close as to justify a conviction of our having at length made a real approach to a geometrical expression of the phenomena. In particular, the singular courses of the variation lines in the Pacific and

Indian Seas, noticed by Mr. Barlow as so characteristic and unaccountable, are made perfectly intelligible as parts of a connected system which would be incomplete without them. The northern magnetic pole too, or point of perpendicular dip given by M. Gauss's formula, coincides, within little more than 200 miles, with its place actually observed, or at least closely approached by Ross in 1832, while the European, African, and Atlantic lines exhibit a correspondence approaching to identity. Some small, but not unimportant, systematic deviations have been pointed out by M. Erman, which a resumption of the calculations with more dependable data will, no doubt, cause to disappear.

A feature we cannot help noticing in this work of M. Gauss, is the uniform predominance of the philosopher over the mere geometer. From his well-known eminence in the latter line, we might have expected undue prominence to be given to methods and artifices, and have looked for displays of formulæ ostentatiously spreading into luxuriance; but, on the contrary, the analysis is everywhere kept subordinate to the physical inquiry, and, though handled throughout with the skill and power of a consummate master, is nowhere suffered to appear as a primary object.

One incidental result of these investigations will appear very striking — astonishing indeed to those whom habit has not familiarized with the enormous numbers which occur when the operations of nature are measured by man's diminutive units.

It is the estimate of the total magnetic power or "moment of magnetism" of the earth, as compared with that of a saturated steel bar one pound in weight. This proportion M. Gauss calculates to be as 8464,000,000,000,000,000,000 to 1; which, supposing the magnetic force uniformly distributed, will be found to amount to about six such bars to every cubic yard.

Besides the secular changes in the magnetic forces which gradually carry the needle far from a fixed direction, according to laws at present unknown, but which at all events act with steadiness and regularity, observation has recognized two subordinate systems of fluctuation to which it is subject, the one, periodical; the other, so far as we can see at present, quite capricious and irregular — in consequence of which the name of magnetic *perturbations* has been assigned to them, as if the needle were disturbed by some external influence of a transitory nature.

The periodical oscillations of the magnetic needle were first observed by Graham in 1722, and have since been studied with much diligence and perseverance by several assiduous and careful observers, among whom our countryman Mr. Gilpin deserves especially to be noticed as having made these observations his constant occupation during the whole period from 1787 to 1806, and having for upwards of sixteen months kept an hourly register extending to twelve hours of every twenty-four, a process by which alone the true laws of such oscillations can be deduced. By these and similar observations by Canton, Wargentin, and Cassini, the existence of periodical movements, both diurnal and annual, has been established. The deficiency of nightly observations has since been supplied by Baron Von Humboldt, who, by investigating the particulars of the nocturnal progress of the oscillation, has completed the outline which Gilpin and others had begun, and enabled us to state with some degree of precision the nature and extent of these periodical changes. The horizontally suspended needle is found to make, each twenty-four hours, two eastward and two westward deviations from its mean position, those which occur in the day time being greater than those taking place in the night. It is curious to remark that this irregularity seems to extend to all similar cases of diurnal fluctuation. In that of the barometer, it is a marked and striking feature; and in the case of the tides, a phenomenon holding a strong analogy to this, called the diurnal inequality, constitutes one of their most singular, and at present mysterious, characters. It is also observed that the extent of excursion differs in summer and winter, as does also the difference between the daily and nightly oscillations. Finally, when the mean places of the needle for each day of a whole year are cleared of the regularly progressive effect of the secular movement, a fluctuation having an annual period is disclosed. Similar periodic changes have of late been traced in the position of the dipping-needle, and there can be no doubt that the total intensity is also subject to periodical increase and diminution.

The periodical oscillations of the needle, then, form a regular and compact system, of which there can be little doubt that the cause is to be sought in superficial changes of temperature, developing electric currents either in the crust of the earth or in the atmosphere. Be this as it may, their general nature and laws may be considered as tolerably well sketched out, though they still require much study in detail. It is otherwise with those irregular and sometimes almost convulsive movements of the needle which constitute the magnetic perturbations, which have of late, and deservedly, attracted great attention by reason of some very extraordinary facts brought to light by their comparison at different and remote stations.

The illustrious Humboldt, to whom every department of science owes so much, and to whom the rare glory belongs of being the first to push onward in so many different lines, gave the forward impulse in this. During the course of those his most memorable voyages and travels in the equinoctial regions of America, in which, all eye, all ear, all thought, he seemed to have received on the expansive *retina* of his mind the picture of universal nature, and to have treasured up its images in the stores of a memory and an intellect worthy of such a prospect, the observation of the magnetic phenomena in all their particulars occupied a large portion of his attention. On his return to Europe, as he informs us in his letter to the Duke of Sussex, he conceived the project of examining the hourly changes of the variation and the perturbations with which the progress of those changes appeared to be effected, on a scale and in a mode not before attempted, and with instruments of superior accuracy. Established in a large garden at Berlin, he observed at the solstices and equinoxes of 1806 and 1807 the changes in the direction of the horizontal needle every half hour, during four, five, or six days, and the intervening nights. The immediate object of this undertaking was the establishment of the nocturnal portion of the daily oscillation already mentioned. But the delicacy of his instrumental means allowing him to appreciate the smallest changes, his attention was excited by the singular and apparently capricious march of the instrument, which appeared agitated by frequent and occasionally sudden and rapid movements, attributable to no accidental or mechanical cause. To these, regarding them as indications of a reaction propagated from the interior of the globe to its surface, he gave the name of *magnetic storms*, in analogy to the sudden changes of electric tension which take place in thunder-storms. In consequence of this discovery M. Von Humboldt conceived the project of procuring magnetic observations to be established to the east and west of Berlin, with a view of tracing the limits and correspondence (if any) of these perturbations. Political events, however, frustrated this project; nor did the subject receive further elucidation till the year 1818, when it was ascertained by a comparison of simultaneous hourly observations by M. Arago at Paris, and M. Kupffer at Kasan, that on making a proper allowance for the difference of longitudes of the stations (no less than 47 degrees), the observed perturbations *were in fact synchronous.* In other words, we are here presented with the surprising phenomenon of an unceasing series of natural signals or pulsations, which, whether propagated from regions deep within the globe, according to Humboldt's first idea, or transmitted down to us from without, as the later discoveries in electrical science seem to indicate, arrive at points of the surface separated from each other by an interval equal at least to the whole breadth of Europe at the same precise moments of time.

A discovery of this magnitude might have been expected to be instantly followed up, yet several years elapsed before any further step was made in this direction; nor was it until 1828–30 that the subject was resumed on a scale of such extent as to secure its successful prosecution. It is again to the indefatigable zeal and great personal influence of Von Humboldt that magnetic science is indebted for this fresh impulse. Taking advantage of his eminent position as a man of science, his free intercourse with persons of rank, power, and official station, and his immense correspondence, and availing himself especially of the opportunity afforded him by his mineralogical visit to Siberia in 1829, he succeeded in procuring the establishment of magnetic observatories not only at Petersburg and at Kasan, but also at Moscow, at Barnaoul, at Nertschinsk, and even at Pekin itself, where the Russian government has constantly supported, by celestial permission, a Greek monastery. These establishments have ever since subsisted, and, as we shall presently see, form important elements in the great system of simultaneous magnetic observation now in progress. At Nicolajeff also, in the mines of Freyberg in Saxony, at Sitka in Russian America, and even in Iceland, the establishment of magnetic stations was solicited and obtained.

The first fruits of this extensive combination appeared in 1830, in the form of a comparison of the hourly observations received from Nicolajeff, Petersburg, Kasan, Freyberg, and Berlin; and by these the synchronism of the magnetic perturbations at these distant localities was placed in full and striking evidence. A confirmation so remarkable of the observations of Arago and Kupffer excited general attention, and led to fresh researches, conducted on a system of maturer concert, and with instruments of far greater precision than had previously been regarded as attainable. As these researches not only embrace the perturbations, but cover the whole ground of magnetic observation, it is necessary to be somewhat particular in our account of them.

It is to M. Gauss that we owe both the new instrumental means employed, the method of reducing their indications to a definite standard, and the establishment of a concerted system of simultaneous observation (having Göttingen for a centre of reference) performed at stated *terms* by observers provided with similar instruments, and dispersed over Europe. The results of the observations made by this "Magnetic Association" at fourteen such terms, and at sixteen stations, extending in latitude from Upsal in Sweden to Catania in Sicily, and in longitude from Petersburg to Dublin, during the years 1836, 1837, and 1838, have been arranged, graphically projected and published by M. Gauss and his indefatigable coadjutor M. Weber, with a full description of the instruments or magnetometers employed, and a complete detail of every particular of their use, in a work entitled, "Resultate aus den Beobachtungen des Magnetischen Vereins." In this system of observation, the perturbations of the horizontal needle (if a bar of steel weighing from four to twenty-five pounds can be called by so familiar a diminutive) are observed both in respect of direction and intensity, not merely at hourly intervals, but at every fifth minute, it having been found that, in proportion as the intervals are narrowed, the coincidence of the projected curves becomes more striking, owing to the great number of momentary fluctuations which escape notice in the longer intervals. Of such coincidence, every sheet of the projections in the work referred to, offers one continuous example. Indeed, so numerous in this improved procedure are the opportunities afforded for fixing on sudden and remarkable movements of the bars, that there would seem to be no difficulty in determining from them, as from any other simultaneously observed signals, the difference of longitudes of the stations. Other distinguishing features of this method are — 1st, the employment of none but telescopic means of measuring the excursions of the bars, the observer never approaching them with his person; 2dly, the maintenance of the bars in a state of continual vibration, owing to their suspension on silk threads, the limits of their excursions and the instants of their attaining those limits being the sole objects of observation; 3rdly, the superaddition of a very ingeniously devised *statical* method of ascertaining the horizontal intensity to the usual dynamical method of observing the time of a given number of vibrations made by the suspended bar. The principle consists in determining the amount of torsion of two parallel fibres, separated by a given interval, used to suspend the bar, which shall suffice to retain it at right angles to the magnetic meridian. The momentary changes of intensity are measured, not by continual fresh adjustments of the torsion, but by noting the limits of excursion of the bar in its vibrations on either side of its original situation.

The instrument destined for this purpose is called by M. Gauss the *bifilar* magnetometer.*

The last and not the least important distinguishing feature in M. Gauss's system of observation is the adoption of a process by which the intensities, concluded from either the statical or dynamical measurement, are freed from the perplexing source of error occasioned by loss of magnetism in the bars employed, and referred to a standard unit verifiable under all circumstances. His work, entitled "Intensitas vis magneticæ," &c., is devoted

* The essential principles of this method — viz.: 1st. The employment of a suspended needle forcibly distorted to a right angle with the meridian; and, 2ndly, the measurement of changes in the directive forces by the fluctuations in its newly-assumed position of equilibrium under such distortion—are of much earlier date, having both been employed by Mr. Christie in his elaborate Memoirs published in the *Phil. Trans.* for 1823, and 1825; papers which have attracted far less attention than their great merit deserves, and which mark a decided epoch in the history of modern refinements in magnetic observation. Mr. Christie used magnets to deflect his needle; but the application of the torsion balance, as a means of measuring the directive force, is expressly suggested by him.—*Ph. Tr.*, 1825, p. 23.—M. Gauss has also been preceded in his ingenious idea of the application of a reflector to his suspended magnet which plays so important a part in his apparatus—at least we know that the idea occurred many years before to Mr. Babbage, though whether applied by him to practice, or even published otherwise than verbally, we are unable to state. (*Note*, 1840.)

Dr. Wollaston, however, was the first to introduce into instrumental practice (in his admirable goniometer) the direction of a reflected ray of light, as the indication of the angular position of a surface too delicate for handling. Mr. Babbage's idea, if our memory serves us, was avowedly an extension of this principle. (*Note added*, 1856.)

to this object, but as the principle of the method—though embraced in formulæ and exemplified in numbers, in that work — is yet nowhere very clearly stated in words, it may not be amiss to explain it. It consists, first, in vibrating a magnet horizontally suspended in the usual manner. By this operation the *product* of the earth's directive force by the magnetic virtue of the bar is obtained. The same bar, in a position at right angles to the magnetic meridian, is then successively presented at given measured distances from the centre of another suspended bar or compass-needle, which it thereby deflects from its position of rest, according to known laws. The angular amount of this deflexion at each distance being observed, gives the *ratio* of the two forces in question, and their product and ratio being thus both known, the forces themselves are determined.

One element, however, is left unprovided for in these arrangements of M. Gauss, viz., the measurement of the vertical component of the magnetic force and its momentary changes, without knowing which, it is impossible to conclude anything as to the real nature, amount, and direction of the perturbative forces. The absolute dip, indeed, may be obtained with much precision, by means well known, but the mode of suspension in ordinary dipping-needles is quite inadequate, in point of freedom and delicacy, to place in evidence, far less to measure the momentary changes of this element. This important desideratum, the only thing wanting to complete our means of observation, has been recently supplied by Professor Lloyd, by the construction of an elegant apparatus termed by him a "Vertical force magnetometer." It is a species of magnetic balance, in which a needle, or magnetized bar, placed in the magnetic meridian, is coerced by the action of small weights, moved by screws, from its natural direction to a horizontal one. This condition renders it possible to rest it, *by knife-edges invariably fixed in and forming a part of it*, on planes of agate, and thus to secure for it in all geographical situations the same delicacy, sensibility, and freedom of motion which belongs to the ordinary weighing balance. Thus coerced, adjusted, and counterpoised, whatever movements take place in it are referable directly to changes in the amount of the vertical magnetic force which opposes, and in its mean situation neutralizes the action of the weights, and, being read off by microscopes and subjected to calculation, afford a measure of the amount of those changes. Mr. Lloyd, we understand, considers that, by the aid of this instrument, a change to the extent of $\frac{1}{40000}$ of the total magnetic intensity may be detected.

To Professor Lloyd we also stand indebted for the geometrical determination of the conditions of situation under which the instruments or magnetometers, destined for observing the three essential elements, can co-exist in one apartment of moderate dimensions without disturbing each other's indications, a consideration of the last importance to the further extension of this system of observation, as diminishing in a most material degree, the cost of erecting a magnetical observatory, and the amount of personal assistance necessary for carrying on the observations. The simple and convenient practical conclusions to which his analysis has led him on this point are given in a paper recently communicated by him to the Royal Irish Academy, to which we must refer.

The effect of these improvements has been to give to magnetic determinations, at least on terra firma, the precision of astronomical observation, while, at sea, the limits of obtainable accuracy, in any moderate weather, have been greatly enlarged by the use of an apparatus recently invented by Mr. R. Were Fox, which also serves to measure the intensity. Armed with such instruments, and in possession of a theory which has proved competent to represent with fidelity all the principal, and many subordinate features of the phenomena, even in the present imperfect state of the data (which, in fact, it reproduces nearly as well as the observations from which they were obtained would probably do could they be repeated), it is clearly impossible longer to rest content with loose or inaccurate determinations, or to sit down in patient expectation that casual visits of travellers or voyagers shall fill in the great *lacunæ* which still subsist in our charts. Voyages and travels especially destined to this object must be undertaken—particular districts traversed and retraversed—stations not only visited but resided in. In a word, the time is evidently arrived for a powerful and united effort, on the part not of individuals but of nations, to place on record the actual state of those data, on a scale, and with an exactness worthy of the subject, and so to render the present epoch a secure point of departure for future ages. Such an effort is now in course of being made, and it will be our object in the remainder of this article to explain the immediate circumstances which have led to it—the nature, aim, and extent of the operations themselves—the leading part which our own country has taken in them, and the general views which ought to guide, and which we conceive to have guided, its promoters in recommending and urging its adoption on their respective governments as a matter of national concern.

The extension of the system of simultaneous observation, ever a favourite object of its original projector, Von Humboldt, was made by him, in April, 1836, the subject of a distinct appeal to the Royal Society, in his Letter to his Royal Highness the Duke of Sussex (then president of that venerable body). In this letter, which contains a brief but lively statement of the history of the magnetic perturbations (from which we have borrowed freely in what precedes); of the progress made and making in the magnetic survey of the globe, and of the chief desiderata of the science as it then stood, he urges the establishment of regular magnetic stations in the British possessions in Canada, Australia, the Cape, and between the tropics, not only for the observation of the momentary perturbations of the needle, but also for that of its periodical and secular movements. Assuredly no nation was ever so favourably situated for such a purpose, nor so strongly called on as a maritime and commercial country for co-operation in a cause directly connected with nautical objects. Nor did this appeal fall on deaf ears. The subject was readily taken up by the Royal Society, and an application to government for a grant of money for the purchase of instruments as readily listened to. The organization, however, of a plan of operations adequate to the ends proposed proved no light or easy matter; nor were the funds thus placed at their disposal by any means sufficient to carry out a large and well arranged scheme. Delays, in consequence, intervened, most fortunate in their event, as giving time for the mature consideration of the subject, and the just appretiation of its magnitude and practical difficulties. While thus in abeyance, a movement from another quarter gave a decisive turn to the whole project, by striking at once an outline so full and sweeping as to meet all the exigencies of the case.

This outline is contained in a series of resolutions adopted by the British Association for the Advancement of Science at their meeting at Newcastle in 1838; and exhibiting, as these resolutions do, a clear view of the general nature and objects of the operations contemplated and now in progress, we cannot do better than extract them from the most authentic reports of that meeting which have hitherto appeared: —

"Resolved 1. That the British Association views with high interest the system of simultaneous magnetic observations which have been for some time carrying on in Germany and various parts of Europe, and the important results to which they have already led, and that they consider it highly desirable that similar series of observations regularly continued in correspondence with, and in extension of these should be instituted in various parts of the British dominions.

"2. That this Association considers the following localities as particularly important: — Canada, Ceylon, St. Helena, Van Diemen's Land, and Mauritius, or the Cape of Good Hope; and that they are willing to supply instruments for their use.

"3. That in these series of observations the three elements of horizontal direction, dip, and intensity, or their theoretical equivalents, be insisted on, as also their hourly changes, and on appointed days their momentary fluctuations.

"4. That the Association considers it highly important that the deficiency yet existing in our knowledge of terrestrial magnetism in the southern hemisphere should be supplied by observations of the magnetic direction and intensity, especially in the high southern latitudes between the meridians of New Holland and Cape Horn; and they desire strongly to recommend to her Majesty's government the appointment of a naval expedition expressly directed to that object.

"5. That in the event of such expedition being undertaken, it would be desirable that the officers charged with its conduct should prosecute both branches of the observation alluded to in Resolution 3, so far as circumstances will permit.

"6. That it would be most desirable that the observations so performed, both at the fixed stations and in the course of the expedition, should be communicated to Professor Lloyd.

"7. That Sir J. Herschel, Mr. Whewell, Mr. Peacock, and Professor Lloyd be appointed a committee to represent to government these recommendations.

"8. That the same gentlemen be empowered to act as a committee, with power to add to their number, for the purpose of drawing up plans of scientific co-operation, &c., relating to the subject, and reporting to the Association.

"9. That the sum of 400*l.* be placed at the disposal of the above-named committee for the above-mentioned purposes."

In consequence of these resolutions, a memorial was addressed to government by the Committee named in them, embodying the chief arguments for taking up the cause as a national concern, and specifying more particularly the objects proposed to be accomplished, and the means of accomplishing them. In this document the memorialists state that—

"In urging the subject on the attention of Her Majesty's Government they wish to be understood as fully recognizing the principle of not resorting to national assistance, except where the object aimed at is of national importance, where private zeal and private means are already in full activity, and exerted to the utmost; and where other nations have set an example which may justly arouse our emulation. 'In this case too,' they add, 'where no private enterprise can accomplish the end proposed.'"

That the full exertion of private effort is a fair criterion of the degree of importance attached in the estimation of the scientific world to any given branch of such pursuits, and one without which it would be quite unreasonable to look for public support in its favour, is, we think, evident enough. But that in the pursuit of great and worthy objects we are coldly to hold back, and wait till foreign nations shall have led the way and roused us by their example, is a doctrine which, as Englishmen, we must repudiate, and which, if acted on by all, would annihilate the principle of national support altogether. And in the case before us, we hold it by no means creditable to have allowed other nations, and Russia in particular, to precede us to the extent to which it must be evident, on a perusal of the foregoing pages, they have done. But let that pass, since a better era is arrived.

"Great physical theories," they go on to observe, "with their trains of practical consequences, are pre-eminently national objects, whether for glory or utility." In effect such they ought to be considered by every nation calling itself civilized; and if we look to consequences, we have only to point to the history of science in all its branches, to show that every great accession to theoretical knowledge has uniformly been followed by *a new practice*, and by the abandonment of ancient methods as comparatively *inefficient* and *uneconomical.* This consideration alone we think sufficient to justify, even on utilitarian grounds, a large and liberal devotion of the public means to setting on foot undertakings, and maintaining establishments in which the investigation of physical laws and the determination of exact data should be the avowed and primary object, and practical application the secondary, incidental, and collateral one. The example of astronomy, on which, as a theoretical science, the sunshine of public support has been hitherto almost exclusively concentrated, may teach us to what extent these collateral benefits conferred on society by such support may go. The perfection of nautical practice, and the establishment of a complete theory, are indeed great social and intellectual results. But we owe more than these to the public recognition of its claims to national support, in the universal impulse given thereby to every other branch of exact inquiry — in the erection everywhere of a higher standard of physical investigation — and in a precision of every determination rendered practicable, and therefore practically insisted upon, which would never else have been dreamed of as attainable.

That the time is now fairly arrived when other great branches of physical knowledge must be considered as entitled to share in that public support and encouragement which has hitherto fallen to the lot of astronomy alone, will, we think, be granted without hesitation by all who duly consider the present state and prospects of science. The great problems which offer themselves on all hands for solution, problems which the wants of the age force upon us as practically interesting, and with which its intellect feels itself competent to deal, are far more complex in their conditions, and depend on data which to be of use must be accumulated in far greater masses, collected over an infinitely wider field, and worked upon with a greater and more systematized power than has sufficed for the necessities of astronomy. The collecting, arranging, and duly combining these data are operations which, to be carried out to the extent of the requirements of modern science, lie utterly beyond the reach of all private industry, means, or enterprise. Our demands are not merely for a slight and casual sprinkling to refresh and invigorate an ornamental or luxurious product, but for a copious, steady, and well-directed stream, to call forth from a soil ready to yield it, an ample, healthful, and remunerating harvest. We may wait, it is true, and consign to centuries to come, the toils, the glories, and the hopes of science, or we may rely on an easy effort distributed over length of years for the accomplishment of much that vigorous exertion might now effect; but we should recollect the admonition of the poet—

> "Nimm die Zögernde zum Rath,
> Nicht zum *Werkzeug* deiner That."

The feeling of the astronomer, labouring under the weight of his vast cycles, patiently watching the slow evolutions of cosmical events, and breathing forth his aspirations after a perfection which he perceives to be attainable in that tone of protracted hope which borders on resignation, has somewhat too much pervaded other sciences. There are secrets of nature we would fain see revealed while we yet live in the flesh — resources hidden in her fertile bosom for the well-being of man upon earth we would fain see opened up for the use of the generation to which we belong. *But if we would be enlightened by the one or benefited by the other, we must lay on power, both moral and physical, without grudging and without stint.*

The presentation of this memorial was backed not only by the personal arguments and representations of its framers, but by similar and even more urgent representations on the part of the President and Council of the Royal Society, who, on this occasion, in a manner most honourable to themselves, and casting behind them every feeling but an earnest desire to render available to science the ancient and established credit of their institution, threw themselves unreservedly, and with their whole weight, into the scale, with immediate and decisive effect. The strong interest taken in the cause by their president, the Marquis of Northampton, on all occasions a warm and zealous friend to science, contributed without doubt not a little to this result.

Science is of no party. Under the government, whether of Whig or Tory, she has often had to complain of the difficulty of making herself heard in recommendation of her objects; but those objects, once recognized by a British government, are taken up in a spirit and with a liberality which ensures success, if success be possible. In the present instance this has been eminently the case. Every point suggested in the above-cited resolutions has been ordered to be carried out into full execution, and every observation recommended provided for in the most ample manner. Ships, buildings, instruments, and, what is of infinitely the most importance, officers and observers selected with care and imbued with the full spirit of their work, have been provided and appointed, while, so far from the general intention being thwarted by lukewarmness or negligence in the execution, every department of the public service concerned in it, or to which it became necessary to apply in the arrangement of details, responded with alacrity to the call.

Of the four observatories recommended, three, viz., those at St. Helena, the Cape of Good Hope, and in Canada, are placed under the direction of the Master-General of the Ordnance, Sir H. Vivian, by whom the necessary orders for their equipment were issued, and every disposition made for their establishment on a footing of complete efficiency, with a promptness indicating no small interest in the success of the undertaking. At the same time, Lieuts. J. H. Lefroy, J. Eardley Wilmot, and C. J. Riddell, of the Royal Corps of Artillery, young officers full of zeal and intelligence, were appointed as directors of those respective observatories and directed to communicate with Major Sabine, R. A., as their immediate military superior. To each observatory are attached three assistants, with a view to the continuance of the observations through the twenty-four hours. Shortly after their appointment, these officers proceeded to Dublin to receive the necessary instructions in the manipulations of the instruments and practice of the new system of observation from Professor Lloyd, who has volunteered the performance of that highly important duty on this and on every subsequent occasion, sparing neither time nor pains in its performance.

The fourth observatory (at Van Diemen's Land) will be conducted by an officer (Lieut. J. H. Kay, R.N.), to be landed with a similar complement of assistants from one of the vessels destined for the antarctic voyage, which also carries out the observers and instruments for the Saint Helena and Cape stations.

One immediate effect of this hearty adoption of the project by the British Government was to call into action the no less hearty and effectual co-operation of the Honourable Court of Directors of the East India Company. That great and powerful body, on every occasion where scientific objects have come recommended to them from quarters which may be held a guarantee for their importance and utility, have shown themselves liberal, even to profusion, in their support — and in this instance, when applied to by the Royal Society to that effect, not a moment was lost by them in complying with the wish expressed by that learned body for the establishment of three (afterwards increased to four) magnetic observatories in their dominions and dependencies, similar and similarly equipped in every respect to those established by government, and destined to a strictly simultaneous and corresponding course of observations. The stations thus ultimately fixed on are Madras — Semla, at an elevation of nearly 8000 feet in the Himalayas — Singapore, as the farthest attainable eastern point, and Aden on the Red Sea, as a point highly important in itself from its position with respect to the magnetic equator which passes nearly through it, as well as from its constituting a link in a chain of stations of high interest, extending in longitude from St. Helena to Singapore.

A basis so extensive, thus afforded for a great combined system of corresponding observation, by which the magnetic state of the whole globe at the present epoch should be, as it were, struck off at a blow, and placed on record for ever, not only justified but demanded that every exertion should be made to procure the co-operation of foreign countries on a regular and concerted plan. In performance of this duty, the Royal Society again bestirred itself by circulars addressed to the various scientific bodies and individuals in its correspondence, by representations to officia authorities abroad, and, where it could be done without a breach of etiquette, to personages in the highest station: and in order that the plan of operations should be so arranged as to consult as far as possible the convenience of Russian and German observers, Professor Lloyd, accompanied by Major Sabine, at the request of the Society, visited Göttingen and Berlin, where being met by M. Kupffer, the director of the Russian magnetic observatories (who for that purpose had undertaken a journey from Petersburg), in personal conference with that eminent and zealous observer, and with Messrs. Von Humboldt, Erman, and Gauss, they were enabled to agree on a scheme of co-operation, which, being subsequently matured by communication with other of the chief European observers, has ultimately been adopted by general consent.

The success of these measures to secure an extensive co-operation may be collected from the following summary of stations, at which it is now certain that magnetic observatories co-operating for the most part to the full extent, but at all events so far as the *personnel* of the establishment will allow, in the proposed plan, and furnished with instruments identical with, or equivalent to, those supplied to the British observatories, are either already established or in immediate course of being so, the instruments being ordered and the observers appointed.

British Stations. — 1. Dublin (Professor Lloyd); 2. Toronto * (Lieut. Riddle, R.A.); 3. St. Helena † (Lieut. Lefroy, R.A.); 4. Cape of Good Hope (Lieut. J. Eardley Wilmot, R.A.); 5. Van Diemen's Land (Lieut. J. H. Kay, R.N.); 6. Madras ‡ (Lieut. Ludlow); 7. Semla (Captain Boileau); 8. Singapore § (Lieut. Elliott); 9. Aden (Lieut. Yule); in addition to which, each ship of the naval expedition, under command of Captain Ross, is provided with a corresponding set of apparatus, to be erected and used in concert wherever opportunity may offer (10, 11).

Russian. — 12. Boulowa; 13. Helsingfors (M. Nervander); 14. Petersburg (M. Kupffer, General-Superintendent); 15. Sitka; 16. Catherinenburg; 17. Kasan; 18. Barnaoul; 19. Nertschinsk; 20. Nicolajeff (M. Knorre); 21. Tiflis; 22. Pekin.‖

Austrian. — 23. Prague (M. Kreil); 24. Milan (Sig. Della Vedova ?).

United States. — 25. Philadelphia (Professor Bache); 26. Cambridge (Professors Lovering and Bond).

French. — 27. Algiers (M. Aimé).

* Substituted for Montreal, originally proposed. This observatory is already in activity, and observations have been received from it.

† Already in activity.

‡ Substituted for Ceylon, originally proposed.

§ Substituted for Bombay, originally proposed.

‖ From Pekin a complete series cannot be expected; but, so far as practicable, the observatory there (already in activity) will co-operate.

Prussian. — 28. Breslau * (M. Boguslawski).

Bavarian.— 29. Munich (M. Lamont, Director of the R. Observatory).

Belgian. — 30. Brussels (M. Quetelet, Director of the R. Observatory).

Egyptian. — 31. Cairo (M. Lambert).

Hindoo. — 32. Trevandrum (Mr. Caldecott, Astronomer to the Rajah of Travancore).

There is every reason to expect that this list will be largely increased within the present year. Indeed six or seven more stations might already be inserted from our knowledge of communications in progress.

The great developement of the Russian system is partly owing to the continuance in activity of the observatories established at the instance of Baron Von Humboldt; partly to the indefatigable zeal and activity of M. Kupffer, on whom their general direction devolves, seconded by representations from England. The occurrence of an Egyptian observatory, established by the extraordinary man who now rules the destinies of that country; and of a Hindoo, one maintained by the liberality of a native prince, and placed under the direction of an English observer, who has already rendered excellent service to magnetic science, are scientific novelties, which will be viewed with interest, as we believe them to be the first instances of potentates, whom European pride regards as semi-barbarous,

* This observatory is supplied with British instruments.

placing themselves so far within the pale of civilization as to co-operate in any scientific proposition.

In casting our eyes over this list we perceive the whole continent of South America unrepresented, though abounding in stations of great interest. We could have wished also to see Otaheite included in the list of primary stations ; for, though aware that measures have been taken to secure *some* observations there, yet its importance well merits for it this distinction. May we not hope that the omission will (before it is too late) be supplied by the missionaries, in whose hands the entire direction of the government and resources of that island may be considered as placed. We know not a point on the surface of the globe so interestingly situated for a physical observatory, or at which, independent of its magnetic interest, the tides, the winds, the barometric oscillations, the habitudes of earth, air, and ocean, all present themselves under aspects so peculiar and so highly deserving to be diligently noted and recorded.

We must now give some account, though necessarily a very succinct one, of the scheme of observations agreed on, which we are enabled to do by the ample and elaborate report of the Committee of Physics of the Royal Society, drawn up on this occasion and forwarded to each station, in which (traced, as we understand, by the able pen of Professor Lloyd) every detail of the construction, adjustment, and use of the magnetometers is clearly explained.

The magnetic apparatus with which each station will be provided consists of three magnetometers, one for the measurement of the declination and its changes, and the dynamical measure of the horizontal intensity; one, on the bifilar construction, for the statical measure of this latter element, and *its momentary changes* (which cannot be obtained by the dynamical process); and one for the measure of the vertical force and its changes on Professor Lloyd's principle; together with a dipping-needle of the best construction, and such astronomical apparatus as is required for ascertaining the time and the true meridian, referring to it the indications of the magnetometers. To these have been also added in each case a most complete and perfect set of meteorological instruments, carefully compared with the standards in possession of the Royal Society, not only for the purpose of affording the necessary corrections of the magnetic observations, but also with a view to obtaining at each station, at very little additional cost and trouble, a complete series of meteorological observations.

Each day is, in the first place, supposed to be divided into twelve equal portions, of two hours each, commencing at all the stations at the same instants of absolute time, which may be called the magnetic hours, viz., 0h. 0m. 0s., 2h. 0m.0s. 3h., &c. of *mean time at Göttingen*, without any regard to the apparent *times of day at the stations themselves*, which will, of course, differ by their differences of longitude, so that the first magnetic hour, which at Göttingen commences at noon, will, at Dublin, for instance (1h. 5m. 8s. west of Göttingen), commence at 10h. 54m. 52s. A.M.; at Madras (4h. 41m. 42s. east of it) at 4h. 41m. 42s. P.M.

At the commencement of every magnetic hour throughout the day and night, of every day (Sundays excepted), the magnetometers are observed, and the meteorological instruments read off. To multiply opportunities for observing remarkable coincidences, the observation at 2h. P.M. Göttingen, MT. is in all cases a triple one, the magnetic readings being thrice repeated in a given order, at intervals of five minutes.

The Göttingen terms, commencing on the Friday preceding the last Saturday in February, May, August, and November, at 10h. P.M. (Gött. MT.) and continued at intervals of two minutes and a half, according to a settled order of the instruments, through the subsequent twenty-four hours, will be observed at all the stations; and moreover, eight additional terms are introduced, viz., on the Wednesday preceding the 21st of each remaining month, commencing at the same hour, and extending to the same series as the other terms.

In this scheme of observation it is easy to see that all the great *quæsita* of magnetic science, so far as they can be at *fixed* stations, are provided for. The continuance of the series for a period of three years, which is contemplated, will afford, by the comparison of mean results, and when the extreme accuracy attainable is considered, abundant data for settling the direction and present amount of the secular variations of the magnetic elements at each station. The subdivision of the entire twenty-four hours into twelve equal portions will furnish more complete and ample data for the evolution of the arguments and co-efficients of every periodical equation; while the simultaneous nature of all the observations, and especially of those of the term days and triplets, cannot but lead us to a knowledge of the nature, laws, and intimate dependencies of the perturbations—with their connexion with meteorological processes, and especially with those which are concerned in the production of the aurora borealis.

Printed forms of registry, drawn up with uncommon care by Captain Boileau, director of the magnetic observatory at Semla, under the advice and inspection of Professor Lloyd, for the magnetic, and others, by Mr. Daniell, at the request of the council of the Royal Society, for the meteorological observations, are adopted in all the British stations, and will no doubt be so at every other, so as to preserve a complete uniformity of registry—a point of great importance, or rather of indispensable necessity, in an immense operation of this nature, the details of which could by no other means be mastered by any one mind. As it is, the comparison of so extensive a collection of data with theory in the developed form it must be expected to assume under their influence, will be a task truly herculean; and we know not which most to admire, the enthusiasm and devotion with which the distinguished individual, whom the universal suffrage of his compatriots declares most competent to the task, has consented to undertake it, or the resources of mathematical skill and practical experience he brings to its execution. The observations, it is understood, are to be continued during three years, and the results, from the British stations, officially forwarded at brief intervals to their proper departments at home, and thence to the Royal Society, which will also become a centre of communication from the foreign stations. Voluminous beyond all former precedent as the mass of data thus accumulated must of necessity be, we trust the whole will be printed (each nation and each department of course providing for the publication of its own). No consideration of economy should be allowed to interfere with the performance of this necessary duty, without which we look upon all that shall be done as virtually thrown away. Highly as we respect the illustrious body above-mentioned, and applaud their selection of the individual into whose hands the results will in the first instance pass; yet their full, fair, and effectual discussion can be secured by no other means than by inviting to it the collective reason of the age, and of all succeeding ones, and affording every one who may think proper to engage in the task, now or hereafter, ample opportunity to do so. To handle so enormous a mass, even in the preliminary and merely mechanical arrangements, is in itself, however, no slight or inconsiderable task, and will demand a well-organized and well-considered system. We have calculated, from the specimens of the registers contained in the report above alluded to, the number of magnetic observations, and such meteorological ones only as are absolutely necessary for their reduction, which will come under discussion, supposing complete series furnished by each of the thirty fixed stations enumerated: and we find them to amount to 1,958,040, a startling sum, and one which, though subject no doubt to large deductions, must still afford matter for serious consideration.

To follow up with full effect the above-described scheme of magnetic observation, it is more than ever desirable that attention should be turned to the subject of magnetic surveys of particular districts, as well in the immediate neighbourhood of the stations as in countries remote from them. In the former, indeed, the necessity of such surveys to connect the stations with the general body of the magnetic lines is so obvious, that we are surprised to see no official provision for it, though the subject is referred to in the memorial already cited in the following terms:—

"In concert with such primary stations, it would be both natural and highly desirable that travellers provided with the requisite instruments, or officers in other stations who may be willing to devote a portion of their time to this service, and who may for that purpose be temporarily provided with the instrumental means, should act. Every such primary station then, supposing such to be established, would henceforth become a point of reference and comparison by which short and desultory series of observations in other localities might be rendered available; including under this head such as might be made in the course of nautical surveys and voyages of discovery, or where, from other causes, it might be impracticable to remain for any considerable time."

If ever magnetic surveys of particular districts can be carried on with advantage, it must be when based on and in concert with a series of regular observations made at stations of reference. We hold it, therefore, to have become the duty of every civilized nation to set on foot and urge to its completion a regular and careful magnetic survey of its own territory and dependencies. For such surveys we have excellent models. Professor Forbes has given us an admirable specimen of this kind in tracing out the course of the isodynamic and isoclinal lines in Switzerland (*Ed. Trans.* xiv.); and for another and very complete example of what such a survey ought to be, we may refer to that of the British Isles, published in the "Eighth Report of the British Association," the joint production of Professor Lloyd and Major Sabine, from the collation of their own observations with those of Captain J. Ross and Messrs. Phillips and Fox. In the chapters of this Report supplied by Professor Lloyd, we are put in possession of every requisite formula of reduction, and with the best and most available mode of combining the observations so as to deduce from them the elements of each magnetic line in its passage through the district under survey, cleared of local irregularities. It is evident that such surveys cannot be considered as complete unless referred to central stations, and unless provision be made for the re-determination, at stated intervals, of the magnetic elements not only at such centres but also at several extreme points, from which to infer the local co-efficients of the secular changes going on throughout each district. It is easy to point out particular fields for such researches. Throughout the whole of North America a wide one exists, which the establishment of a Canadian station renders particularly desirable should be entered upon immediately; the deficiency of trustworthy magnetic observations in all that vast region being lamentable. In the United States, at least, there is no lack of individual spirit and enterprise for the task, and it is with pleasure we learn that a private association, comprising the most distinguished names in American science, Bache, Bartlett, Henry, Locke, Loomis, Renwick, Rogers, &c., is already preparing to distribute that country among them for survey, each taking his share. In Southern Africa, too, a magnetic survey, in correspondence with the proposed station at the Cape, would be most desirable; that vast colony being in this respect at present a mere blank. The difficulties presented by the nature of the country and the mode of travelling to the transport of instruments might easily be overcome; and among the multitude of wealthy, intelligent, and enterprising Indians who resort thither for health, and to whom mere active locomotion in that favoured climate, is, literally speaking, the breath of life, it may not be too much to hope that some may be found to whom the determination of a magnetic dip or intensity may have as high attraction and offer as good sport as a long shot at a lion or an antelope. In India itself an excellent example has been already set by the surveys of Messrs. Taylor and Caldecott, of the southern part of the peninsula, which we trust to see extended to every part of the Anglo-Indian territory, in connexion with those central stations which the liberality of the East India Company is on the point of establishing there.

In Van Diemen's Land and New South Wales especially, the subject is of crying and urgent practical importance, and indeed in every new settlement where the allotment of land is going on, and where, as a matter of necessity, the compass must be appealed to for the direction of boundary lines.

The consideration of magnetic surveys naturally leads us to the second branch of this great public undertaking, the naval expedition which has lately left our shores. This expedition, under the command of Captain J. Clerk Ross, consists of two ships, the Erebus of 370 tons, and Terror of 340, the latter commanded by Captain F. Crozier, an old and long tried shipmate of Captain Ross, and bound to him by strong ties of mutual attachment and esteem, a circumstance of no small importance on the long, dangerous, and difficult service contemplated. As a winter near the South Pole is among the contingencies to which they may be subjected, and at all events much exploration among the frozen seas which surround it, is inevitable, the vessels are strengthened for their conflict with the ice by every means which the art of the shipwright could devise, nor has any arrangement or contrivance for the warmth, comfort, and accommodation of their inmates, which previous experience could suggest, nor any imaginable resource in case of accident (such as subdivision of the vessels into distinct water-tight compartments, &c.) been omitted in their fitting up. The crews are all picked volunteers on double pay, and both officers* and men animated with the finest spirit. In the choice of a commander the expedition has been singularly fortunate—Captain Ross, to say nothing of his many excellent qualities as

* *Erebus*—Captain J. C. Ross; Lieutenants E. J. Bird, J. F. L. Wood, J. Sibbald; Master, Charles Tucker; Surgeon, R. Maccormick; Purser, T. R. Hallett; Assistant-Surgeon, J. D. Hooker. *Terror*—Captain F. R. M. Crozier; Lieutenants A. Mac Murdo, C. Phillips, J. H. Kay; Master, P. P. Cotter; Surgeon, J. Robertson; Assistant-Surgeon, D. Lyall; Clerk, G. H. Mowbray.

an officer and a man, having already signalised himself in the extraordinary voyage undertaken by his uncle, Sir J. Ross, in search of the wreck of the Fury, as much by his conduct and resource as by the actual discovery of the northern magnetic pole, and having ever since his return been engaged as a matter of taste and private pursuit, in magnetic observations both at home and abroad, and being perfectly familiar with the principles of the new methods.

The object of this expedition is, emphatically, the collection of magnetic observations in the southern hemisphere, and more especially in those regions, which, owing to their high south latitudes, are little accessible, and unlikely to be visited for purposes of commercial intercourse or enterprise, and in which, from the analogy of the northern hemisphere as well as from the general configuration of the magnetic lines, so far as the existing charts can be trusted, there is reason to believe the most interesting points and inflexions of those lines are situated, such as the southern magnetic pole or poles and the points of maximum intensity. To the former of these points, considering it as probable that only one exists, M. Gauss has assigned, by the interpretation of his general formula, a probable situation in latitude 66° S., longitude 146° E., or on the meridian, nearly, of Hobart Town. On the correctness of this conclusion Captain Ross's observations will of course enable us to decide; but it ought to be borne in mind that, owing to the great deficiency of antarctic observations, this theoretical position can only be regarded as a first approximation, open to large corrections. By a singular and most fortunate coincidence, an island or islands have been recently discovered, nearly in this latitude, and so situated in respect of longitude as to afford a station, certainly on one side, and possibly also on the other, of the point in question. Should this discovery be verified to its full extent, a base will be afforded, the convergence of the needle at whose extremities will hardly fail to point out nearly the situation of the pole, should direct access to it prove impracticable. We say nearly, for it is a mistake to suppose, as is commonly done, that the magnetic pole or point of perpendicular dip is a *precise* point of convergence to the needle in its neighbourhood. The probable situations of the points of greatest intensity are in latitudes 47° and 60° S., longitudes 130° and 235° E., respectively, and are both, therefore, accessible. To traverse the isodynamic ovals which surround these points, in their immediate neighbourhood, will be also an object of prominent interest. In fact, however, there is no point in those unexplored or imperfectly explored seas which surround the South Pole at which magnetic observations will not be of extreme interest. Wherever it may be practicable to land and observe, especially on the polar ice, the determinations, being there obtained with perfect precision and free from all local influence, will possess the highest value, especially in those cases where it may be practicable to erect the magnetometers with which also the expedition is furnished, and observe for the diurnal changes and perturbations.

The Erebus and Terror, having taken in the officers and assistants for the establishments at St. Helena, the Cape, and Van Diemen's Land, with the instruments for the equipment of those observatories, dropped down the river and sailed from Margate on the 30th September, 1839; a day for ever memorable in the annals of British science. After touching and observing at Madeira, Porto Praya, St. Paul's Rock, Trinidad, &c., and crossing the magnetic equator in 14° 2′ S., 30° 30′ W., they made St. Helena on the 31st January, 1840, where they remained only so long as was necessary for landing Lieutenant Lefroy and his party, and selecting a favourable site for their establishment. The point selected is one calculated to give rise to reflections of no ordinary interest on "the various turns of fate below," being no other than Longwood, a spot in every respect except one admirably calculated for the purpose, and in that one (viz., in the extraordinary amount of local magnetism) no worse than the rest of the island, which, being entirely of volcanic and basaltic formation, is, in fact, a magnetic nucleus. This circumstance, however, though fatal to absolute determinations of the elements on it, no way interferes with the principal objects of its selection as a station, their secular, periodic and perturbative changes being quite as well deduced in the presence as in the absence of local attraction. Meanwhile the great amount of such attraction at this island is understood to have given occasion for several very interesting and important observations on board the ships, producing singular anomalies in their results, assuming the form of discordancies between them, which were only obviated by quitting the anchorage and standing off to sea, out of the reach of the local influence; while, on all other occasions, the observations on board both ships manifested the most satisfactory accordance, the dips often agreeing to the same minute, and being seldom more than a few minutes apart.

The establishment at Longwood being satisfactorily arranged through the hearty co-operation of the insular authorities, who seemed bent on emulating those at home in removing everything like an obstacle, the expedition proceeded, on the 9th of February, in its voyage to the Cape, where it arrived on the 17th of March, having traversed in its course the system of isodynamic ovals surrounding the point of least intensity in the South Atlantic, passing as nearly as was practicable over that important point itself, and thus accomplishing satisfactorily, it may be presumed, one of the objects pointed out in the instructions furnished to Captain Ross, by procuring data for settling with accuracy its true situation, and ascertaining the amount of the *absolute minimum* of magnetic intensity at present existing on the globe. Arrived at the Cape, Lieutenant Wilmot and his party were landed, and a site selected for them close to the superb astronomical observatory maintained there by the British government, where, aided by the same prompt attention on the part of the colonial government, and the scientific assistance and local knowledge of the distinguished and public spirited director of the observatory (Mr. Maclear), the latest accounts we have been favoured with a sight of left them in full and satisfactory progress towards the completion of their establishment.

In the establishment of the Canadian observatory (it may here be mentioned) delays equally unforeseen and unavoidable occurred. The party under Lieutenant Riddell, having landed at New York after narrowly escaping shipwreck, and still more narrowly the destruction of all their instruments in the confusion of lightening their vessel by throwing overboard all its heavy stores, proceeded to Montreal, the point originally pitched on for the station. It proved, however, so objectionable, by reason of local magnetism, as to render it advisable to alter the locality to Toronto, a situation apparently quite free from this annoyance; but before a proper site could be selected and the preliminary arrangements made for building, the setting in of winter had suspended all proceedings of that nature, which could not be resumed till April, but are probably by this time complete. Meanwhile, the activity and resource of Lieutenant Riddell supplying the want of every convenience, the observations, so far as their nature would permit, were commenced and are in progress — the first *term* having been observed in March, as agreed on.

The expedition quitted the Cape on the 4th of April, since which, of course, no account of its further progress can have been received. The advanced state of the season must preclude any attempt to penetrate southwards, as originally proposed, during the voyage to Van Diemen's Land, so that the exploration of the land discovered by Kemp and Enderby will necessarily be left for another season. The establishment of the Van Diemen's Land observatory being a point of primary importance, the voyage thither will probably be direct, taking in the way only those few points of land which offer stations of interest, such as Prince Edward's and the Croxet Islands, and the desolate shores of Kerguelen's Land, where, should time and circumstances permit, a magnetic term will be observed. Arrived at Hobart Town, the party landed and settled, and the instruments erected, preparations will be made for a push to the southward with the earliest return of the warm season, in search of the magnetic pole or poles, and in prosecution of the general objects of the voyage. Ulterior to this, the circumnavigation of the southern pole, the magnetic exploration of every accessible point of land in the polar basin, and the observation of terms in strict correspondence with those in the fixed observatories at every station where the vessels may remain long enough, will be distributed over the remaining duration of the enterprise, in such order as shall seem most practicable to its able and experienced commander.

Although, as has been said, the object of Capt. Ross's voyage is emphatically the collection of magnetic data, yet it must not be supposed that the many other important scientific objects attainable in such a voyage have been anywise overlooked or disregarded either in its plan or in the provision made for its execution. Never, on the contrary, we believe, has an expedition of discovery left our shores so largely provided with apparatus of every description for physical research, and with instructions embracing so many points of scientific interest, and so distinctly and expressly stating the desiderata which it may supply, and the most available means of supplying them. These instructions have been furnished in the form of reports by the several scientific committees into which the Royal Society has of late thought proper to break up its line of battle, each in its own department; but of these reports, one alone, that of the committee of physics (including meteorology), has been hitherto published for general circulation. It is not our intention very minutely to criticise this report. Were it so, we might object to the ambitious form of its title, assuming as it does a generality and a unity of design which neither its contents nor its original purpose warrant. We know how difficult it is for two or three, much more for a committee of thirty, acting under the subsequent revision and remodelling of a council of twelve, to indite and publish a connected work. Accordingly, to such a title the work before us has no claim, being in fact rather remarkably the reverse. Nevertheless it is full of interest to the voyager and traveller. It abounds with pertinent and useful suggestions relative to every species of physical observation, such as magnetism, the tides, the measure of the force of gravity, the distribution of temperature over sea and land, the depth and currents of the ocean, refraction, eclipses, variable stars, meteors, aurora borealis, &c. &c. The instructions relative to terrestrial magnetism are especially full and precise as the occasion required, and are accompanied with abstracts of the forms of registry intended to be followed in the magnetic observatories. Those relating to meteorology amount in effect almost to a practical treatise on meteorological instruments, the management of a meteorological observatory, and the systematic registry of its observations, and, coming now from authority entitled to so much confidence, we do hope they will have the effect of inducing something like order, system, and unity of co-operation into this most important, but at the same time most straggling, disjointed, and imperfect science.

The requisitions for information relative to the depth, constitution, temperature, and currents of the ocean, are both numerous and calculated to excite a lively interest. The explanation of the oceanic currents can never be complete till we know the elements which affect the density of the water at different depths, and the seat of action of the forces which produce the disturbance of its equilibrium of density and pressure. Those elements are the temperature, saltness, and compression of the sea water; the two former of which are determinable by direct observation—the latter by calculation from the depth. As regards the seat of action of the motive forces, it is justly remarked in the "Report," that the order of the phenomena is precisely the reverse of what obtains in the atmosphere. In the sea, the sun's rays are totally absorbed at the surface or within a few fathoms of it, and having no tendency to penetrate deeper by conduction, and but little liability to be carried down by superficial agitation, are merely, as it were, *floated* on the surface without any tendency whatever to produce *ascensional currents*, such as arise in the atmosphere from the heated surface of equatorial continents or seas. On the other hand, as the density of sea water goes on increasing by cold to its freezing point, it follows that there must be constantly in action, in the two polar basins, but chiefly in that where winter prevails, a *descensional* force producing subaqueous currents radiating outwards from the poles, which in their progress towards the equator, are of course modified by the earth's rotation in analogy with the trade-winds, whenever the form of the bottom or the depth and extent of the channels by which the deeper seas communicate, will permit. The depth and form of these channels therefore, and of the subaqueous basins which they connect—or, in other words, the configuration of the subaqueous mountains and valleys, enters as a most material element into the problem, and adds greatly to the geological interest attached to deep sea soundings. On this head we understand that Captain Ross has already arrived at some very remarkable results, having so completely overcome the great difficulty which attaches to this operation, as to have procured soundings at a depth beyond all former experience, and in one instance especially to have attained a depth below the surface exceeding the altitude of the summit of Mont Blanc above it!—and that too with a facility and certainty which promises to afford a speedy solution of the long agitated question of the mean and maximum depths of the ocean. In fact, we may already fairly conclude from these experiments a *general* depth of sea far exceeding the *general* elevation of the continents, since it is extremely improbable *either* that the deepest, or nearly the deepest, *region* should have been the scene of the few trials yet made—*or* that within the particular region attempted, precisely the deepest *points* should have been those which have now, for the first time, received the lead.

Appended to this report are two highly interesting communications from Baron von Humboldt and M. Erman, respectively suggesting a multitude of observations and experiments, in addition to those recommended in the body of the report,* and which being by this time, as well as the report itself, in the hands of all the parties concerned, will of course receive every attention. Independent of the very great value of many of these suggestions, this proof, among an infinity of others which have occurred of the lively interest this great scientific operation has excited and is exciting abroad, cannot but be most welcome. Though we may not perhaps entirely coincide in the great stress laid by M. von Humboldt in the document emanating from him on the precise tracing out of the course of the magnetic equator, and the line of no declination in preference to precise determinations spread over a wider range, yet it is impossible not to agree with him in the strong view which he appears disposed to take of the extreme value of the present conjuncture for securing observations in all parts of the ocean, by taking advantage of every practicable opportunity, by a liberal supply of instruments and by every sort of encouragement

* While these pages are going through the press, additional proofs of this interest are afforded in the form in which it is most desirable it should be exhibited, that of active co-operation on the part of foreign governments. Of such co-operation on the part of Holland we are now assured—a point of the utmost importance by reason of her colonial possessions in the East, where two observatories at least will be established. Report also speaks of observatories at Kremsmunster, Cadiz, Bologna, &c.

and inducement held out to those who are willing and competent to use them.

We cannot close this imperfect sketch of the great combination thus happily set on foot, and we trust to be as happily brought to a conclusion, without remarking one peculiarity attending the expedition under Captain Ross's command, which cannot but be most encouraging and satisfactory to those who have embarked in it as well as to all who have had any share in recommending its being undertaken. If it return at all, after covering any considerable extent of the antarctic seas, it cannot return otherwise than successful. It is hardly conceivable that the existence and situation of the actual magnetic pole or poles should escape detection by observations made in the course of an antarctic circumnavigation, though the points themselves may prove inaccessible; nor is there any one *geographical* point to be pushed for in preference to another on which the success of the enterprise can be said to be in any way staked. The harvest of discovery will be reaped alike either at sea, on land, or on ice indifferently. No insuperable barrier interposed by nature between our brave countrymen and the object of their toils can frustrate their exertions. They will gather as they go, and whatever they collect is sure to be of value. That the actual circumstances in which they must be occasionally placed in the prosecution of their objects will here, as on every other such occasion, call forth the manifestation of those great and glorious as well as most endearing qualities of the British seaman which have shone so conspicuously on former similar occasions we cannot doubt, nor that the public sympathy will be as warmly excited on this as it has been on any such occasion in favour of those who are thus leading the forlorn hope in the siege which science lays to the strongholds and fastnesses of Nature.

WHEWELL ON THE INDUCTIVE SCIENCES.

History of the Inductive Sciences from the earliest to the present Times. By the Rev. WILLIAM WHEWELL, M. A., Fellow and Tutor of Trinity College, Cambridge, President of the Geological Society of London. 3 vols. 8vo. 1837.

The Philosophy of the Inductive Sciences founded upon their History. By the Rev. WILLIAM WHEWELL, B.D., Fellow of Trinity College, and Professor of Moral Philosophy in the University of Cambridge, Vice President of the Geological Society of London. 2 vols. 8vo. 1840.

(FROM THE QUARTERLY REVIEW, No. 135.)

IF the moral and intellectual relations of man have ever been justly regarded as transcending in importance all other subjects of human interest, the necessary dependence of his duties and responsibilities on his natural faculties must render it impossible to appretiate or define the one without entering into a close investigation and analysis of the other. And if, in the course of this inquiry, it appear, by reference to history and experience, that there exist in the intellectual constitution of our species springs of power and capacities of intelligence which have been but rarely drawn upon, and which have lain, as it were, torpid and dormant during long portions of history and among vast masses of population, it will become not less our interest than our duty to study with the most earnest solicitude the conditions under which the vigorous development and worthy employment of that power and those capacities can subsist.

That man is a speculative as well as a sentient being, searching in every thing for connexion and harmony, the perception of which mixes itself with his choicest pleasures, is what we need not to be reminded of. To call up their images, even transiently, in his mind, the powers of his imagination and fancy are continually tasked, while to trace them through the realities of universal nature constitutes at once the noblest and most delightful, but, at the same time, the most arduous exercise of his reason. Chained, however, to the ground by his material wants, and solicited unceasingly by his passions, which tax to the utmost all his faculties for their gratification, man has been found in every age but too ready to forget this lofty privilege, and, degrading reason from its highest office, to employ it, now as the laborious drudge of his appetites, and now as the subservient instrument of his designs. The experience of all history has shown that the gratification arising from the exercise of the purely intellectual faculties is especially apt to be postponed to almost every other, and in its higher degrees to have been as unduly appretiated by the many as it has been rarely enjoyed by the few who are susceptible of them. The mass of mankind, too happy in a respite from severe toil and bitter contention, are well content with easy pleasures which cost them little exertion to procure and none to enjoy. To the poor and overwrought, a mere oblivion of care and pain; to the rich and refined, luxurious ease and pleasing objects and emotions presented in rapid succession, and received and enjoyed without effort, offer a paradise beyond which their wishes hardly care to roam. The most robust and vigorous constitutions only, whether of mind or body, find a charm in the ardour of pursuit, and feel that inward prompting which excites them to follow out great or distant objects in defiance of difficulties. Even these, for the most part, require the stimulus of external sympathy and applause to cheer them on their career: and great indeed, and nobly self-dependent, must that mind be, which, unrepressed by difficulty, unbroken by labour, and unexcited by applause, can find in the working out of a useful purpose, or in the prosecution of an arduous research, attractions which will lead him to face, endure, and overcome the one, and to dispense with or despise the other. The sympathies of mankind, however, have rarely been accorded to purely intellectual struggles. Men seldom applaud what they do not in some considerable degree comprehend. The deductions of reason require for the most part no small contention of mind to be understood when first propounded, and if their objects lie remote from vulgar apprehension, and their bearing on immediate interests be but slender, the probability is equally so that they will experience any other reception than neglect. And thus it has happened that, in so many cases, the impulse of intellectual activity even when given has failed of propagation. The ball has not been caught up at the rebound and urged forward by emulous hands. The march of progress, in place of quickening to a race, has halted in tardy and intermitted steps, and soon ceased altogether.

The consequences of these and similar hinderances which have operated at every period of history and in every state of humanity against the effective exercise of our reason in its pure and proper field, and on those high objects with which it has been found competent to grapple, will appear, if we look for its results among the more ancient monuments of human thought and action. As a conquering, contriving, adorning, and imaginative being, the vestiges left by man are innumerable and imperishable; but, as a reflective and reasoning one, how few do we find which will bear examination, and justify his claim! How few are the conclusions drawn from the combined experience and thought of so many generations which are worth treasuring as truths of extensive application and utility! How rarely do we find in the writings of antiquity or of the middle ages any general and serviceable conclusion respecting things that be—any philosophical deduction from experience beyond the most obvious and superficial on the one hand, or the most vague, loose, and infertile on the other—any result fairly reasoned out, or any intelligible law established from data afforded by observation of phenomena; whether material, having reference to the organization of the system around us, or psychological, bearing on the inward nature of man?

But from the epoch, comparatively so recent, when man began to consider himself not merely as the denizen, but as the interpreter of nature, and, warmed and inspired by the noble prospects opening on him from this exalted point of view, to speculate on her laws, less in the spirit of an interested occupant than of an admitted and privileged spectator, humbly but diligently seeking to unravel some of the lowest of her mysteries, and catch thereby a glimpse, however dim and distant, of the designs of her glorious Author—since this inspiring note has been sounded in our ears, and found its responsive chord in innumerable bosoms, how different is the scene which has opened! Instead of barren and effete generalities—of vague and verbal classifications—of propositions promising everything to the ear, but performing nothing to the sense—of maxims grounded on pure assumption, and argument dogmatically taking its stand on the appeal to our irremediable ignorance, we find that it has been practicable for human faculties to attain a knowledge of truths based on a foundation co-extensive with the universe, yet applicable to the closest realities. And while thus exercising our faculties in these their primary essays within the narrower and safer circuit of material laws (which yet, opening out in vista after vista, seem to lead onwards to the point where the material blends with, and is lost in, the spiritual and intellectual), may we not look forward with no presumptuous hope to the attainment of a position from which, with an eye schooled and disciplined by such experience and with a mind thoroughly familiarized with the characters of truth as it presents itself to us in these passionless researches, we may follow out its traces and recognize its features through the mist of interest or in the storm of emotion, when engaged in those far more difficult subjects of inquiry which the social and intellectual world afford? It is a hope long deferred and often damped, but never utterly extinguished; springing afresh in youthful and ardent bosoms in perpetual aspiration, and which finally to dismiss would be to deprive philosophy of its most sacred object, and of its only abiding charm.

With the indulgence of such hopes, and with the steadily increasing conviction of the possibility of their ultimate realization, which every fresh advance in science affords, arises a necessity of occasionally, and indeed frequently, passing in review both the assemblage of the results obtained and the mode in which they have been obtained; with a view not only to the duly estimating the real value of our actual acquirements, and the direction in which further progress appears most immediately practicable, but to the deducing from our experience of the progress already made, maxims and principles available in our future career. Science itself thus comes to be considered as an object not simply of philosophical interest, but of inductive inquiry. If we cannot succeed in laying down rules which shall conduct us infallibly to the discovery of unknown truths, we may at least expect to ascertain, by thus passing in review the history of science, what have been the stages and conditions of society in which its greatest acquisitions have been made; what symptoms have been their usual precursors; what tendencies have arrested them in their development; what is that attitude of mind which affords the most favourable condition for the occurrence of discovery to individuals, and that state of public feeling and general occupation and interest which contributes to make one age or one nation more distinguished than another for their magnitude and frequency. Grave questions these, since, as we have already remarked, there are duties and responsibilities, individual and social, attached to their discussion.

But not only has the philosophy of science this practical object, it has its speculations as well as its applications, its theories as well as its maxims, which constitute it a philosophy; and these, it must be confessed, lie among very thorny, difficult, and abstruse considerations, which is no wonder, seeing that it is occupied with the grounds of human belief, the reality of human knowledge, nay, the very nature of truth itself, and the competency of the human faculties to its perception; all subjects of the utmost obscurity, and which involve us, at its very outset, in the most intricate and puzzling discussions of metaphysics. What is the nature of general and of universal propositions? Are all true universal propositions *necessary* truths, or is any truth, or all truth, necessary? What is the act or series of acts of the mind in constructing general propositions, and when constructed, in what manner do we rest in them as expressive of truth? Is it that we simply admit them as results of experience, until habitual acquiescence and unbroken verification renders dissent first difficult, next impracticable, and finally, inconceivable? Or do we recognize in them but the echo of a voice within our own bosoms, which for the first time we have learned to interpret, and whose announcements we receive as revelations? In other words, whether any, and what portion of our knowledge be innate, or whether the whole be a mere collection of deductions from experience, systematized by the act of the mind, continually reviewing and arranging its acquisitions, and moulding them into forms of its own, whether merely adapted for ready use and recollection, or as essential to their recognition as parts of a whole, or as subject-matter for high and abstract meditation. Do we apply to the objects of our reasoning, ideas of which we have a perception, and propositions of which we have a conviction antecedent to experience (and which may therefore be regarded as impressed on our intellectual nature by the Author of our being), linking them together by their appropriateness to form subjects of these innate propositions in the way of special application, and by the conformity of the perceptions connected with them to these innate fundamental ideas? Or do we simply distribute all the phenomena of the world around us, and of our own minds, into groupes, according to the analogies of the impressions they make on our perceptive faculties, whether bodily or mental (the perception of such analogies being itself one of the primordial faculties of our minds); and do we then, by a peculiar and irresistible impulse of our intellectual nature, which we term generalization, attribute to all the members of such groupe, not only those with which we have become familiar, but also all those which we do or can conceive in our minds as appertaining to it, the same attributes, properties and relations, according to their special natures, which we have observed to belong to any one of them, and especially that which has served as the ground of analogy and the motive for so connecting them?

These at first sight appear widely different, and indeed almost diametrically opposite views of the philosophy of knowledge; and we are thus, at the very outset of the subject, presented with two schools of such philosophy—that, which refers all our knowledge to experience, reserving to the mind only a high degree of activity and excursiveness in collecting, grouping, and systematizing, its suggestions—and that, which assumes the presence of innate conceptions and truths antecedent to experience, intertwined and ingrained in the very staple and essence of our intellectual being, and commanding as with a divine voice, universal assent as soon as understood. The author of the very striking, profound, and in many important respects, original works of which we have undertaken to give some account, belongs to the latter of these schools; and, indeed, appears disposed to press its doctrines and assumptions to a very far greater extent, and to place them in an infinitely bolder prominence than we have been at all aware of having been before done, except perhaps in the writings of some of the later German metaphysicians. We confess in ourselves a leaning, though we trust not a bigoted one, to the other side. And this it is as well to notice at the outset, as it will occasionally tend to place us involuntarily in the apparent position of objectors to the form in which the matter of these works is propounded and treated; while yet we are impressed with a most hearty conviction of their substantive value and importance, and a most genuine admiration of the extraordinary talent and boundless command of resources displayed in their conduct. And after all, it seems far from certain that this opposition of views is anything more than apparent; for among the infinite analogies which may exist among natural things, it may very well be admitted that those only are designed, in the original constitution of our minds, to strike us with permanent force, to embody around them the greatest masses of thought and interest, to become elaborated into general propositions, and finally to work their way to universal reception, and attain to all the recognizable characters of truth, which are really dependent on the intimate nature of things as that nature is known to their Creator, and which have relation to their essential qualities and conditions as impressed on them by Him; so that the power bestowed on the mind of seizing on those primordial analogies, and its impulse to generalize the propositions which their consideration suggests, on the one view of the subject are equivalent to its endowment with a direct recognition of fundamental ideas and relations not derived from experience, and the evolution from those ideas of necessary truths equally independent of experience in the other. And, perhaps, with this explanation both parties ought to rest content, satisfied that, on either view of the subject, the mind of man is represented as in harmony with universal nature; that we are consequently capable of attaining to real knowledge; and that the design and intelligence which we trace throughout creation is no visionary conception, but a truth as certain as the existence of that creation itself.

We must, however, proceed to our analysis of the works before us, which, though separated by a considerable interval in the times of their publication, stand nevertheless to each other so essentially in the relation of parts of one continuous whole, that they cannot be rightly appretiated otherwise than in connexion—the first of them, or the "History," being so constructed, while passing in chronological review the several steps of progress in each department of physical science, as to bring forward in especial salience those features and epochs of scientific discovery in which general principles have been contained and comprehensive views elicited, in such a manner as to lay bare the workings not only of the inventor's mind but of that of his age. From such a review, the "Philosophy" of the subject is not simply left to be collected—it is pointedly led up to; and it is by their combination that we can alone expect to have at length presented to us, in the Philosophy of Inductive Science, what Horace has so clearly and happily indicated as the one great desideratum in that of Life and Morals—

> "*Respicere exemplar* vitæ morumque jubebo
> Doctum imitatorem, et VIVAS HINC DUCERE VOCES."

A work which professes to present a history, so philosophically arranged, of physical discovery in all its departments, and afterwards (passing that history in view—examining it in its various lights—comparing its parts with each other, and from each deriving its appropriate lesson) to deduce therefrom a body of philosophy based on legitimate inductions—to trace out the nature and sequence of the intellectual processes which have led and must continue to lead to discovery—and not only to do this in a general way, but to show by reference to the history of each science that these processes have actually been followed out in its particular case, and to point out in what special mode the application has been made—and all this with the professed ulterior object of deducing from the greatest body of assured and dispassionate truths which the world has yet seen collected, guides and rules, hints and warnings, to aid us in our future researches after truth in more mixed and agitating inquiries;—a work conducted on such a plan, and having such objects, if in any way answering to its design, must deserve to be considered, and must take its rank accordingly, among the most important contributions which have ever been made to the philosophy of mind; nor can it fail to exercise a powerful influence on the future progress of knowledge itself in all its branches.

Mr. Whewell appears on all occasions to be fully alive to the extent of these pretensions and the consequent importance and dignity of his task. There is, however no arrogance in the tone in which they are put forward—and, so far as we can perceive, no partiality in the bias, and assuredly no levity in the temper, of his decisions on the many delicate and difficult points on which, as an historian and a philosopher, he has to pass judgment—not merely as to simple personal questions of priority but as to the substantial merits and value of inductions and discoveries themselves. His own words, in which he states his views and feelings on these essential points, deserve to be cited in illustration of the spirit in which he writes:—

"It is impossible not to see that the writer of such a history imposes upon himself a task of no ordinary difficulty and delicacy; since it is necessary for him to pronounce a judgment upon the characters and achievements of all the great physical philosophers of all ages and in all sciences. But the assumption of this judicial function is so inevitably involved in the functions of the historian (whatever be his subject) that he cannot justly be deemed presumptuous on that account. . . And if I may speak my own grounds of trust and encouragement in venturing on such a task, I knew that my life had been principally spent in those studies which were most requisite to enable me to understand what had been done; and I had been in habits of intercourse with several of the most eminent men of science in our time, both in our own and other countries. Having then lived with some of the great intellects, both of the past and present, I had found myself capable of rejoicing in their beauties, of admiring their endowments, and, I trusted also, of understanding their discoveries and views, their hopes and aims. I did not therefore turn aside from the responsibility which the character of the historian of science imposed upon me. I have not even shrunk from it when it led me into the circle of those who are now alive and among whom we live. . . I trusted, moreover, that my study of the philosophers of former times had enabled me to appreciate the discoveries of the present, and that I should be able to speak of persons now alive with the same impartiality and in the same spirit as if they were already numbered with the great men of the past. . . With all these grounds of hope, it is still impossible not to see that such an undertaking is in no small degree arduous, and its event obscure. — *Pref. Hist.* vol. i.

"I rejoice on many accounts to find myself arriving at the termination of the task which I have attempted. One reason why I am glad to close my history is, that in it I have been compelled to speak as a judge respecting eminent philosophers whom I reverence as my teachers in those very sciences on which I have had to pronounce, if indeed the appellation of pupil be not too presumptuous: but I doubt not that such men are as full of candour and tolerance as they are of knowledge and thought; and if they deem, as I did, that such a history of science ought to be attempted, they will know that it was not only the historian's privilege but his duty to estimate the import and amount of the advances which he had to narrate: and if they judge, as I trust they will, that the attempt has been made with full integrity of intention and no want of labour, they will look upon the inevitable imperfections in the execution of my work with indulgence and hope. There is another source of satisfaction in arriving at this point of my labours. If after our long wandering through the regions of physical science we were left, with minds unsatisfied and unraised, to ask 'Whether this be all?' our employment might well be deemed weary and idle. If it appeared that all the vast labour and intense thought which had passed under our review had produced nothing but a barren knowledge of the external world or a few arts ministering merely to our gratification; or if it seemed that the methods of arriving at truth, so successfully applied to these cases, aid us not when we come to the higher aims and prospects of our being; — this history might well be estimated as no less melancholy and unprofitable than those which narrate the wars of states and the wiles of statesmen. But such is not the impression which our survey has tended to produce. At various points the researches which we have followed have offered to lead us from matter to mind — from the external to the internal world; and it was not because the thread of investigation snapped in our hands, but rather because we were resolved to confine ourselves for the present to the material sciences, that we did not proceed onwards to subjects of a closer interest." — *History*, vol. iii. p. 62.

This is excellent; but in illustration of the general spirit in which the work is written, we must yet cite a few more sentences: —

"Bacon's purpose was that his new organ should produce material as well as intellectual profit—works as well as knowledge. That the study of the order of nature does add to man's power, the history of the sciences since Bacon has abundantly shown; but though this hope of derivative advantages may stimulate our exertions, it cannot govern our methods of seeking knowledge without leading us away from the most general and genuine forms of knowledge. The nature of knowledge must be studied in itself and for its own sake before we attempt to learn what external rewards it will bring us. I have not therefore aimed at imitating Bacon in those parts of his work in which he contemplates the increase of man's dominion over nature as the main object of natural philosophy; being fully persuaded that, if Bacon himself had had unfolded before him the great theories which have been established since his time, he would have acquiesced in their contemplation, and would readily have proclaimed the real reason for aiming at the knowledge of such truths to be—that they are *true*."—*Philosophy of the Ind. Sci. Pref.* xiii.

"As we have already said, knowledge is power, but its

interest for us in the present work is—not that it is power, but that it is knowledge."—*Philosophy*, vol. ii. p. 576.

This is a chord which we rejoice to hear sounded: science has scattered her material benefits so lavishly wherever she has been in presence, that no small number of her followers — and all the multitude — have left off gazing on the resplendency of her countenance in their eager scramble for her gifts. From those who frequent her courts with such views she veils her brightness and withdraws her spirit, leaving them to grovel, poring like Mammon on the golden pavements of her mansion, while their ears are deaf to its celestial harmonies, and their nostrils closed to its breathings of paradise. Our age and our nation, we grieve to say it, too often need to be so reminded.

In presenting the History of the Sciences, Mr. Whewell pursues a course not a little novel, and which gives a picturesque or rather epic interest to his narrative, while it secures the eminent advantage of concentrating attention on the most important and characteristic epochs. These, to which he attaches the epithet "inductive epochs," or those "in which the inductive process by which science is formed has been exercised in a more energetic and powerful manner," are each, in his mode of presenting the subjects, considered as led up to, and ushered in by, a *prelude*, during which "the ideas and facts on which they turned were called into action; were gradually evolved into clearness and connexion, permanency and certainty; till at last the discovery which marks the epoch seized and fixed for ever the truth which had till then been obscurely and doubtfully discerned."

"And again, when this step has been made by the principal discoverers, there may generally be observed another period, which we may call the *sequel* of the epoch, during which the discovery has acquired a more perfect certainty and a more complete development among the leaders of the advance; has been diffused to the wider throng of the secondary cultivators of such knowledge, and traced to its distant consequences. This is a work, always of time and labour, often of difficulty and conflict."

Every such epoch in short we may look upon as the hunger, the meal, and the digestion of one intellectual day; or, if we prefer a less ignoble simile, the muster, the victory, and the pursuit of each decisive intellectual struggle; though, perhaps, our author's idea of the *sequel* may be better illustrated by the occupation and settling of the country under the dominion of the conquerors, quelling the insurrectionary movements of ignorance and prejudice under the new régime, and partitioning out the land in provinces and domains.

In presenting Scientific History under this form, Mr. Whewell has been led almost unavoidably to assign to each of the most active inductive epochs its hero, on whom all the strong lights of his pictures are thrown — its Protagonist, on whom the highest interest of the drama is concentered. Thus we have the inductive epochs of Hipparchus and of Copernicus in formal, and of Newton in physical astronomy — of Galileo in mechanics — of Young and Fresnel in Photology — that of Stahl, of Lavoisier, and of Davy and Faraday in chemistry, &c. It may perhaps be objected to this course, that it can hardly be pursued without throwing into comparative shade, and so far lightly treating, characters of great eminence, to whom Science is deeply indebted, who have either pioneered the way before, or beaten it after the passage of those triumphal cars in which the more fortunate leaders receive our homage. Provided the selection, however, be duly made, and merit be always accorded in other cases where merit is really due, we see no injustice in this. It must be remembered that the History of Science is the History of the Mind — of that which is most essentially and emphatically personal. The thoughts of a philosopher, and his incursions into the realm of unexplored truth, are far more strictly his personal exploits than the victories of the general or the combinations of the statesman. Every step in the higher theories has been an achievement in which the *spolia opima* have fallen to the leader's prowess, and in falling have decided the day, however the masses may have then rushed in and secured the conquest. It is too much the present fashion to ascribe all progress — at least all modern progress — in inductive science, and indeed in every department of human thought and action, to "the Age," as if there were some magic in the word, and as if by its use it were possible to elude or abate down the acknowledgement of individual pre-eminence. True it is that in the collection of facts, and in those subordinate inductions by which classes are established and laws evolved — in all that is the province of mere experiment and observation, and in much that conduces to their right understanding — the great command of means and leisure enjoyed by multitudes of clever men, and the spirit of open-eyed inquiry which pervades all the educated part of society, will do, and is doing, much to facilitate those last steps of the inductive processes which terminate in *established theories.* But no merely *clever man* ever struck out a great theory, and it remains no less true that these steps are in all cases gigantic strides, in which a gulf is passed, a barrier overleaped; and that, from the advance so gained, all precursory knowledge suddenly assumes an aspect of novelty, and may be said almost to have been at that moment entirely rediscovered, so effectually is it summed up in its new form of enunciation. Nor is it less certain that this final and consummating step is in all cases an impossibility to any mind but one which grasps and controls the sum of what is known with a force capable of crushing it into condensation and moulding it into a form congruous with yet more general harmonies. And—what in a philosophical point of view is of chief importance — these, to use the language of Bacon, are the "glaring instances" (*instantiæ ostensivæ*) in which the phenomena of the inventive faculty stand out in their strongest and most eminent form, and whose study promises to lead by the nearest induction to a knowledge of the laws and conditions of this faculty. It is precisely these steps which it is of most importance to contemplate, both as the most difficult in themselves and as leading to the widest consequences. The following very striking passages from Mr. Whewell's Reflections on the Epoch of Newton, and the doctrine of Universal Gravitation, will put our readers in possession of his views on this subject, which appear to us to have both truth and originality: —

"Such then is the great Newtonian doctrine of Universal Gravitation, and such its history. . . . Any one of the five steps into which we have separated the doctrine would of itself have been considered an important advance; would have conferred distinction on the person who made it and the time to which it belonged. All the five steps made at once formed not a leap but a flight — not an improvement merely but a metamorphosis — not an epoch but a termination. . . . The requisite conditions for such a discovery in the mind of its author were, in this as in other cases, the idea, and its comparison with facts; the conception of the law, and the moulding this conception in such a form as to correspond with known realities. . . . In the mere conception of universal gravitation Newton must have gone far beyond his contemporaries both in generality and distinctness; and in the inventiveness and sagacity with which he traced the consequences of this conception he was, as we have shown, without a rival, and almost without a second. . . . It is not easy to anatomize the constitution and the operations of the mind which makes such an advance in knowledge. Yet we may observe that there must exist in it, in an eminent degree, the elements which compose the mathematical talent. It must possess distinctness of intuition, tenacity, and facility in tracing logical connexion, fertility of invention, and a strong tendency to generalisation. . . . Newton's inventive power appears in the number and variety of the mathematical artifices and combinations which he devised, and of which his books are full. If we conceive the operation of the inventive faculty in the only way in which it appears possible to conceive it — that while some hidden source supplies a rapid stream of possible suggestions, the mind is on the watch to seize and detain any one of these which will suit the case in hand, allowing the rest to pass by and be forgotten — we shall see what extraordinary fertility of mind is implied by so many successful efforts: what an innumerable host of thoughts must have been produced to supply so many that deserved to be selected. And since the selection is performed by tracing the consequences of each suggestion, so as to compare them with the requisite conditions, we see also what rapidity and certainty in drawing conclusions the mind must possess as a talent, and what watchfulness and patience as a habit." — *History*, ii. 180, *et seq.*

The personal character of Newton, and the painful interval of suspension in which, at one period, his mental faculties appear to have been held, in consequence of excessive fatigue and over-excitement, have been of late so much discussed, that we must be pardoned if we prolong this extract beyond what is immediately necessary to our present purpose, by a few sentences bearing more directly on his individual character and habits. He has been represented as in some degree deficient in the loftier and more powerful elements of moral, as distinguished from intellectual character. We deem otherwise; and that, had circumstances, unhappily for mankind, forced the development of his faculties in some other line, he would have shown the same ascendency of a determined purpose — the same predominance over difficulties and obstacles — the same profound and perseveringly executed plans, that characterized the scientific career which consumed the vigour of his best years. Mr. Whewell would seem to have formed a similar estimate.

"The stories which are told of his extreme absence of mind probably refer to the two years during which he was composing his 'Principia,' and thus following out a train of reasoning the most fertile, the most complex, and the most important which any philosopher had ever to deal with. The magnificent and striking questions which, during this period, he must have had daily rising before him, the perpetual succession of difficult problems, of which the solution was necessary to his great object, may well have entirely occupied and possessed him. He existed only to calculate and to think. Often, lost in meditation, he knew not what he did, and his mind appeared to have quite forgotten its connexion with his body. His servant reported that in rising in a morning he frequently sat a large portion of the day half dressed on the side of his bed; and that his meals waited on his table for hours before he came to take them. *Even with his transcendant powers, to do what he did was almost irreconcilable with the common conditions of human life, and required the utmost devotion of thought, energy of effort, and steadiness of will —the strongest character as well as the highest endowments which belong to man.*" — *Hist.* ii. 185-6.

It is not our purpose to enter into any minute analysis of the historical part of Mr. Whewell's work. Admirable as it is, and justly as it might claim a more detailed criticism, the far higher interest of the philosophical volumes demands our chief attention. The field into which it would be necessary to enter, were we disposed to pursue a different course, is so wide, that a separate article, and that of no ordinary extent, would be required to convey an adequate impression of its merits. A general sketch of its arrangement and conduct will be, however, necessary for the understanding of what follows, and must suffice for our present purpose.

It is among the Greeks that we are to look for the first dawn of inquiry into the causes and principles of natural events and the constitution of the world — the first at least of which any distinct knowledge has descended to us. Their versatile and inquisitive character led them by no cautious or measured steps into the most obscure and abstract, as well as in the most obvious and tempting paths of speculation. Mind and matter, moral and physical relations, seemed spread before their eager gaze, rather as a flowery field where brilliant discoveries and general truths, freely offered in spontaneous growth, might be gathered up with little effort, than as (what it really is) a tangled region of dark and thorny enigmas to be resolved by patient thought no less than by happy divination. Their early philosophers, therefore

"entered upon the work of physical speculation in a manner which showed the vigour and confidence of the questioning spirit, as yet untamed by labours and reverses. It was for later ages to learn that man must acquire slowly and patiently, letter by letter, the alphabet in which Nature writes her answer to such inquiries. The first students wished to divine, at a single glance, the import of the whole book."

The signal and complete failure of every attempt of the early Greeks to establish any sound principle in Physics, contrasts remarkably with their brilliant successes in abstract mathematics. But whence this failure? The question is one of great importance in the outset of a Philosophical History of Science, and accordingly is made by Mr. Whewell the subject-matter of his first book. We may condense in a few words his solution of this curious problem. The founders of the Greek School Philosophy, sought, it is true, the elements of their inductions in the phenomena of nature, but sought them not in a careful and philosophical analysis of facts, but rather in a minute examination of the *words* and *forms of language* in which those facts are expressed by superficial observers in the crude and commonplace parlance of everyday life. Were language a true picture of nature, a perfect *daguerreotype* of all her forms, this proceeding might be pardonable. Half the labour of the modern inductive philosopher is to construct a language which shall be such. But common language is a mass of metaphor, grounded, not on philosophical resemblances, but on loose, fanciful, and often most mistaken analogies. From studying such language as the representative of Nature, no pure and fundamental classification of facts, such as legitimate Induction requires, can result; but, on the contrary, the greater the acuteness and the broader the induction, the wider will be the departure from sound philosophy. "In Aristotle," says Mr. Whewell, "we have the consummation of this mode of speculation. The usual point from which he starts in his inquiries is, that *we say* thus or thus in common language." And this he exemplifies in various instances. Hence the doctrine of contrarieties, a most fertile source of Aristotelian confusion, in which

"it was assumed that adjectives or substantives which are in common language, or in some abstract mode of conception, opposed to each other, must point at some fundamental antithesis in nature which it is important to study."

Thus, for example, *light* came to be considered as the opposite to *heavy*, not as its inferior degree, to the utter vitiation of the Aristotelian statics and dynamics.

We see, then, that in the Greek School Philosophy facts *were* appealed to, but facts as they stand distorted and falsified in vulgar language, not as they really existed in nature; still less as subjected to any process of just analysis. Hence, in their classifications, though they had in their possession both facts and ideas, the ideas, to use Mr. Whewell's pointed form of expression, were neither *distinct* nor *appropriate to the facts;* without which there can be no science.

"It will appear from what has been said," says Mr. Whewell, "that there are certain ideas or forms of mental apprehension which may be applied to facts in such a manner as to bring into view fundamental principles of Science; while the same facts, however arranged or reasoned about, so long as their appropriate ideas are not employed, cannot give rise to any exact or substantial knowledge."

We call the reader's attention to this passage, because "the forms of mental apprehension" to which he alludes in it play a very conspicuous part in his philosophical views. The obvious sense of the passage to those who are familiar with what has previously been written on this subject would seem to be that there are both appropriate and inappropriate *Heads of Classification*, under which facts may be grouped, and that, if grouped under the former, *causes* (whether proximate or ultimate) or laws fitted to form elements of higher inductions, will *ipso facto* be suggested—if under the latter, nothing but vague and fallacious inductions will be raised, while the true principles will elude our grasp. But this is not *all* Mr. Whewell's meaning, as will abundantly appear in the sequel.

Archimedes alone among the Greeks succeeded in obtaining clear hold of one, and that the most important, of these fundamental ideas, viz.—force or pressure as a *measurable* quantity, and as measured by the conditions of its equilibrium with other forces assumed as known. A "*glaring in-*

stance," drawn from vulgar experience, furnished the axiom which served him to render a true account of the property of the lever, viz.—that the weight of a body or collection of bodies, or its pressure on the point of its suspension, is not altered by moulding the body into different forms or by changing the arrangement of the individuals of such collection. "The weight of a basket of stones is not altered by shaking the stones into a new position." Now it must be observed that the "*instance*" in question is a general, not an individual one. It is in the strictest sense an *inductive* proposition, drawn not from a single case but from the unbroken experience of all mankind. That which makes it fertile in Philosophy is, that the individual facts which have gone to make up this general one were grouped by Archimedes under their appropriate head, *i. e.*, *Total pressure regarded as the sum of partial pressures.* That which can be variously subdivided, and yet always summed up into the same total, must be quantitatively measurable, susceptible of precise numerical relations and capable of affording a handle to exact mathematical reasoning. Mr. Whewell's comment on this induction is remarkable. The general fact, he says,

"is obvious, when we possess in our minds the ideas to apprehend it clearly. When we are so prepared, the truth appears to be manifest, *independent of experience*, and is seen to be a rule to which experience *must* conform."—*History*, book ii. p. 93. (The italics are our own.)

Here we have the first instance of that erection of a standard of *physical*, as distinct from logical truth, yet wholly *within the mind*, a standard different from and paramount to experience, and so far, therefore, antecedent to it, which forms, as we have before observed, so distinguishing a feature of Mr. Whewell's Philosophy. We cite it thus early as it occurs, to show how entirely it pervades every part of his speculations, and how integrant a portion it constitutes of them.

We owe to Archimedes also the discovery of the fundamental principles of hydrostatics. The character of this philosopher offers many points of close resemblance to that of Newton. We trace in him the same paramount development of the mathematical faculty—the same tendency to apply it to physical subjects—the same acute perception of really important and essential features, such as admit of general and abstract statement, and are thereby fitted to become axioms in science—the same fertility of resource in the creation of new geometrical methods when the powers of the old ones proved inadequate to his objects; methods which in effect, and as involving the passage from the finite to the infinite, contained the germ of the fluxional or differential calculus, and enabled him to resolve problems which peculiarly and essentially belong to the domain of that calculus. We find in him, too, the same habits of intense, continued, and abstracted thought, nay, even the same tendency to mechanical constructions and optical improvements; in a word, the only combination the history of mind has offered which we can believe capable, if placed in Newton's position, of accomplishing what Newton did. When Archimedes perished, in the wreck of his nation, a light was extinguished which, had it been suffered to shine, might have accelerated by a thousand years the maturity of the inductive philosophy.

The formal astronomy of the Greeks forms the subject of the third book of Mr. Whewell's "History," and both in that work and in the "Philosophy" affords room for much valuable and instructive remark. The earlier stages of this science, the determination, with some degree of exactness, of the relation between the year, the month, and the day—the establishment of cycles expressive of this relation, and of others adapted to the prediction of eclipses by their periodic recurrence—the recognition of the earth's sphericity, &c.; these are matters which involve little theory, and draw but little on the inventive faculty. On these, however, Mr. Whewell observes that

"the familiar act of thought exercised for the common purposes of life, by which we give to an assemblage of our impressions such a unity as implied in the above notions and terms, a month, a year, ... and the like, is in reality an *inductive* act, and shares the nature of the processes by which all sciences are formed."—*Hist.* b. i. p. 109.

If the term inductive, applied here to this very important mental act, be understood in that technical sense in which it is commonly used, when speaking of physical discoveries, viz., as the concluding of something more general by the assemblage of particulars of a less general kind, we must demur to this remark; but if it be intended to designate every inductive act of the mind, as an instance of the exercise by it of that peculiar constructive or plastic faculty in virtue of which out of the assembled perception of qualities, it constitutes an object — out of extension, figure, resistance, colour, smell, a body — out of a series of dots an outline, &c.; then we not only agree with the assertion, but regard it as expressing a full and complete theory of induction itself, and of the mode in which our minds not only form to themselves conceptions of numerical aggregates by the contemplation of units, but *construct general propositions themselves from the contemplation of particulars, and attribute to them a universality which experience alone is incapable of warranting.* When by repeated verifications of its assertion in individual cases, the course of a general proposition is, so to speak, *dotted out* before the mind, and when the particulars are brought so close, that the attention glides easily, and is, as it were, conducted from one to the other, so as to suggest a law of connexion, there requires no more to induce the mind to fill up by its own act the intervals between them. Urged by a powerful and ready impulse of which we can give no account but that it *is so*, but which would seem to be a modification of the influence of habit (if it be not itself the origin of that influence), *we assume a continuity where we find none*, and in this manner are led to believe the cases where we have no experience, on the evidence of those in which we have. We are far from imagining, however, that Mr. Whewell would be disposed to acquiesce in this view of the inductive *nisus*. His views assume something yet more active and independent in the operation of the mind in such a case. According to his conception of the matter, the mind supplies much more than the mere completion of continuity. It spins from a store within itself that thread on which, and on no other, the pearls shall be strung. It finds, already self-traced on its own tablets, that subjective line to which the *dots* of experience only give the semblance of an objective reality. Experience, according to him, only exemplifies, cannot prove a general proposition. Its truth stands on the higher and independent ground of *inherent necessity*, and is recognized to do so by the mind so soon as it becomes thoroughly familiarized with the terms of its expression.

The hero of the inductive epoch of the Greek astronomy is Hipparchus, having for his forerunners in its prelude Eudoxus and Calippus; the epicyclic theory, its matter of induction; and the development of this by Ptolemy and his successors down to Aboul Wefa and Tycho, its sequel. This theory, though clumsy as a physical hypothesis, and consistent only with a part of the facts of the system it undertakes to explain, and we may add, assuredly not believed in as a mechanism by its devisers, was yet a bold and fine conception for the embodying a large assemblage of facts, and one which, as regards those facts which it does include, has continued, under a very different aspect, to maintain and even to extend its ground in modern theory, being in effect a shadowing forth of the now demonstrated principle of the sufficiency of circular functions of the time to represent all the phenomena of the planetary motions. We have here, then, a case of very high philosophical interest. The general proposition of the epicyclic theory remains true, though stated in the language of falsehood, and though arrived at by fanciful analogies and untrue assumptions. "We thus see," observes Mr. Whewell,

"how theories may be highly estimable, though they contain false representations of the real state of things; and may be extremely useful, though they involve unnecessary complexity. In the advance of knowledge, the value of the true part of the theory may much outweigh the accompanying error, and the use of a rule may be little impaired by its want of simplicity." — *Hist.* b. iii. p. 181.

"The principles which constituted the triumph of preceding stages of science may appear to be subverted and ejected by later discoveries, but in fact they are (so far as they are true) taken up into the subsequent doctrines and included in them. They continue to be an essential part of the science. The earlier truths are not expelled but absorbed, not contradicted but extended; and the history of each science which may thus appear like a succession of revolutions is, in reality, a series of developments." — *Introd. Hist.* b. i. p. 10.

The discoveries of Copernicus and Kepler, which complete the history of formal astronomy (thenceforward to be merged in the more extensive views of its physical theories), form the subject of Mr. Whewell's fifth book. But before entering on this theme, his narrative is suspended, to afford opportunity for a general view of the state of science in the middle ages, or, as he terms it, the stationary period, in which,

"along with the breaking up of the ancient forms of society, were broken up the ancient energy of thinking the clearness of idea, and steadiness of intellectual action. This mental declension produced a servile admiration for the genius of better times, and thus the spirit of commentation. Christianity established the claim of truth to govern the world; and this principle, misinterpreted and combined with the ignorance and servility of the times, gave rise to the dogmatic system: while the love of speculation, finding no sure and permitted path on solid ground, went off into the regions of mysticism." — *Hist.* i. 355.

These several heads, therefore, viz., the indistinctness of ideas — the commentatorial spirit — the mysticism — and the dogmatism of the middle ages — furnish matter for four admirably written chapters of the book devoted to the history of this period, while a fifth, replete with interest, is assigned to the progress of the arts in those ages, in so far as that progress can be said to have any bearing on science. We regret that our limits will not allow us to cite several of the many striking passages with which these chapters abound, and one in particular on the revival of architecture in the twelfth and succeeding centuries (a subject which appears to have occupied much of our author's attention), by reason of the ingenious manner in which it connects the curious and original views of Mr. Willis on the character and formation of the Gothic style with the revival of sound mechanical ideas.

The Copernican or heliocentric doctrine of the planetary system is so familiar to us, and so entirely identified with the ideas we have received as elementary, that perhaps it may startle some of our readers to be told that the Epicyclic theory formed an essential part of Copernicus's views—so much so, indeed, that his chief, nay his only merit, in the revival of this ancient doctrine, and the only ground on which we can justifiably continue to attach his name to it, is, that he demonstrated the applicability to the heliocentric system of this theory, which had been previously found efficacious in embodying all the then known parts of the geocentric.

In discussing the reception and diffusion of the theory of Copernicus, Mr. Whewell is necessarily led to the subject of the persecutions of Galileo for their advocacy. In his observations on these transactions, and on the general subject of the scientific interpretation of scriptural expressions, there is a right-mindedness, a tolerance, and a moderation, which we would recommend to the especial notice of all who venture on the bitter and troubled waters of religious controversy: —

> "The meaning," he observes, "which any generation puts upon the phrases of Scripture depends, more than is at first sight supposed, upon the received philosophy of the time. Hence, while men imagine that they are contending for revelation, they are in fact contending for their own interpretation of revelation, unconsciously adapted to what they believe to be rationally probable. And the new interpretation which the new philosophy requires, and which appears to the older school to be a fatal violence done to the authority of religion, is accepted by their successors without any of the dangerous results which were apprehended. When the language of Scripture invested with its new meaning has become familiar to men, it is found that the ideas which it calls up are quite as reconcileable as the former ones were with the soundest religious views. And the world then looks back with surprise at the error of those who thought that the essence of religion was involved in their own arbitrary version of some collateral circumstance." — *Hist.* i. 403.

The philosophical character of Kepler is admirably drawn; the quest in which this most garrulous and amusing writer, but at the same time most ardent and truth-loving man, set forth in the heavens, has much analogy to that of Columbus on earth. Each was urged by a strong inward conviction that there *must be* a body of truth capable of detection, a new realm to be laid open in that particular direction in which his researches tended. Each made its discovery the object of his entire devotion—pursued it with a dogged, and what might be thought a desperate perseverance, and not content with partial success when attained, renewing the attempt again and again, and always with increasing good fortune. In all that regards the tone of personal character there cannot be a stronger contrast than between the grave and stately bearing of the noble Genoese and the mercurial vivaciousness and *naïve* self-exposure of his astronomical parallel, but in the earnest devotion of each to his dominant idea, and the magnificent disclosures with which that devotion in each case was rewarded, the parallel is close.

Kepler was indefatigable in framing and trying hypotheses, and many of those which he did try, and which proved unsuccessful, have been since censured as visionary and fanciful, while some have felt scandalized that *any* perseverance in a mere system of guesses should have been so brilliantly rewarded. But, in the first place, it is difficult to say, among mere guesses, in the absence of all sound principle, that those which proved successful were to be deemed less fanciful than those which failed: and in the next place, it must be remembered that almost all Kepler's guesses were grounded on what he considered as physical assumptions. "In making many conjectures which on trial proved erroneous, Kepler was not more fanciful or unphilosophical than other discoverers have been. Discovery is not a 'cautious' or a 'rigorous' process in the sense of abstaining from such suppositions." Kepler's guesses, Mr. Whewell goes on to say, "exhibit to us the usual process, somewhat caricatured, of inventive minds—they rather exemplify the rule of genius than, as has been hitherto taught, the exception." (*Hist.* i. 412.)

"This is the spirit in which the pursuit of knowledge is generally carried on with success: those men arrive at truths who eagerly endeavour to connect remote points of their knowledge, not those who stop cautiously at each point till something compels them to go beyond it."—*Hist.* vol. i. p. 423.

"Kepler's talents were a kindly and fertile soil which he cultivated with abundant toil and vigour, but with great scantiness of agricultural skill and implements. Weeds and grain throve and flourished side by side almost undistinguished, and he gave a peculiar appearance to the harvest by gathering and preserving the one class of plants with as much care and diligence as the other."—*Hist.* vol. i. p. 415.

The sixth and seventh books of Mr. Whewell's History contain a condensed, but well arranged and philosophical summary of the completion of the science of dynamics, and its triumphant application to physical astronomy, in the inductive epochs of Galileo and Newton, with all their noble train of consequences. This is beaten ground, and admitting of little novelty in the mode of traversing it. In that which Mr. Whewell has chosen, and which was necessary to his plan, the chronological order of discovery in the general science and in its application is pursued separately, a condition which gives rise to some confusion in details, inasmuch as the creation of new methods in dynamical science, and the generalization of its conceptions were mainly consequent on and directed to the solution of those great problems which the system of the world involves, and which have stamped their own character on the larger portion of the general science.

Until the laws of mechanical action were discovered and applied through the intermedium of mathematical analysis to the explanation of natural phenomena, all physical science might be considered as groping in the dark. In no previous instance had speculation been able to lead up to a clear perception of efficient causes, far less to an exact apprehension of their mode of action, so as to trace them into their effects. In the broad daylight which the discoveries of Newton and his followers poured over every part of the system of nature, men saw with astonishment in how wondrous a complication of reciprocal actions and influences its frame subsists; and in attempting to carry their newly-acquired principles into all its details, they beheld, developing themselves as corollaries and dependencies on each particular point of those discoveries, branches of science either altogether new, or receiving from the new light thrown on them such novelty of aspect and such vast and rapid accessions as may justify us in regarding them of modern creation. Moreover, it speedily became evident in the endeavour to give a purely mechanical explanation of phenomena, that whatever forces act to produce certain classes of them, must be conceived to act through the medium of some organization or mechanism, different according to their nature, and so imposing peculiar characters on their explanation. And we may now further add, on a review of those classes and of the phenomena which later research has brought to light, that although, undoubtedly, all sensible changes and movements of matter are *directly* referable to acting *forces*, and are therefore the *immediate* results of mechanical effort; yet in the explanation of innumerable phenomena, it is impossible to limit our views to such effort even as an *ultimate physical* cause. We have to ascend a step higher, and to assign—or if not to assign, to seek—if not to seek, at least to recognise as admissible, an ulterior cause (as something distinct from a *motive* or a *reason*) for the exertion or development of force itself under the circumstances; nay, to admit the possible agency of more than one such cause, giving rise to the development of forces under a variety of different but definite aspects. In a word, we seem on the verge of obtaining a glimpse of causes, which, though strictly physical, are yet of a higher order than force itself, and of which this latter is one of the direct or indirect effects. Such a cause we think we recognize as an object of consciousness, in that effort (accompanied with fatigue and exhaustion) which intervenes between the mental act of mere volition and the muscular contraction which moves our limbs.* Such causes, too, may possibly lie at the root of chemical affinity, of electric and magnetical polarity, and thence, by no remote analogy, of gravitation itself, and of all those material forces whose action is not merely temporary or occasional, but permanent and continuous.

But not to plunge deeper at present into a line of speculation which is very forcibly suggested by several passages in Mr. Whewell's work, and to which we shall probably again be led in our further remarks on it,—it is clear, meanwhile, that the multitude of branches into which, from the Newtonian epoch downwards, the path of science has been constantly diverging—renders it necessary to define and classify them in order to follow out their history with anything like distinctness and with any regard to philosophical views in its treatment. The classification which Mr. Whewell adopts, though not unexceptionable, is perhaps, in the present state of human knowledge, as convenient for his especial purpose as any which could have been made. Under one general head (" The Secon-

* On this subject see " Cabinet Cyclopædia," Astronomy, § 370, and the note thereon. The appeal is to the consciousness of those who will very carefully attend to their own sensations and mental acts. Disease, by retarding and disturbing processes which in health are performed almost unconsciously, will often enable us to analyse phenomena that common observation regards as simple. In Dr. Holland's " Medical Notes and Reflections," p. 504, a work replete with profound philosophy, we find cases recorded strikingly in point to the idea in the text.

dary Mechanical Sciences"), he includes acoustics, optics, and thermotics, because "in these, phenomena are reduced to their mechanical laws and causes in a secondary manner," or by the intervention of a *medium*. Under the "Mechanico-chemical" sciences he classes electricity, magnetism, and galvanism, or voltaic electricity, with its new appendage of electro-magnetism. Chemistry itself is classed as "The Analytical Science;" mineralogy as the "Analytico-classificatory," constituting a sort of link between the science of pure analysis and those which he regards as purely classificatory, such as botany and zoology. Under "Organical Sciences," we have physiology (or, as he terms it subsequently and more properly, biology) and comparative anatomy; while geology forms the nucleus of a class of especial and novel interest under the title of "Palætiological Sciences," "whose object it is to ascend from the present state of things to a more ancient condition from which the present is derived by intelligible causes."

It must be quite obvious that this enormous bill of fare, if taken in detail, can, by no conceivable process of intellectual cookery, be brought within the compass of a single meal; nor within our limits, and with the deeper interest of the philosophical volumes yet soliciting our attention, can we undertake even to condense a quintessence, or select a leading flavour from each course. The fact is that the eleven books, of which the remainder of Mr. Whewell's history consists, must rather be regarded as philosophical epitomes of their several subjects —outlines struck with a large and free hand, and destined to fix attention on leading features (though traced with perfect mastery and with consummate skill) — than as digested histories of the above enumerated branches. To have made them such, would not only have been impracticable within thrice the compass to which the work extends, but would have utterly overlaid and defeated the author's objects in writing it as we have above stated them. Accordingly, he expressly disclaims any such intention. (*Hist.* vol. ii. p. 293.) Regarding as we do, both in the remarks we have already made and in those we are about to offer, the merely historical as quite subordinate to the philosophical interest of the subject, we entirely approve of this mode of proceeding, though we could perhaps have wished that, by some modification in the title, the particular scope and limits of the work itself had been more pointedly expressed.

Of these books we find most to admire and approve in those which treat of the purely Classificatory and Palætiological Sciences, while on the other hand, that on the "Analytico-Classificatory Science," or Mineralogy, though apparently laboured with more care than any of the rest, strikes us as somewhat less successful, not from any want of perfect and intimate acquaintance with the subject, but rather, on the contrary, from a too intimate perception of its weakness as a science. Mineralogy, indeed, is of all sciences perhaps the least satisfactory; nay, we are even disposed to question whether it ought not rather to be struck out of their list, or degraded from an independent rank. A mineral which is neither a definite chemical compound, nor a recognizable crystalline aggregate, must assuredly stand low as an object of scientific attention and inquiry, though as a deposit it may interest the geologist, or as a material the artist. To dignify the science itself Mr. Whewell is obliged to generalize it.

"We have seen," he says, "that the existence of chemistry as a science which declares the ingredients and essential constitution of all kinds of bodies, implies the existence of another corresponding science which shall divide bodies into kinds, and point out, steadily and precisely, what bodies they are which we have analysed. But a science thus dividing and defining bodies is but one member of an order of sciences, different from those which we have hitherto treated, viz., the Classificatory Sciences. Mineralogy is the branch of knowledge which has discharged the office of such a science so far as it has been discharged; and indeed has been gradually approaching to a clear consciousness of its real place and whole task."—*Hist.* vol. iii. pp. 188. 190.

This is assuredly very ingenious. But it amounts to merging the science of *Mineralogy* in that higher and purer branch which Mr. Whewell has the great merit of here, for the first time, distinctly pointing out, and which has for its objects the classification of chemical elements and combinations in general by their crystalline and optical relations and mechanical and external qualities, and thus connecting the sciences of chemistry, optics, and crystallography, and perhaps many others, by the most important fundamental relations of polar forces. Classification in such a case is only another word for the announcement of general laws, the results of inductive observation: results, that is to say, of a more elevated order than those which depend on a mere remarking of general resemblance, or even on the specification of particular arbitrarily selected points on which the logical proof of such resemblance can be rested. Accordingly, in so far as in this last sense of the word, mineralogy is to be regarded as a classificatory science, its history offers only a succession of failures. Perhaps the most remarkable of these are precisely those in which the specified points of resemblance are the most distinct and systematic, viz., those of Berzelius and Mohs, both which Mr. Whewell condemns, and we think justly.

In geology our author is a catastrophist, or rather an anti-uniformist.

"*Time*," he says, "inexhaustible and ever accumulating his efficacy, can undoubtedly do much in geology: — but *Force*, whose limits we cannot measure, and whose nature we cannot fathom, is also a power never to be slighted: and to call in the one to protect us from the other is equally presumptuous to whichever side our superstition leans." — *Hist.* vol. iii. p. 616.

This is sensibly as well as pointedly stated. The most strenuous advocate for the exclusion of paroxysmal epochs will not contend for *perfect* uniformity so long as earthquakes are not of daily occurrence and calculable intensity: and the question as to what is and what is not paroxysm, — to what extent the excursion from repose or gentle oscillation may go without incurring the epithet of a catastrophe, is one of mere degree, and of no scientific importance whatever. Geology as a body of science has been always too much divided by antagonist doctrines and by the opposition of rival schools. The eagerness of the combatants in the Plutonic and Neptunian controversy surpassed the bounds of amicable discussion, and decidedly retarded the progress of sound theory: and now that these rival divinities have sacrificed their exclusive claims and agreed to act in unison, the cataclysmal and uniformitarian systems, though advocated in a far better spirit, are yet, we think, rather too deeply tinging the views of modern geologists and biasing their course of speculation. Mr. Whewell by mooting the question as to what *is* uniformity, has afforded the antagonist schools a point of approximation where they may merge their differences and unite their efforts.

Though we are glad to observe that a small part only of these chapters is devoted to controversial points, yet we were hardly prepared to expect so decided an undervaluing of Dr. Hutton's really important contributions to geological science as we find in Mr. Whewell's section "On Premature Geological Theories," where his "Theory of the Earth" is simply mentioned to be condemned as such, and in which Playfair's fascinating "Illustrations" of that theory, a work which we cannot but believe to have exercised a most important influence on the science generally, by showing the complete untenability of a simple aqueous doctrine, and the absolute necessity for admitting heat at least to a share in its explanations, is passed unmentioned. But, on the other hand, the chapters on "Systematic Descriptive Geology," and those on "Geological Dynamics," are not only excellent as historical compendiums, but so abundant in philosophical views, and present so graphic a picture of the science, that we cannot recommend to the student of that science a better guide to his reading, and key to its speculative difficulties, than he will find in their perusal. In particular we would recommend a careful perusal of the section headed "Question of Creation as relating to Science," and that which follows it as admirably calculated to infuse a spirit of sobriety and caution into all future speculations on the subject of the gradual introduction and extinction of species, a subject doubtless the most startling and bewildering which has ever yet gained admission within the pale of legitimate physical inquiry.

Mr. Whewell divides the "Philosophy of the Inductive Sciences" into two parts; the first treating of "Ideas," the second of "Knowledge;" divisions which, for our purpose, and perhaps also as respects the probable influence of the work on the progress of science, it will be proper to regard as the theoretical and practical departments of this philosophy. The subject of Ideas, which occupies somewhat more than one of the two volumes of which the work consists, is subdivided into ten books. The first, "Of Ideas in General," being devoted to the explication of metaphysical views on the nature of scientific truths, the grounds of our knowledge of them, and the analysis of those mental acts by which we attain and recognize them. The remaining nine books exhibit the application of these general views and principles to the philosophy of each of the great subdivisions of science adopted in the historical work, *seriatim*; with the superaddition, however, of a preliminary book on the philosophy of the pure sciences (the mathematics). These our author has excluded from his history, on the ground of their not being *inductive sciences*. "Their progress," he says, "has not consisted in collecting laws from phenomena, true theories from observed facts, and more general from more limited laws, but in tracing the consequences of the ideas themselves," which lie at the root of them, viz., space and number. As a matter of philosophy, we think this distinction untenable, on grounds we shall presently state, though there can be no doubt that the *inductive part* of these sciences, so far as it has yet been carried, offers no historical points, furnishes no matter of history. Their highest axioms have been quickly and readily arrived at; and it is only on their *deductive* part that any great amount of intellectual effort has been expended. It is on this very ground, however, that we perceive the greater propriety in their occupying a prominent place in the philosophy of inductive science, in which we hold them to exemplify what Bacon would term clandestine instances, a class always replete with instruction.

As it is in the first of these books that Mr. Whewell developes and distinctly lays down those peculiar *à priori* views to which we have before alluded, and to which, as already said, we feel unprepared to yield entire assent, it will be necessary for us to examine rather in detail this part of his work, at the risk, it may be, of some degree of tedium to our non-metaphysical readers; though we shall endeavour, as far as possible, to divest our observations of technical metaphysical phraseology, which, sooth to say, we do not think that very obscure and imperfect science yet sufficiently advanced to indulge in otherwise than sparingly, and as it were *emphasis gratiâ*.

Mr. Whewell's general aim in this book is to show that there exist "certain fundamental ideas or forms of mental apprehension," which, whether by reason of their simplicity, clearness, facility of suggestion, or otherwise, but more especially by reason of their *appropriateness* to the subjects, are peculiarly fitted to become, and have accordingly become, as of necessity they must, the leading features of particular branches of science, and the bases of all sound knowledge in those branches. That these ideas, or some of them (according to their appropriateness), are, in virtue of the activity of the mind, superinduced on, or in some intellectual manner combined with our perceptions, and thus bind together in a certain unity, and according to a certain mode of apprehension, — first, all those sensible perceptions, which, simultaneously affecting the mind, impress it with the conception of *a fact;* secondly, all those facts which, when contemplated together, appear to have a certain relation fitting them to be so united or bound together by one or other of these fundamental ties. These facts, when so bound together, constitute facts of a more general kind, or theories: which, when confirmed by long experience, rendered perfectly familiar by habit, and adopted into common language, come to be regarded as facts, and spoken of and referred to as such (as when, for instance, we speak of the earth's rotation on its axis, or its revolution in an ecliptic orbit round the sun, as facts).

This aggregation, or rather intellectual *cementation*, of facts into theories, is, however, usually performed, not by the direct intervention of the fundamental idea appropriate to each theory, such idea being frequently of an order too elevated and remote for that purpose, but commonly by the intervention of certain "modifications and limitations of the fundamental idea," which may be termed "ideal conceptions." Thus an ellipse is an ideal conception, a *modification* of the fundamental idea of space; genus, a modification or *limitation* of the fundamental idea of resemblance, and so forth.

Were we to express this in ordinary language, we should say that we rise by steps only to the highest degree of abstraction and generality, and in working our way upwards in that direction, we employ terms and phrases more or less abstract, according to the degree of generality which we feel ourselves competent to attain. The line, therefore, between the fundamental idea and the ideal conception appropriate to each step of advancing science, and to each scientific theory, is necessarily indefinite, and accordingly we observe that throughout the work Mr. Whewell uses the one term for the other with little hesitation. The formation of a theory out of facts, and the nature of the inductive process itself, are thus well and clearly described: —

"When we have become possessed of such ideal conceptions as those just described, cases frequently occur in which we can, by means of such conceptions, connect the facts which we learn from experience, and thus obtain truths from materials supplied by experience. In such cases the truth to which we are thus led is said to be collected from the observed facts by induction." — *Phil.* vol. i. p. 42.

After giving examples of this, Mr. Whewell proceeds: —

"And in like manner in all other cases, the discovery of a truth by induction consists in finding a conception, or combination of conceptions, which agrees with, connects, and arranges the facts. Such ideal conceptions, or combination of conceptions, superinduced upon the facts, and reducing them to rule and order, are theories." — *Ib.* vol. i. p. 43.

"The act of the mind, by which it converts facts into theories, is of the same kind as that by which it converts impressions into facts. In both cases there is a new principle of unity introduced by the mind, an ideal connexion established: that which was many becomes one: that which was loose and lawless becomes connected and fixed by rule. And this is done by induction, or, as we have described this process, by superinducing upon the facts, as given by observation, the conception of our minds."—*Phil.* vol. i. p. 44.

"Thus it appears that, understanding the term *induction* in that comprehensive sense in which alone it is consistent with itself, it is requisite to give unity to a fact no less than to give connexion to a theory."— *Ib.* vol. i. p. 45.

It is impossible to express with more precision than Mr. Whewell has done in the passages above extracted, or in a more luminous manner, the true nature of the inductive processes, as regards facts and theories. Two important points, however, remain to be decided: first, the origin within the mind of these ideal conceptions or fundamental ideas themselves; and, secondly, whether, and in what manner, we are justified in extending *theories* so framed, or propositions so concluded, beyond the limits of the individual facts on which our conceptions have been superinduced.

There can be no doubt that the origin of all induction is referable to that plastic faculty of the mind, which assigns an unity to an assemblage of independent particulars.* But in order to carry

* On this subject we will merely refer the reader to Mr. Douglas's excellent work on the Philosophy of the Mind, (Ed. 1839), p. 182. *et seq.*

out this idea to its entire meaning, it is necessary to extend the field in which this faculty exerts itself to every description of impression of which the mind is susceptible. Thus, from the impression it receives from its own acts, states, and faculties, which are never for two consecutive instants the same, or equally exerted, so inductively bound together, the ideas or conceptions of personal existence and identity, time, and mental power arise within it. Again, from those which it receives directly (and antecedently to all *other* experience), from its connexion with the body, it is led to form in a similar way its conceptions of space and mechanical force, which are therefore, we apprehend, in the most complete and absolute sense *suggested* by experience—by the experience, that is to say, of certain *peculiar mental sensations* (if we may coin a word for the purpose) which distance, direction, and force, when perceived, excite within us. Then again, from that mixed multitude of impressions received through the bodily senses, it frames to itself, by a similar induction, the conception, fact, or theory, as we please to call it, of an independent external world. Moreover, from the impressions it receives on contemplating these external relations (which, besides bringing back on it, confirming, and elucidating in innumerable modes, all those more original and simple conceptions, furnish in a thousand ways that which is the true "fundamental idea" of all science, viz., harmony, regularity, or law), it rises by a constantly extending and unbroken chain of experience to the *law of continuity*, which is perhaps the highest inductive axiom to which the mind of man is capable of attaining, and, as one of the most important results of this law, to the perception and admission of general truths, on the ground of particular verifications.

By contemplating our own faculties of attention, recollection, and other similar processes, whereby the mind continually influences the succession of its own thoughts, or rather, in the same instant that we experience that peculiar *mental sensation* which is connected with the exercise of these faculties, we come to have suggested the notion of mental power. By dwelling on the *effort* whereby we put our limbs into motion, the conception of vital effort as expended in the production of mechanical force is in like manner suggested; and by dwelling on the only feature these remarkable phenomena have in common, viz., *change*, predictable beforehand, as sure to be consequent on their *voluntary* exercise, we attain to an abstract conception of *cause* as the origin of *all* change, a conception which once so originated within our minds by this, our highest form of experience, personal consciousness, is reflected back, and verified by all external experience, though in forms far less pure and unadulterated than that in which it is presented to us by these internal phenomena. Lastly, by the experience of our own intentions as capable of being carried out into execution by material or moral combinations, we have suggested to us the notion of *design* or final cause, and by that of our emotions as dependent on the result of our designed acts, the conception of *motive* and of moral responsibility.

Mr. Whewell, however, puts a most decided and unhesitating negative on the claims of experience to the origination of these ideas. We must, therefore, examine the argument by which he supports this negative:—

"We have seen," he says, "that there are propositions which are known to be necessarily true, and that such knowledge is not and cannot be obtained by mere observation of actual facts. It has been shown also that these necessary truths are the results of certain fundamental ideas, such as those of space, time, number, and the like. Hence it follows inevitably that these ideas and others of the same kind are not derived from experience. For these ideas possess a power of infusing into their developments that very necessity which experience can in no way bestow. This power they do not borrow from the external world, but possess by their own nature. Thus we unfold out of the idea of space, the propositions of geometry, which are plainly truths of the most rigorous necessity and universality. But if the idea of space were merely collected from observation of the external world, it could never enable or entitle us to assert such propositions: it could never authorise us to say that not merely some lines but all lines not only have but must have those properties which geometry teaches. Geometry in every proposition speaks a language which experience never dares to utter, and indeed of which she but half comprehends the meaning. Experience sees that the assertions are true, but she sees not how profound and absolute is their truth." — *Phil.* i. 71.

The necessity of geometrical truths has never, we believe, been questioned, nor is it our disposition to do so now. It is not, however, with their *necessity* that we are just now concerned. All true propositions about realities are necessarily true, provided their subject-matter be necessarily such as it is, since every reality must be consistent with itself. Whether space be, as we conceive it to be, a substantive reality independent of our minds, and whether capable of being directly contemplated by them or not, or as Mr. Whewell, adopting the Kantian doctrine, maintains it to be, a *real condition* of the perception of our own and all other existence—if it be a *necessary* reality, or a *necessary* condition, then are the expressions of its properties, in geometrical language, necessary truths. The truths of geometry *exist* and are verified in every part of space, as the statue in the marble. They may depend on the thinking mind for their conception and discovery, but they cannot be contradictory to that which forms their subject-matter, and in which they are realized, in every place and at every instant of time.

But it is with the universality, not the necessity, of its truths that we are concerned, or rather with the nature and grounds of our conviction of their universality:—

"Experience," says Mr. Whewell, "must always consist of a limited number of observations; and however numerous these may be, they can show nothing with regard to the infinite number of cases in which the experiment has not been made. . . . Truths can only be known to be general, not universal, if they depend upon experience alone. Experience cannot bestow that universality which she herself cannot have, nor that necessity of which she has no comprehension." — *Phil.* i. 60, 61.

Now we conceive that a full answer to this argument is afforded by the nature of the inductive propensity — by the irresistible impulse of the mind to generalize *ad infinitum*, when nothing in the nature of limitation or opposition offers itself to the imagination—and by our involuntary application of the law of continuity to fill up, by the same ideal substance of truth, every interval which uncontradicted experience may have left blank in our inductive conclusions. What we contend for is, not that the propositions of geometry are other than necessary and universal, but that space being a reality (or a real condition), the mind, applying itself to that reality, discovers its properties by such application, which is experience, and embodies the results of that experience in axiomatic propositions. For what, we may ask, *can* impress us with a sense of truth other than a clear perception of *meaning?* And what *is* a perception of meaning other than *an intellectual experience of the real qualities and relations of the objects of our thoughts, as exemplified in special cases?*

And after all, the truths of geometry are summed up and embodied in its definitions and axioms. The definitions we need not consider, but let us turn to the axioms, and what do we find? A string of propositions concerning magnitude in the *abstract*, which are equally true of space, time, force, number, and every other magnitude susceptible of aggregation and subdivision. Such propositions, where they are not mere definitions, as some of them are, carry their inductive origin on the face of their enunciation. Of those which expressly relate to space, the axiom which declares magnitudes equal which exactly fill the same space, is clearly only a rule of interpretation declaring how the word equal is to be understood when space is the object of reference, and how the measurement of space is to be executed, and is only the ordinary practical process of measurement embodied in words. Those which declare that two straight lines cannot enclose a space, and that two straight lines which cut one another cannot both be parallel to a third, are in reality the only ones which express characteristic properties of space, and these it will be worth while to consider more nearly. Now the only clear notion we can form of straightness is uniformity of direction, for space in its ultimate analysis is nothing but an assemblage of distances and directions. And (not to dwell on the notion of continued contemplation, *i. e.*, mental experience, as included in the very idea of uniformity; nor on that of transfer of the contemplating being from point to point, and of experience, during such transfer, of the homogeneity of the interval passed over), we cannot even propose the proposition in an intelligible form, to any one whose experience ever since he was born has not assured him of the fact. The *unity of direction*, or that we cannot march from a given point by more than one path *direct to the same object*, is matter of practical experience long before it can by possibility become matter of abstract thought. We cannot attempt mentally to exemplify the conditions of the assertion in an imaginary case opposed to it, without violating our habitual recollection of this experience and defacing our mental picture of space as grounded on it. What *but* experience, we may ask, can possibly assure us of the *homogeneity* of the parts of distance, time, force, and measurable aggregates in general, on which the truth of the other axioms depends ? As regards the latter axiom, after what has been said, it must be clear that the very same course of remarks equally applies to its case, and that its truth is quite as much forced on the mind as that of the former by daily and hourly experience.

We have considered the perception of space, in its ultimate analysis, as resolvable into perceptions of distance and direction; into line and angle; but it may be urged that our ideas of superficial and solid space involve something more than these elements—that surface and solidity are not in their essence resolvable into *mere* distance and direction. It is here that we trace, as we conceive the matter, the result of the mind's plastic faculty, by which, out of the assemblage of simple perceptions, it forms to itself a *picture*, or *conception*, or *idea* (call it what we will) in which those perceptions are mentally realized, but which seems to us to be something more than those perceptions—what the Lockian school terms, in short, *substance;* and which we consider to be no other than the mind's *perception of its own active effort* in this process. The conception of solid extension stands, we apprehend, to these simple elementary perceptions of distance and direction in the same relation as that of *body* to the perceptions of resistance, extension, colour, figure, &c., which are all that common experience affords us of *matter;* and this is the only sense in which we can agree with, or indeed attach any distinct meaning to, a remarkable passage in Mr. Whewell's chapter "On the Idea of Space": —

"By speaking of space, as an idea, I intend to imply that the apprehension of objects as existing in space, and of the relation of position, &c., which thus prevail among them, is not a consequence of experience, but a result of a peculiar constitution and activity of the mind which" [*i. e.*, the activity] "is independent of experience in its origin, though constantly combined with it in its exercise." — *Phil.* b. ii. p. 81.

But when he goes on to declare, in the next page, "that space is not a notion obtained from experience," and in addition to the argument from the universality and necessity of its properties which we have already considered, supports this doctrine by such arguments as these: —

"Experience gives us information concerning things without us, but our apprehending them *as* without us, takes for granted their existence in space. Experience acquaints us with what are the form, position, magnitude, &c., of particular objects, but that they have form, position, magnitude, presuppose that they are in space."—*Phil.* i. p. 82.

we cannot avoid placing on record our dissent from the conclusion, and our inability to perceive the cogency of the reasoning. The reason, we conceive, why we apprehend things as without us is, that they *are* without us. We take for granted that they exist in space, because they *do* so exist, and because such their existence is a matter of direct perception which can neither be explained in words, nor contravened in imagination; because, in short, space is a *reality* and not a matter of mere convention or imagination. Still less can we attribute the smallest force to such reasons as those in p. 86, where it is denied that space "exists as a thing," because "that thing is infinite in all its dimensions, and moreover is a thing which, being nothing in itself, exists only that other things may exist in it." We might meet such reasoning in its own spirit, by declaring that that which has parts, proportions, and susceptibilities of exact measurement, must be "a thing."

The philosophy of the pure sciences involves not merely the idea of space, but of magnitude in the abstract. It is common indeed to represent, in elementary books, such magnitudes by geometrical lines and areas, and thus to demonstrate the truths which serve as the bases of the sciences of arithmetic, algebra, &c. But this is only legitimate, because the axioms of abstract magnitude are verified among such lines and areas in the same manner as they are verified among the various other objects to which they apply, and *by induction from which* they have been concluded to be generally true. That equals added to equals produce equal aggregates is true of equal times, equal weights, equal numbers, as well as of equal spaces. Were we to grant (which we do not) that the truth of the proposition in each of these forms is a direct result of simple intuition involving no induction—no consideration of particular cases, *i. e.*, no experience—still the combination of all these separate truths into one general expression equally applicable to all the forms, must surely be allowed to be an act of inductive generalization. To maintain the contrary, is to maintain that the mind conceives and reasons on the abstract in anticipation of the concrete, on the general before the particular, which is in fact Platonism, and to which indeed, in many respects, and as purified of its more extravagant features, Mr. Whewell's theory closely approximates. A remarkable instance of this is afforded by his reasoning respecting time:—

"Since all particular times are considered as derivable from time in general, it is manifest that the notion of time in general cannot be derived from the notion of particular time. The notion of time in general is therefore not a general conception derived from experience."—*Phil.* i. p. 124.

This is as if any one should argue that as there is but one material universe, of which all particular bodies are necessarily parts—therefore our notion of the material universe is not a general conception derived from our experience of individual bodies. The fact is, that if we were to select an idea which must more emphatically than another be derived from experience, it would be that of time; for what is it which excites in us the perception of its lapse, but the internal comparison of our mental state at the beginning and end of each instant, which *is* experience, if the word have any meaning. The lapse of the instant is a *reality;* a very obscure and mysterious one, no doubt; and our notion of it (the result, or perhaps we should rather say the perception, of the active effort of the mind to connect its present and past state) is that substantive conception which may be considered as bearing the same relation to the reality of time, whatever that be, as our substantive conception of space bears to the realities of distance and direction.

As respects number, Mr. Whewell has adopted a mode of considering it which has lately grown much in vogue, but which we regard as, to say the least, very problematic; viz., that it is a mere modification of the idea of time. Now things may be repeated in space as well as in time, and though it may be perfectly true (though of that we have some doubts) that the attention at each instant is so wholly absorbed in the contemplation of one object, that every other is absolutely *unperceived,* and is to us, to all intents and purposes, as if it existed not; yet this would only go to show that, owing to the imperfection of our faculties, time is necessary as a *mean* to enable us to *count* number, but not that it enters otherwise than as a mean into an idea of any particular number, as two. Two horses are two horses, whether we require time to count them or not, and whether counted or uncounted. On precisely the same principle, time might be declared an element in our conception of figure, and indeed of space itself. Number, therefore, we cannot help regarding as an abstraction, and *consequently* its general properties or its axioms to be of necessity inductively concluded from the consideration of particular cases. And surely this is the way in which children *do* acquire their knowledge of number, and in which they learn its axioms. The apples and the marbles are put in requisition, and through the multitude of gingerbread nuts their ideas acquire clearness, precision, and generality. And it is so impossible for us to divest ourselves, either as respects number, or any of those primary relations, as space, time, &c., of the bias given to all our notions by the unbroken influence of an experience which commenced with our birth and perhaps even before it, that we may well be excused if we more than hesitate in our assent to a doctrine, which requires us so entirely to unmould and unbuild the whole structure of our mental habits and acquirements, as does that of the non-suggestion of ideas of this class, and the non-establishment of their axioms by experience; including always, be it observed, in our notion of experience, that which is gained by contemplation of the inward picture which the mind forms to itself in any proposed case, or which it arbitrarily selects as an example—such picture, in virtue of the extreme simplicity of these primary relations, being called up by the imagination with as much vividness and clearness as could be done by any external impression, which is the only meaning we can attach to the word *intuition*, as applied to such relations.

Into the philosophy of the abstract sciences the notion of cause does not explicitly enter; relations, not events, being the subject of inquiry in these sciences. But in those where phenomena come to be explained, the reference of these to their causes, and the development of the processes by which the action of such causes is carried out through a chain of intermediate effects, till they result in the phenomena observed, is our sole, at least our ultimate, object of inquiry. Now it deserves especial notice that most of the phenomena which nature presents are cases of indirect causation. Conceptions of *cause* suggested by such phenomena can hardly be other than crude, imperfect, and even perhaps erroneous. For example, invariable antecedence of cause and consequence of effect is laid down by writers on this subject as an essential feature of this relation. But this must be understood in reference to the *state of things*, historically speaking, which precedes and that which follows that indivisible instant of time in which action takes place, as the two portions of a line separated by a point are necessarily the one on one side, the other on the other of that point. If the antecedence and consequence in question be understood as the interposition of an interval of time, however small, between the action of the cause and the production of the effect, we regard it as inadmissible. In the production of motion by force, for instance, though the effect be cumulative, with continued exertion of the cause, yet each elementary or individual action of the force is, to our apprehension, *instanter* accompanied with its corresponding increment of momentum in the body moved. In all dynamical reasonings, no one has ever thought of interposing an instant of time between the action and its resulting momentum; nor does it appear necessary. The process has more the character of a simple transformation of force into momentum, without gain or loss. The cause (this particular cause) seems to be neither destroyed nor enfeebled, but absorbed, and transformed into its effect, and therein treasured up. In this view, which seems quite as tenable as any other which has yet been taken of the relation of physical cause and effect, the time lost in cases of indirect physical causation is that consumed in the movements which take place among the parts of the mechanism set in action, by which the active forces so transformed into momentum are transported over intervals of space to new points of action, the motion of matter in such cases being regarded as a mere carrier of force. So also, when force is directly counteracted by force, their mutual destruction must be conceived, we think, as instantaneous. It appears to us, therefore, well worthy of consideration, *whether*, in deriving any part of our abstractions of cause and effect from external phenomena, *we be not misled in assuming sequence as a necessary feature in that relation*, and whether sequence, when observed, is not rather to be held as a sure indication of indirect action, accompanied with a movement of parts. Certain it is, that the higher we ascend in the scale of physical causation the more inconceivably rapid do we find the *propagation* of action. The *play* of the mechanism (if we may borrow a metaphor) seems less, and the approach to perfect fitting and contact of its parts more near.

The direct personal consciousness of causation which we have when we either exert voluntary force or influence the train of our own thoughts, has been much and singularly lost sight of by many writers on this subject. Whatever be the essential nature of that relation (or whether even it be in all cases the same), we are no more left in doubt of its being a real relation, when we experience this consciousness, than we are of our own reality, or of that of an external world. When once suggested (as we conceive it to be) by such experience, as a kind of mental sensation, it is seized and dwelt on with a force and tenacity which strongly indicates its real importance to our knowledge and well-being. The energy and assurance with which it is generalized, or rather universalized, and extended to all the events of nature, must be held as another indication in the same direction. Nothing can be imagined more different than the two lines of experience by which this consciousness of effective action is impressed. They agree in nothing but in change consequent on or simultaneous with voluntary effort, and predictable beforehand, as sure to accompany such effort. Yet this point of analogy is seized and made the basis of a *universal theory* with an invariable verification by experience, and a decisive acknowledgment of its irresistible cogency, which proves it to be one of those grand primordial analogies alluded to above (p. 152.); an analogy by which the physical and intellectual world are brought into inseparable contact, by establishing the influence of *will* over both.

There are, no doubt, other lines of experience in which we also receive, but more obscurely, and as it were conversely, through the medium of effect, the idea of cause. But from the very diversity of these modes of suggestion it follows that this idea is, as Mr. Whewell admits it to be, an abstraction. And from this consideration alone it seems to us imperatively to follow that whatever axioms (if there be any) belong to this idea, must be inductively concluded from their verification in each of those several particular lines of experience in which we recognize and insulate the peculiar mental sensation of *causality*. It must be very clear, for instance, that an axiom which, though verified in one form of causation, is yet unmeaning or incorrect in another, cannot be an axiom of causation in the abstract, or must be inadequately worded as such. And the same must surely be the case with axioms requiring limitations and conditions dependent on the *kind* of cause.

These considerations seem to us essential to forming a right understanding of the metaphysics of Mr. Whewell's three books on the Philosophy of the Mechanical Sciences. For the basis of these he takes the fundamental idea of cause—not that this relation is not to be considered equally involved in other sciences, but emphatically, because in these we have succeeded, in those not, in tracing phenomena up to one of those causes of whose existence our own consciousness assures us, viz., force. In pursuance of his general plan of ascribing a necessary universality to physical as well as to every other class of general truths and deriving this necessity and this universality from the assumed *à priori* origin within the mind of whatever abstract principles are involved in their enunciation, he lays down three axioms of causation as flowing not from experience, but from our fundamental idea of that relation, viz.:—1. Nothing can take place without a cause. 2. Effects are proportional to their causes, and causes are measured by their effects. 3. Reaction is equal and opposite to action. Of these the first in our view of the matter is the mere generalization of our internal consciousness in the two distinct lines of experience above mentioned—a generalization cogent doubtless in the highest degree, as all such impulses of the generalizing instinct are when the mind feels no obstacle, and finds itself contradicted by no opposing experience. The second axiom presents only a vague, if any, meaning where causes are unsusceptible of numerical addition or conjoined agency—and where they are so susceptible Mr. Whewell admits that "there may be circumstances in the nature of the cause which may further determine the *kind* of effect which we must take for the measure of the cause." But it is clear that we are now discussing the relation of causes to their *direct* effects, and that consequently we are allowed no latitude of choice. We are not to range about the results of their action till we find some one, be it direct or remote, by which our rule shall be saved. We are to take the direct effect as we find it, viz., that which is separated from the action of the cause by no interval of time and by no intermedium of mechanism; and if with *this* for an effect the axiom be verified, all is well.

On the third axiom Mr. Whewell reasons as follows:—

"The reaction is an effect of the action, and is determined by it. And since the two, action and reaction, are forces of the same nature, each may be considered as cause and as effect, and they must therefore determine each other by a common rule. But this consideration leads necessarily to their equality: for since the rule is mutual, if we could for an instant suppose the reaction to be less than the action, we must by the same rule suppose the action to be less than the reaction." — *Phil.* i. p. 175.

"Like our other axioms, this has its source in an idea, viz., the idea of cause under that particular condition in which cause and effect are mutual." — *Ibid.*

We trust Mr. Whewell will believe that we speak in all sincerity, and not without diffidence in our own impressions, when we declare that this is a modification of the idea of cause, which we can no how bring ourselves to conceive. It seems to lead direct to the conclusion, with no escape, that a cause can cause itself. For if A be the cause of R, and R, by the rule of mutuality, the cause of A; then is A the ultimate and R the proximate link in a chain of causation by which it is derived from itself. This, it may be said, is a verbal quibble. But if it be (which we think it is not), it is one that inevitably forces itself on the thoughts on the bare mention of such a proposition, as that cause and effect can in any case be justly regarded as mutual. If indeed we admit the doctrine of sequence as a general feature of causality, and suppose ever so small an interval of time interposed between cause and effect, the rule of mutuality is evidently impossible. This doctrine, however, as already said, we regard as untenable; and from a single, short, and insulated sentence in p. 252., which seems to have called up when written no further mental remark, it appears that Mr. Whewell herein agrees with us.

It would not be difficult, however, so to word this axiom as to render it applicable and intelligible in every form of causation, and at the same time to avoid introducing the term Reaction, which though highly convenient, and therefore readily admissible in dynamical reasonings, ceases altogether to present any distinct meaning when used in reference to other than mechanical cause. The axiom, for instance, taken as a general proposition, deduced from and verified by experience in every form of causality, may be held to assert the limitation of a finite amount of cause to the production of a finite amount of immediate effect, in consequence of which limitation the total effect must be such as to leave no part of the energy of the cause outstanding and applicable to the production of further effect. In other words, it must be such as to exhaust, or absorb, or transform into itself, as the case may be, *the whole cause*. Dynamically interpreted, this leads to the law of reaction, while physiologically, it expresses merely fatigue or exhaustion, which every one is conscious of on bodily or mental exertion. For it must be observed, (and the remark appears to us of great moment,) that in the production of voluntary motion we do not conceive the mind or will as directly exerting force on, and so producing motion in matter. Were such indeed the case, we might reasonably ask what becomes of reaction where mind is at one end of the rod and matter at the other? Here we recognize the importance of that intermediate link in the chain of causation, that physiological effort dependent on the will, but yet distinct from mere volition, already before alluded to. Of the nature of this effort it seems impossible to frame any other conception than this—that without being itself force it evolves or creates force having all the characters of molecular attractions and repulsions, either among the contiguous particles of the muscles directly; or else indirectly in them, through a chain of polar arrangements among those of the nerves — a cause, in short, of a higher order than force, and which, for anything we can know to the contrary, may be in action even among the particles of inanimate matter, whenever force is exerted, though whether in all cases under the immediate control of a directing will, transcends of course our faculties to decide on physical grounds. However convenient it may be in common language, or in dynamical reasoning, to speak of force as the action, of one body upon another, and as accompanied with a reaction of the other back upon the first, it is far more consonant with this view of voluntary action, and indeed with the mass of facts in other sciences, to regard it as a cause or disposition to motion, originating indifferently *between* them, and *manifesting itself by an effect which has always a twofold or polar character*, *i. e.*, the production (unless counteracted) of equal momenta in opposite directions at either extremity of its line of action: the sum of such momenta being (as in all cases of polar action) equal to zero.

Mr. Whewell, in his chapter "On the Origin of our Conceptions of Force and Matter," traces them simply to our sense of muscular action and resistance, but without distinguishing, as we have done, between the effort and the action, and of course without drawing from that distinction the consequences which we have above suggested, and which seem to us so important. He then proceeds to treat, at great length in separate chapters, of the establishment of the principles of statics and dynamics. These chapters are extremely valuable. We perceive in them the results of great labour and a long series of intense and persevering thought bestowed on their subjects, the fruits of which have from time to time appeared in several previous works*, and are here brought together as in a focus. Of these works it is but justice to say that we know of no treatises extant which afford so complete and philosophical a view of the principles of these sciences, and of the steps by which they have acquired their ultimate development and demonstrative character. Though assuredly not the most brilliant of the many gems which adorn our author's wreath of merit, their sterling value

* "Elementary Treatise on Mechanics." Cambridge, 1819. —"On the Free Motion of Points, and on Universal Gravitation." Cambridge, 1832.—"The First Principles of Mechanics." Cambridge, 1832. — "The Mechanical Euclid." Cambridge, 1837.

will secure them an estimation superior even to that of many original discoveries.

In these chapters, as well as in the works alluded to, the whole of mechanical science is made to depend on a few simple propositions of axiomatic self-evidence—and with this, as regards systematic and logical deduction, we can have, of course, no quarrel. It is when we find it put forward that these axioms owe their evidence and universality solely to our fundamental and abstract idea of causation, to the general axioms thence derived, and in no way to experience, that we demur. As we admit no such propositions, *other than as truths inductively collected from observation*, even in geometry itself, it can hardly be expected that, in a science of obviously contingent relations, we should acquiesce in a contrary view. As we conceive matter to have been created, and to admit of annihilation, we can of course conceive the non-existence of force, and if so, it certainly does appear a violent inroad on the liberty and power of thought to maintain that we may not, or cannot, conceive the laws of force to have been otherwise established than as we find them. But let us take one of these axioms and examine its evidence: for instance, that equal forces perpendicularly applied at the opposite ends of equal arms of a straight lever will balance each other. What but experience, we may ask, in the first place, can possibly inform us that a force so applied will have any tendency to turn the lever on its centre at all? Or that force can be so transmitted along a rigid line *perpendicular to its direction*, as to act elsewhere in space than along its own line of action? Surely this is so far from being self-evident that it has even a paradoxical appearance, which is only to be removed by giving our lever thickness, material composition, and molecular powers. Again, we conclude that the two forces, being equal and applied under precisely similar circumstances, must, if they exert any effort at all to turn the lever, exert equal and opposite efforts: but what *à priori* reasoning can possibly assure us that they *do* act under precisely similar circumstances? that points which differ in place, *are* similarly circumstanced as regards the exertion of force?—that universal space may not have relations to universal force—or, at all events that the organization of the material universe may not be such as to place that portion of space occupied by it in such relations to the forces excited in it, as may invalidate the absolute similarity of circumstances assumed? Or we may argue, what have we to do with the notion of angular movement in the lever at all? The case is one of rest, and of quiescent destruction of force by force. Now how is this destruction effected? Assuredly by the counter-pressure which supports the fulcrum. But would not this destruction equally arise, and by the same amount of counteracting force, if each force simply pressed *its own half* of the lever against the fulcrum? And what can assure us that it is not so, except removal of one or other force and consequent tilting of the lever?

The other fundamental axiom of statics, that the pressure on the point of support is the sum of the weights, is derived by Mr. Whewell, from the principle of reaction. "If it be not an axiom," he asks, "deriving its truth from the fundamental conception of equal action and reaction, which equilibrium always implies, what is the origin of its certainty?" Equilibrium implies, however, not merely equal action and reaction, which law subsists whether equilibrium take place or no, but equal action and *counter*-action. The pressure on the fulcrum *is not destroyed by the reaction* of the fulcrum, for that would subsist were the fulcrum pushed from its place by the pressure. If it be destroyed at all, it must be destroyed by a *counter-acting* force *applied* for that purpose, and the question is, what is the amount of the force that must be so applied. Were the pressure on the fulcrum ten times the sum of the weights, its *reaction* would still be equal to that pressure. Such reaction, in our view of the nature of force, is simply the simultaneous and opposite manifestation of its polar action, and can in no case afford an available measure of its intensity. Force can only be measured by motion produced, or by amount of force *elsewhere originating* necessary to prevent motion.

What then, it must of course be asked, *is* the origin of our certainty of the axiom? We reply, simple experience. It is merely a scientific transformation and more refined mode of stating a coarse and obvious result of universal experience, viz., that the weight of a rigid body is the same, handle it or suspend it in what position or by what point we will, and that whatever sustains a body sustains its total weight. Assuredly, as Mr. Whewell justly remarks—

> "no one probably ever made a trial for the purpose of showing that the pressure on the support is equal to the sum of the weights. Certainly no person, with clear mechanical conceptions, ever wanted such a trial to convince him of its truth, or thought the truth clearer after the trial had been made."

But it is precisely because in every action of his life from earliest infancy, he has been *continually* making the trial and seeing it made by every other living being about him, that he never dreams of staking its result on one additional attempt made with scientific accuracy. This would be as if a man should resolve to decide by experiment whether his eyes were useful for the purpose of seeing by hermetically sealing himself up for half an hour in a metal case.

In making these remarks on Mr. Whewell's *à priori* doctrines, we are most anxious to be understood as limiting our disapproval strictly to the point of view from which he has contemplated his subject. In its handling there is every thing to admire, nor are we aware that we have ever in the same compass encountered such a mine of recondite thought, searching inquiry, and pointed and brilliant illustration. But to these views he recurs again and again, and always with increasing decision, *vires acquirit eundo*, as if their force had grown upon him in their contemplation. Thus, even in the midst of his mechanical applications, he suspends his argument to insert a chapter on "the paradox of universal propositions obtained by experience," a paradox in which, however, we see nothing that strikes us as paradoxical. If there be necessary and universal truths (which we unconditionally admit) expressible in propositions of axiomatic simplicity and obviousness, and having for their subject-matter the elements of all our experience and all our knowledge, surely these are the truths which, if experience suggest to us any truths at all, it ought to suggest most readily, clearly and unceasingly. If it were a truth, universal and necessary, that a net is spread over the whole surface of every planetary globe, we should not travel far on our own without getting entangled in its meshes, and making the necessity of some means of extrication an axiom of locomotion.

The only tests of abstract truth are entire consistency in itself, and accordance with its exemplification in particulars. A mingled host of individual relations is suggested to our understandings by every object and event. It is *consistency of suggestion* by many particular events and objects which leads us to make any abstract propositions at all, since without such consistency we must for ever remain not merely passive but bewildered percipients. But, on perceiving this consistency, we are not simply led, but urged to make them by the most irresistible of all our mental impulses — the generalizing or inductive *nisus*. "We do not," as Mr. Whewell most justly remarks, "acquire from mere observation a right to assert that a proposition is true in all cases." But that we do possess the *propensity* is clear from this, that we generalize the abstract suggestion of mistaken relations, if of frequent occurrence, as readily as of true ones, nor ever dream of abandoning our conclusions till their inconsistency with further observation stares us in the face.

There is, therefore, nothing paradoxical, but the reverse, in our being led by observation to a recognition of such truths, as *general* propositions, co-extensive at least *with all human experience*. That they pervade all the objects of experience, must ensure their continual suggestion *by* experience; that they are true, must ensure that consistency of suggestion, that iteration of uncontradicted assertion which commands implicit assent, and removes all occasion of exception; that they are simple, and admit of no misunderstanding, must secure their admission by every mind.

Necessity and universality are large words — perhaps somewhat too large for the human understanding fairly to handle. Mr. Whewell himself does not "venture absolutely to pronounce whether the laws of motion, as we know them, can be rigorously traced to an absolute necessity in the nature of things;" though "some of the most acute and profound mathematicians have believed that for these laws of motion, or some of them, there was a demonstrable necessity compelling them to be such as they are, and no other." Such phrases, after what has been stated of his views, might give occasion to much remark—the only one they suggest to us is the nicety of the line in such matters between belief and demonstration, between belief spontaneous and belief compelled.

The moment we get out of particulars, we get into abstractions, out of real into logical relations. The test of truth by its application to particulars being laid aside, nothing remains but its self-consistency to guide us in its recognition. But this in axiomatic propositions amounts to no test at all. *It is the essence of such propositions to stand aloof and insulated from each other.* One abstract proposition can only be shown to be consistent with another in two ways — either by both being verified in one particular, or concrete as the logicians call it, or by the one being logically derivable as a necessary consequence of the other, in which case one or other ceases to be axiomatic. Axioms, rigorously such, can admit of no meaning in common. *Their mutual compatibility, as fundamental elements of the same body of truth, can only be shown by experience* — by the observed fact of their co-existence as *literal truths* in a particular case produced.

A truth, necessary and universal, relative to any

object of our knowledge, must verify itself in every instance where that object is before our contemplation, and if, at the same time, it be simple and intelligible, its verification must be obvious. The sentiment of such a truth cannot, therefore, but be present to our minds whenever that object is contemplated, and must therefore make a part of the mental picture or idea of that object which we may on any occasion summon before our imagination. If that sentiment be wanting, the picture is unfaithful: it is, in fact, no picture at all. It is, therefore, impracticable for us to frame any logically true and consistent proposition concerning such object, in which that sentiment is not at least implicitly involved, much less one in which it is explicitly contradicted. All propositions, therefore, become not only untrue, but *inconceivable*, if necessary axioms be violated in their enunciation.

It is requisite, also, to bear in mind, in this argument, the prerogatives of experience. The mind cannot give to arbitrary combinations of its own that impress of reality and unity which it acknowledges when it contemplates realities. It cannot imagine to itself, for example, a being in which time is solidified, space set in motion, matter invested with the property of being in two places at once, &c. It may jumble the ideas, or conceive them in succession, but finds them always incoherent, and can no-how educe from its own stores the substantive conception of a being or reality in which they shall co-exist. In the case of space, if the axioms of geometry be not present to our minds directly or by implication, when we think of it, there is nothing left for us to think of—for these axioms express its whole essence. If we try to frame a conception of space in which they shall not be verified, or shall be replaced by others essentially different, we find it impracticable, and this is our criterion of their necessity. Some such notion the Hibernian must have formed of space, when he declared that if all the people were in the hall, the hall would not hold them. Again, in the case of matter, if inertia be not present to our minds in any act of reasoning, it is no longer matter about which we reason, but that which may subsist, if inertia be absent; for instance, moveable and coloured extension, which we can no-how figure to ourselves as a "*a thing.*" And, if we admit into our conception an idea contradictory to those suggested by experience as belonging to it, such as immobility, then again it is not matter about which we reason, but a new creature, such as experience has never presented. Such a being, if it exist, must exist according to its own laws, but they cannot be the laws of matter and motion, which remain therefore unaffected by the supposition. Relations which pervade all human experience, and all human power of conception grounded on that experience, we may call necessary relations without much violence to language or reason.

It may, however, be alleged, that one criterion

of abstract truth remains unconsidered—its direct recognition *in the abstract* without mental reference to *any* particular case, to *any* example, to *any* experience. How truth may or may not impress conviction in other minds, it is doubtless presumptuous to assert, for which reason we have dwelt only on the received *tests* of truth, as conveyed from mind to mind by the intervention of language. If there be those who can persuade themselves that they are yielding a rational assent to the terms of an abstract proposition on the mere jingle of its sound in their ears, while refusing to test it by calling up in their minds those images with their attributes which experience has inseparably associated with its words, they have certainly a very different notion of logical evidence from our own.

That our success in abstract and physical research may aid us in extending our views to what may be called the social sciences, it is of primary importance in our choice *if choose we must*, between a logical and an empirical philosophy, that we should be well aware how far and with what restrictions and humiliating conditions the former is possible or practicable. The citadel of truth equally vindicates its altitude whether we measure it by toil and upward struggle, or by throwing ourselves headlong from its battlements. It is then that we are taught caution and reserve when observation presents us its axioms in a form inextricably involved, and when experiment is fraught with hazard to our own happiness and that of others. A logical philosophy in such sciences which shall start from necessary and universal formulæ can only be safe when human history shall be complete and the book of events on the point of closing for ever. Logically speaking, we may indeed so limit the acceptation of our terms as to make our axioms, if other than barren truisms, intelligible only when empirically true. Yet what is this but to bind our philosophy for ever in the leading-strings of experience, and declare it, with the aspirations of maturity in a ceaseless state of pupilage? Mr. Whewell's good sense, which may always be trusted, whatever be the phase under which his excursive intellect delights to manifest itself, has led him direct to this conclusion, a conclusion which draws the teeth of the general doctrine and renders it perfectly innocuous. Speaking of the laws of motion — but in language generally applicable — he says they

"borrow their form from the idea of causation, though their matter be given by experience; and hence they possess a universality which experience cannot give. They are certainly and universally valid; and the only question for observation to decide is, how they are to be understood. They are like general mathematical formulæ which are known to be true even while we are ignorant what are the unknown quantities which they involve. It must be allowed, on the other hand, that so long as these formulæ are not interpreted by a real study of nature, they are not only useless but prejudicial, filling men's minds with vague general terms, empty maxims, and unintelligible abstractions, which they mistake for knowledge. Of such perversion of the speculative propensities of man's nature, the world has seen too much in all ages. Yet we must not on that account despise these forms of truth, since without them no general knowledge is possible. Without general terms and maxims and abstractions, we can have no science, no speculation; hardly, indeed, consistent thought or the exercise of reason. The course of real knowledge is to obtain from thought and experience the right interpretation of our general terms, the real import of our maxims — the true generalizations which our abstractions involve." — *Phil.* i. p. 242.

In such a spirit we may trust the philosopher, let him take what ground he will. The high *priori* Pegasus, so curbed and guided, is a noble and generous steed who bounds over obstacles which confine the plain matter of fact roadster to tardier paths and a longer circuit. There is no denying to this philosophy, for one of its distinguishing characters, a *verve* and energy which a merely tentative and empirical one must draw from foreign sources, from a solemn and earnest feeling of duty and devotion in its followers, and a firm reliance on the ultimate sufficiency of its resources to accomplish every purpose which Providence has destined it to attain.

The distinction between the primary and secondary qualities of bodies has given some trouble to metaphysicians. We are not quite sure that this distinction, as usually taken, is tenable. All sensible qualities of material objects, not excepting even their extension and figure, are manifestations, *by multitude*, of powers, arrangements, mechanisms, and movements, in particles individually imperceptible. We have not the shadow of a proof that the particles of bodies are extended. The contrary seems to us all but demonstrable—and if not, then are extension and figure merely dotted outlines which the mind, acting according to the law of continuity, fills up and unites. Primary qualities, therefore, can only be received by us as provisionally such (like the undecomposed elements in chemistry), while such as can be referred to a traceable mechanism ought assuredly not to be so considered. But these again may be advantageously subdivided according to the mode of their manifestation to our senses, and the line which Mr. Whewell has drawn, by classing under one head those which depend for their perception on the intervention of a medium between the bodies in which they originate and our organs of sensation, is at once natural, and convenient as a ground of classification. The idea or conception of *a medium*, therefore, is made by him the bases of those sciences, as acoustics, photology, and thermotics, which relate to such qualities.

On the other hand, there is a class of sciences in which the powers of matter, whether primary or derivative, manifest themselves in their action only incidentally on us as percipients, but immediately in the production of visible movements and modifications, permanent or transitory, of the material agents themselves. Such are those which relate to the intimate construction and mechanism of matter, and which, so far as yet developed by chemical, optical, and electrical research, all agree in bringing forward, in a more or less prominent form, that which Mr. Whewell has pitched upon as the "fundamental idea" of these sciences: viz., *polarity*—or, as he abstracts and generalizes it (not finding it ready made in our minds), the con ception of "opposite properties in opposite positions." Thus generalized speculations on the ultimate identity of all the forms in which it occurs throughout nature appear no longer extravagant or fantastic, and can hardly even be considered premature, when, as in Mr. Whewell's chapter "on the Connexion of Polarities," we find these manifestations so closely linked, two by two, as to form an unbroken chain pervading all nature. Thus we have, first, magnetic brought into immediate relation with electrical polarities, by the great discoveries of Oersted and Ampere; electrical with chemical, by those of Davy and Faraday; chemical with crystallographical, by those of Haüy and Mitscherlich; and these, again, with optical polarities, by the striking experimental researches of Brewster, and the grand dynamical generalizations of Fresnel. We have certainly never seen the case so strikingly put. The main link in this wonderful chain of connexion—and we may add, too, a link inferior to none in the clearness and steadiness of thought and refinement of experiment, demanded for its establishment—is that supplied by the recent electro-chemical researches of Dr. Faraday, to whose transcendent merits as a philosopher we are delighted to find Mr. Whewell here, as on all occasions, doing full and cordial justice. Not a little pleased also are we to find him, in this chapter, dealing out equal justice, though of a very different kind (not, however, without a leaning to the side of mercy) to the ravings of Hegel and Schelling on the subject of magnetic and optical polarizations; thereby separating himself in the most decided manner from that exaggerated *à priori* school of metaphysical speculation which finds in "the Absolute," or in the proposition "A=A," the totality of all existence and all knowledge discovered or discoverable!

The fundamental ideas assumed for the philosophy of chemistry are "affinity" in that sense in which it is understood by chemists, and "element" as a modification of the idea of material "substance"—the indestructibility of which is laid down as an axiom of universal, undisputed authority, on the somewhat singular ground (for an axiomatic proposition) of its *opposition* to the common course of our experience, and its apparently paradoxical air (vol. i. p. 391.) when proposed. As we are quite sure that it is not Mr. Whewell's intention to maintain the necessary and eternal self-existence of matter, we would recommend him, in the next edition of his work, to modify the expressions in the passage alluded to, which go to place the idea of material substance in this respect on a par with those of space and number. The general notion of substance is applied to chemistry, by the additional axiom that a body is equal to the sum of its ponderable elements; which excludes the phlogistic theory, on the ground of its assuming a *negative* element, and gives occasion for the assertion, as a general maxim, that "imponderable fluids are not to be admitted as chemical elements of bodies"—nay, that such fluids are to be regarded as incapable of being affected by mechanical impulse and pressure—which is in effect to deny them altogether the properties of matter (vol. i. p. 400, note). We are hardly prepared for so sweeping a conclusion, though we may admit that impulse and pressure must be conceived in a very refined way when dealing with such subtle agents.

The atomic doctrine is treated in this and the next book, "On the Philosophy of Morphology," as applied especially to crystallography, in which we find enunciated a principle whose importance is best felt on a contemplation of its utter neglect by all who have attempted to frame distinct conceptions of the intimate atomic structure of chemical compounds. The principle is this: "that all hypotheses concerning the arrangement of the elementary atoms of bodies in space must be constructed with reference to the general facts of crystallization." We cannot help believing that this principle will prove a fertile one, and that by admitting the *particles* of bodies to consist — not as has been done hitherto, by Dalton, Wollaston, and Ampere — of a few only, but of great multitudes—

of thousands perhaps, or millions — of *atoms;* not only may the facts of crystallography be represented, but much light thrown on many obscure points in the theory of the absorption of light, the colours of bodies, and their power of conducting heat. The great stumbling block of the atomic chemistry — the occasional necessary subdivision of an "atom" — would at once disappear under such a mode of considering ingredients.

The "Philosophy of the Classificatory Sciences" is full of interest and instruction. The fundamental idea of resemblance traced into assemblages of items and adjuncts, variously associated and differing in degree in different kinds — the unity of object emerging from the multiplicity of such particulars — the substitution of type for definition, of central grouping for determining limit — the important office of terminology in such sciences, and the conditions under which terms must be applied "so as to make general propositions possible" (an apophthegm which merits to be regarded as the axiom of systematic terminology) — are all admirably treated. Our limits leave no room but for a single and somewhat garbled extract, where the conditions of our perception of an object as an individual are stated. And here we must observe, once for all, that Mr. Whewell, of all authors we have read, is perhaps the most difficult to *extract* briefly. The copiousness of his illustration and the point of his language are such that it is scarcely possible to draw a line, or to omit; we are led on from sentence to sentence, from image to image, from point to point — all adding to the general effect of the picture, and none capable of being sacrificed without real detriment. It is a flowing and embroidered robe, but which sits so well to the person that it will not bear to be trimmed or curtailed.

"*Condition of unity.* — The primary and fundamental condition is, that we must be able to make intelligible assertions respecting the object, and to entertain that belief of which assertions are the exposition. A tree *grows*, *sheds* its leaves in autumn, and *buds* again in spring, *waves* in the wind, or *falls* before the storm. And to the tree belong all those parts which must be included in order that such declarations, and the thoughts that they convey, shall have a coherent and permanent meaning. . . The permanent connexions which we observe — permanent among unconnected changes which affect the surrounding appearances — are what we bind together as belonging to one object. This permanence is the condition of our conceiving the object as one. . . . We may therefore express the condition of the unity of an object to be this: — that *assertions concerning the object shall be possible;* or rather, we should say, that the acts of belief which such assertions enunciate shall be possible."

The application of this principle is wider than the domain of natural philosophy — it applies in literature, and especially to the unity of dramatic, nay, even of historical and national character; and will often serve as a criterion of truth in assertions relative to such characters.

The application of the axioms and principles of resemblance to natural history, with especial reference to mineralogy, finishes this volume. Our author here returns to the charge, in advocacy of the extension of mineralogy to the classification of chemical products, and inorganic bodies in general, whether natural or artificial, by sensible qualities, and on some principle of graduated resemblance. Some of the widest and deepest questions, as he justly remarks, of the philosophy of classification are here brought under consideration. The most essential is, what we are to understand by individuals and species, where life and reproduction are absent. Mr. Whewell's definition of a mineralogical individual is at least precise. It is "*that portion* of any mineral substance which is determined by crystalline forces acting to the same axes," a definition which applies, in the absence of all natural faces, and makes the individual co-extensive with the reasons which determine it to be one body rather than another — *so far at least as crystalline polarities include those reasons.* As regards species, these must be determined, here as elsewhere, by *the predominant principle of the existence of the object;* and, the principle of reproduction being absent, the forces which make the individual permanent and its properties definite must stand in place of those which preserve the race where individuals are generated and die; and thus we are of necessity led to make the crystallization of bodies, on both grounds, the basis of arrangement, and in cases where, owing to pulverulence, or the liquid or gaseous state of a body, this character cannot be *observed*, it must be *con-*

cluded, provisionally, from its chemical, electrical, or other habitudes. Mr. Whewell has certainly made out so strong a case for the admission of this new science on our list, that we earnestly desire to see the work of constructing it fairly undertaken, whatever denomination, whether External chemistry, Mineralogy, or the Natural history of inorganic bodies, may appear best suited to it.

In applying the fundamental idea of resemblance to natural history, we are of course led to the consideration of natural families; of their object in nature, as means to an end, or whatever else we may interpret as the *philosophical import* of such families; and of the criteria by which, among positive arbitrary arrangements, such families may be recognized. These last are of the utmost importance, and they resolve themselves into one which is, in fact, the criterion of all true induction, viz., what Mr. Whewell terms "the consilience of inductions." "The maxim," he says, "by which all systems, professing to be natural, must be tested is this, that the arrangement obtained from one set of characters coincides with the arrangement obtained from another set." That such families do exist among animals and vegetables is not a matter which can now be called in doubt — but the part they play in nature is no way to be understood without reference to a deeper and more mysterious philosophy — the Philosophy of Life and of Final Cause. These, accordingly, form the subjects of Mr. Whewell's consideration in the next or ninth book.

That the idea of Life, of which we are all conscious, should be so obscure as to render it even in a high degree difficult to say in what life consists, may well seem strange; but the wonder vanishes if we reflect that it is only of our bodily sensations and mental acts that we have that consciousness which makes them objects of direct attention. Of the principle of life within us, and the means by which the nourishment, and actions of our organs are maintained —nay, even of most of the functions they are continually performing — we have absolutely no consciousness whatever — the whole process going on without our knowledge and without the concurrence of our will. There is a profound mystery cast about the whole subject, which all attempts to explain by mere reference to chemical affinities and changes on the one hand, or to mechanical movements of particles on the other, have utterly and miserably failed. The notion of a vital fluid, conducted along the nerves and consumed or changed in its operation on the organs, offered a better promise. Electrical action *is* so communicated, and *does*, to a certain extent, produce effects simulating some of the manifestations of life. But however abstract our conception of such transferable agent, the question still arises, whence the supply, and whence the organization by which it is conveyed and acts at its point of destination. Mr. Whewell seems disposed to lean to the conception of an animal soul, or ultra-material agent (to which we know not why he should have hesitated in applying the word *life*, in its simplicity, and as applicable alike to plants and animals), a "soul," however, from which all the higher attributes which that term involves are utterly and carefully excluded. The *psychical* theory (which is as old as Aristotle), he observes —

"not only gives unity to the living body, but marks more clearly than any other the wide interval which separates mechanical and chemical from vital action, and fixes our attention upon the new powers which the consideration of life compels us to assume. It not only reminds us that these powers are elevated above the known laws of the material world, but also that they are closely connected with the world of thought and feeling, with will and reason. . . . The psychical school are mainly right in this, that, in ascribing the functions of life to a soul, they mark strongly and justly the impossibility of ascribing them to any known attributes of body." — *Phil.* ii. 29.

We pass over the various definitions which have been given of life — the attempts which have been made, with more or less success, to break up the general conception of it into an assemblage of separate (and possibly independent) ones of vital forces or powers — nay, even the curious and interesting speculations of Mr. Whewell on that marvellous subject — animal instinct — to extract some passages from his chapter on Final Causes, which (albeit our limits begin to press) appear to us indispensable to conveying a fit impression of that earnest yet right-minded, that strong and solemn yet sober feeling with which our author contemplates and powerfully induces and persuades his reader to contemplate all those dispositions, intellectual and material, which tend to lead the mind from the frame of nature to its Eternal Author. The argument of design has never been more pointedly, more irresistibly urged than in this chapter — and that chiefly from being made to rest on its main point of strength — *organization* as distinct from *law*. "An organized product," says Kant, "is that in which all the parts are mutually ends and means," and it is therefore not without reason that the idea of final cause is here introduced in an especial manner: —

"It has been objected that the doctrine of final causes supposes us acquainted with the intentions of the Creator, which, it is insinuated, is a most presumptuous and irrational basis for our reasonings. But there can be nothing presumptuous or irrational in reasoning on that basis which if we reject we cannot reason at all. If men really can discern and cannot help discerning a design in certain portions of the works of creation, this perception is the soundest and most satisfactory ground for the convictions to which it leads. The ideas which we necessarily employ in the contemplation of the world around us afford the only natural means of forming any conception of the Creator and Governor of the universe, and if we are by such means enabled to elevate our thoughts, however inadequately, towards Him, where is the presumption of doing so? or rather, where is the wisdom of refusing to open our minds to contemplations so animating and elevating and yet so entirely convincing. The assertion appears to be quite unfounded that, as science advances from point to point, final causes recede before it and disappear one after the other. . . . We are rather by the discovery of the general laws of nature led into a scene of wider design, of deeper contrivance, of more comprehensive adjustments. Final causes, if they appear driven further from us by such an extension of our views, embrace us only with a vaster and more majestic circuit: instead of a few threads connecting some detached objects, they become a stupendous network which is wound round and round the universal frame of things." — *Phil.* ii. 92. *et seq.*

On these extracts, and on the whole of this admirable chapter, we shall only add one remark. Cause, design, and motive are, as *we* conceive them, abstractions drawn from observed analogies of which our own personal and conscious experience supplies the chief materials. It is by these primordial analogies that we are led upward from creation to Creator, and animated by the prospects of our own immortal destiny. And these are precisely the analogies which by the original constitution of our minds we seize and generalize with the strongest impulse and fullest reliance. In such a constitution, no less than in our physical organization, we trace *design*, but a design as much loftier in its ends as our minds excel our bodies in worth and dignity — and pointing, as its origin, to a *motive* of which, whatever is good and great in humanity is only a dim and feeble adumbration.

In the "Philosophy of Palætiology," Mr. Whewell pushes on his frontier to the verge of all that is dark, awful, and overwhelming in antiquity. Every trifling pedantry and consecrated puerility of grammar and history, the tales of seanachies, and the dreams of cosmogonists, shrink and die away before the profound and solemn but shadowy images which this subject calls up, as the light Nymphs of fountains and Dryads of the woods before the fabled throne of ancient Night and Demogorgon. Yet the darkness which rests on that vanishing point to which every line, though broken, converges, is far different from the gloom of elder and despairing mythology—it is the palpitating reaction of an effulgence ineffable and intolerable, before which our gaze is sealed and our faculties prostrated. We will not injure the effect of this book by extracts.

The remaining books of this philosophy, constituting its second part, treat of *knowledge*—of the construction of science. To this all that has gone before is, properly speaking, subordinate and preparatory—in that sense, however, in which the base of a pyramid is subordinate and preparatory to its apex. Whatever be the origin of our fundamental ideas, and whatever the nature of the faculty by which we frame out of them ideal conceptions applicable to the explanation or connecting of phenomena, it is clear that, possessing such ideas, and the faculty of framing such conceptions, every step in our knowledge must consist in bringing them to bear upon facts, and binding together the latter in ideal connexion by means of them. Those processes, therefore, by which the ideas appropriate to particular classes of facts are brought into view and rendered more clear, and by which conceptions involving such ideas are made to fit and bind together the facts more closely, are those by which science is constructed. The former of these Mr. Whewell terms the explication of conceptions, the latter the colligation of facts: terms which strike us as particularly neat and well chosen, and which will doubtless henceforward become part of the fixed nomenclature of the subject. To the former belong almost all scientific controversies and discussions, which are thus seen to be anything but vexatious and injurious (as often thought) to the true interests of science, however too often fatal to the happiness of the disputants. They are the struggles by which thinking men emerge from darkness into day, and in trying to convert or confute their adversaries get to understand themselves. All battle, it has been well remarked, is misunderstanding, and all victory *terminating in permanent conquest* has been said to have right in some form or other on its side. The latter maxim, though we deem it profoundly false in history and politics, is yet certain in science. When controversy terminates, the defeated party is not suppressed, but extinguished. The inconsistency of its tenets becomes unfolded into self-contradiction, and they are thenceforward regarded not only as false, but as inconceivable.

The battle, as Mr. Whewell justly observes, is often one of definitions—for these are not, as is too commonly supposed, arbitrary. On the contrary, in science *their office is to embody in precise terms the very conception which is to serve as a key to the whole subject.* Hence a definition is always followed by a proposition of more or less generality dependent on it for its truth, and which expresses the manner in which many facts are intelligibly bound together by the conception it involves. In geometry, for example, the definition of a straight line is immediately followed by the axiom that two such lines cannot include a space; on which all geometrical truth depends. "In many cases, perhaps in most, the proposition which contains a scientific truth is apprehended with confidence, but with some vagueness and vacillation, before it is put in a positive, distinct, and definite form." Definition is here of essential service by compelling the propounder to give clearness and body to whatever was shadowy and indefinite in his conception. Still, in this shadowy state, it must exist in the mind of him who first perceives that facts *can* be so availably connected. The sagacity of him who frames a sound and pregnant definition must be preceded by the equal or superior sagacity of those who from the assemblage of facts are led to perceive what are the ideas and what the nature of their modifications which the definition ought to embody.

The ideas must be appropriate to the facts; but in discerning what ideas *are* appropriate lies one of the difficulties of inductive discovery — in modifying them into a suitable conception another, and usually a far greater. For these processes no rules can be given, nor does Mr. Whewell attempt it. In the analysis which he gives of the inductive process into three steps, which he describes as "the *selection of the idea*, the *construction of the conception*, and the *determination of the magnitude*," he says, "No general method of evolving such ideas can be given: such events appear to result from a peculiar sagacity and felicity of mind—never without labour — never without preparation ; yet with no constant dependence upon preparation, upon labour, or even entirely upon personal endowments." (vol. ii. p. 553.)

The true *idea*, it is to be observed, in Mr. Whewell's sense of the word, often presents itself almost spontaneously. Accident, by throwing before the most careless observer a "glaring instance," or vulgar experience of the mutual dependence of phenomena, has, in innumerable cases, done for us this part of the work. Reference of facts to the right fundamental idea generally takes place in what Mr. Whewell calls the prelude of an inductive epoch. One age proposes a problem in terms, referring facts to a right principle — a subsequent age resolves it by applying the principle according to a right conception. This step is always the result of sagacity, labour, and intimate acquaintance with the subject. The other may, but this can never be accidental.

Let us now consider the colligation of facts.

All facts, as we have seen, are theories — all true theories, facts, according to the position from which we contemplate them. Sensations (mental as well as bodily) inductively bound together, make *things* and (as we conceive the matter) *ideas;* things and ideas, *facts:* facts and ideas, theories or *general facts;* and so on. In binding together our fagot of facts, therefore, it is impossible to exclude from them ideas — they form an essential part of the bundle; indeed, the most essential of all, for its strength and coherence depends upon them. It is not, however, a collection, but an assortment that we aim at making. Our facts, therefore, have to be examined and *decomposed,* so as to bring into view the elementary ideas which they involve, with a view to the exclusion, or at least disregard, of all which are unsusceptible of scientific precision or otherwise inappropriate to inductive inquiry. Of the latter class are all which refer to emotions of wonder or terror, to passion or interest. Science is essentially abstract, passionless, and disinterested. Results are to be accepted for their truth alone: joy and fear have no part in their approval or disapproval; and the facts on which it depends must be selected in this view of its character; the precise, the abstract, and the measurable, being the grounds of their selection.

Hypotheses must of all things be framed — not loose and incapable of being exactly tested by following them into consequences, like those which Newton proscribed in his celebrated "hypotheses non fingo," — but such as can be so tested by reference to number, time, quantity, &c.; such as refer rather to modes of action of known causes than to the assumption of unknown, or (if that be necessary) which point out an intelligible and traceable line of connexion between the cause assumed and the results observed. Our facts may be homogeneous and well assorted—nay, they may have an obvious disposition to lie side by side and fit well together, yet be incoherent for want of the bond which is to unite them. For this we have to search, and the search consists in framing hypotheses and testing them by their legitimate results. Kepler constructed no less than nineteen for representing the apparent motion of Mars before that of an elliptic orbit about the sun suggested itself to his mind, which proved the true one, and the simplest of them all.

The rule of referring phenomena to known rather than to unknown causes (which is what Newton meant by his *vera causa*), is no doubt a good one. Like a new element in chemistry, a new *cause* must not be resorted to till all known causes are proved at fault. Nevertheless, seeing, as we do in the actual state of science, far beneath the surface of things, having acquired as it were new senses in the powerful agents we employ, new causes *may* work their way into evidence — may mark their peculiarities in so many lines of inquiry as to render it impossible not to admit them into the list of *true* causes, or those which are understood among philosophers to be available for explanation. The rule of the *vera causa* Mr. Whewell, as we understand him, very justly limits in its acceptation to this sense, and with equal justice and force of argument combats that dry and unsatisfactory philosophy which declares *laws*, not causes, to be the legitimate objects of human research. To proscribe the inquiry into causes is to annihilate science under shelter of "that barren caution which hopes for truth without daring to venture in quest of it."

It is of great moment to distinguish the characters of a sound induction. One of them is its ready identification with our conception of facts, so as to make itself a part of them, to engraft itself into language, and by no subsequent effort of the mind to be got rid of. The leading term of a true theory once pronounced, we cannot fall back even in thought to that helpless state of doubt and bewilderment in which we gazed on the facts before. The general proposition is more than a sum of the particulars. Our dots are filled in and connected by an ideal outline which we pursue even beyond their limits, — assign it a name, and speak of it as *a thing*. In all our propositions this *new thing* is referred to, the elements of which it is formed forgotten; and thus we arrive at an inductive formula; a general, perhaps a universal proposition.

Another character of sound inductions is, that they enable us to predict. We feel secure that our rule is based on the realities of nature, when it stands us in the stead of more experience; when it embodies facts as an experience wider than our own would do, and in a way that our ordinary experience would never reach; when it will bear not stress but torture, and gives true results in cases studiously different from those which led to its discovery. The theories of Newton and Fresnel are full of such cases. In the latter, indeed, this test is carried to such an extreme, that theory has actually remanded back experiment to read her lesson anew, and convicted her of blindness and error. It has informed her of facts so strange as to appear to her impossible, and showed her all the singularities she would observe in critical cases she never dreamed of trying.

Another character, which is exemplified only in the greatest theories, is the *consilience of inductions*, where many and widely different lines of experience spring together into one theory which explains them all, and that in a more simple manner than seemed to be required for either separately. Thus in the infinitely varied phenomena of physical astronomy, when all are discussed and all explained, we hear from all quarters the consentaneous echoes of but one word, GRAVITATION. And so in optics, each of its endless classes of complex and splendid phenomena being interpreted by its own conception; when these conceptions are assembled and compared, they all turn out to be translations into their peculiar language of the single phrase TRANSVERSE UNDULATION. Mr. Whewell has given us, as examples of the "logic of induction," what he terms "inductive tables" of each of these noble generalizations, which form not the least interesting feature of the work, enabling us, as they do, to trace, as in a map, the separate rills of discovery flowing at first each in its own narrow basin, thence confluent into important streams, which, uniting at length into one grand river, bear downwards to an ocean of truth beyond our tracing.

The theory of the construction of science being thus reduced to an analysis of these three processes —the decomposition of phenomena, the explication of conceptions, and the colligation of facts — the important question of course arises, how far the theory avails us in the practice; what progress it enables us to make to an *art of discovery?* and if, as Mr. Whewell acknowledges, such an art be, strictly speaking, impossible, what benefit do we derive from thus breaking up and reviewing its principles? The reply is clear: whatever we do, it is desirable at least to know fully *what is to be done*, and to be familiar with every facility and every method by which particular parts of the process have been ascertained to be materially aided or shortened. Thus the measurement of phenomena being an essential part of the process by which facts are rendered precise and strictly comparable with theories, *methods of observation* come to be considered with a view to the detection of general causes of error, the means of obviating them, and the establishment of maxims and habits which shall afford the inexperienced observer the benefit of his predecessor's failures and successes. An art of observation at least is possible, though an art of invention is not. Again — the research of causes is of necessity preceded by that of laws, which to be useful as tests of hypotheses must be quantitative, and involve precise numerical data. In the discovery of these much trouble may be saved, and much clearer insight gained by regular systematic methods of grouping and combining observations. Four such methods are laid down by Mr. Whewell in his chapter on this subject — those of "curves," "of means," "of least squares," and "of residues." Of these, the method of curves depends on the very principle on which we have metaphorically explained the nature of inductive generalization itself, the power which the mind possesses of connecting a series of dots, by a continuous outline — in virtue of which it has the especial and invaluable quality of detecting and eliminating casual errors. Mr. Whewell has exhibited the principle of this very powerful method with much clearness, and carefully traced the limits of its applicability. We may add, too, that nowhere will be found more beautiful instances of its systematic application than in his own elaborate and most successful researches on the tides. The methods of means and of least squares, which are properly one and the same, depend on the laws of probability—a subject which we are somewhat surprised to find slightly, or not at all, alluded to in any part of these works. That of residues is susceptible of far wider than mere quantitative application, and is in fact one of the most fertile and certain means of discovery that we possess.

A very large space is devoted by Mr. Whewell to a "review of opinions on the nature of knowledge, and the methods of seeking it," from Plato and Aristotle downwards. It is curious to observe the grand antithesis between an ideal and an empirical philosophy propagating itself onwards from these great masters to the present day, with little or no approach to a decision. Mr. Whewell, in the work before us, gives a masterly specimen of what may be done to make Platonism a solid and compact body of philosophy, while the views we have attempted to advocate (we are but too conscious how inadequately) are fundamentally Aristotelian, strange as it may seem to find the Stagyrite, of all philosophers, figuring as the father of induction.

Among the "innovators of the middle ages" brought into especial notice by Mr. Whewell in this review, Roger Bacon claims the first rank—a rank scarcely, if at all, inferior to that which the universal suffrage of posterity has vindicated to his great namesake Francis. The way in which he "sticks fiery off" from the general darkness of his era is indeed something marvellous; nor is the marvel diminished when we come to compare his ideas, as delivered in the "Opus Majus," with those of his illustrious successor, in the "Novum Organum." The resemblance indeed is so close as to be more than a mere resemblance—it is all but identity. When reading his exposition of the four general causes of human ignorance, his animated and impatient recalcitration against the authority of Aristotle (as then understood, or rather misunderstood, but at all events supreme in the schools), and his urgent and eloquent recommendation of mathematics and experiment, as the only true roads to knowledge, we fancy ourselves transported over the broad gulph of four centuries, and communing with the spirit of the great reformer. In one respect he far surpassed his successor, having been quite as remarkable for successful research in the practice of physical and experimental inquiry as the latter was unfortunate in every attempt to apply his principles to practice.

But science, as a body, has its aids and modes of progress, which may be considered in general, and without reference to the ways in which it may be advanced in detail. In this, as in many other cases, the whole may be advantageously considered as something different from the sum of its parts. The great value and importance of scientific truths as conducive not only to the physical, but, as we firmly believe, to the moral well-being of man, justifies us in regarding it as *a duty inseparable from our claim to civilization, to push forward the frontier of sound and well-established knowledge in every possible direction and by every form of individual and national effort.* Herein we conceive to consist one of those grave responsibilities consequent on acquisitions made, and powers ascertained, which we have alluded to in the commencement of this article. Already the public mind is beginning to be awakened to the sense of these responsibilities, nor was there ever a period in the history of mankind in which the sober and well-weighed judgments of men earnest in the cause, and competent to the task of suggestion, were listened to with more deference, and acted on with more readiness and sequence. We feel therefore grateful, and listen with doubly-excited attention when one who has shown himself in so decided a manner and on so many occasions a leader in the van of science, and whose influential position in one of our great universities enables him to carry out into practice his own suggestions in a field where they are sure to be productive of immediate effect, places before us the results of his thought and experience on the subject of intellectual education as a means of securing the spread and general reception of clear scientific ideas. "The period," he says, in a short but important chapter on this subject, which we most earnestly recommend to the attentive perusal of all who have anything to do with public education,

"appears now to be arrived when we may venture, or

rather when we are bound to endeavour, to include a new class of fundamental ideas in the elementary discipline of the human intellect. This is indispensable if we wish to educe the powers which we know it possesses, and to enrich it with the wealth which lies within its reach." — *Phil.*, vol. ii. p. 512.

The ideas to which Mr. Whewell especially alludes in this passage, in addition to those of space and number, which form the basis of a purely mathematical discipline, are those of *force* and *definite resemblance*, as the grounds of instruction in the principles of mechanics and natural history, the latter more especially being introduced as a corrective, and, we must say, as appears to us, a very valuable one, of those habits of thoughts and reasoning from mere definitions and axioms which a too extensive attention to mathematics is sure to generate. "The lessons afforded by this study," he says,

"are of the highest value with regard to all employments of the human mind; for the mode in which words in common use acquire their meaning approaches far more nearly to the *method of type* than to the method of definition. The terms which belong to our practical concerns or to our spontaneous and unscientific speculations, are rarely capable of exact definition. They have been devised in order to express assertions often very important, yet very vaguely conceived, and the signification of the word is extended as far as the assertion conveyed by it can be extended by apparent connection and analogy." — *Phil.*, vol. ii. p. 518.

In Mr. Whewell's recommendation also of "a

continued and connected system of observation and calculation," imitating the system which has been found so efficacious in astronomy, and extended to other branches of science, we cordially join. Such a system is commenced on a scale worthy of our nation in the magnetic and meteorological observations recently set on foot by the British Government and the East India Company, and though only intended in their origin for a temporary purpose, we entertain little doubt that the results they will furnish will prove of such importance as to induce their continuance.

The great length to which this article has extended prevents our giving any account, as we had originally intended, of a highly elaborate dissertation on the language of science, *i. e.*, on nomenclature and terminology, which, under the form of aphorisms, illustrated and explained, Mr. Whewell has prefixed to his Philosophy: the more so as the subject itself, though important, being far from inviting, and the pages assigned to it being kept as it were in a perfect foam of unpronounceable Greek, Latin, and German technical terms, it is not unlikely to be passed over by readers anxious to become acquainted with the substantial matter of the work. It is full, however, of valuable instruction, the great need of which, arising from the absence of general and distinct views on the subject among those who invent and use new terms, is much to be deplored. The *ultimatum* of unintelligible and unmanageable nomenclature, however, seems at length to have been reached *, since we can hardly conceive it possible in those respects to go beyond the system lately adopted by the French chemists for the designation of organic compounds.

Of the style of Mr. Whewell's work it may be expected that we should say something, the extracts above given having been selected rather with a view to their matter than their manner. Its chief characters are a remarkable occasional point and felicity of expression, and the almost systematic adoption, as a mode of illustration, of a great assemblage and variety of metaphorical allusion, much greater indeed than we should like to see adopted by an author less thoroughly imbued with his own meaning, and less capable of curbing the exuberance of a brilliant fancy into an entire subordination to his reason. We say systematic — for we have no doubt that it is intentional; and the object, moreover, is attained, the convergence of illustrations from so many different quarters rendering it perfectly impossible to mistake the point to which they are directed. Among our author's various and brilliant accomplishments not one of the least remarkable is his poetical talent, of

* Not quite. The British school has outdone the French, and Garagantua's mouth must be put in requisition. Witness such names (*Philosophical Transactions,* 1851, p. 380.) as Methylethylamylophenylammonium (H. 1857.)

which we have specimens in the mottoes prefixed to the several books of his "History," and in the following perfect little *bijou* from Goëthe, with which, as with a sweetener after such a dose of bitter metaphysic as we have been forced to inflict upon our readers, we shall endeavour to win them back to smiles and good humour:—

"Thou, my love, art perplexed with the endless seeming confusion
Of the luxuriant wealth which in the garden is spread.
Name upon name thou hearest; and, in thy dissatisfied hearing,
With a barbarian noise one drives another along:—
All the forms resemble, yet none is the same as another.
Thus the whole of the throng points at a deep-hidden law,—
Points at a sacred riddle. Oh! could I to thee, my beloved friend,
Whisper the fortunate word by which the riddle is read!"

HUMBOLDT'S KOSMOS.

Kosmos, Entwurf einer Physischen Weltheschreibung, VON ALEXANDER VON HUMBOLDT. Ersten Band, Stuttgart und Tübingen. J. G. Cotta'scher Verlag, 1845.

Cosmos. Sketch of a Physical Description of the Universe. By ALEXANDER VON HUMBOLDT, Vol. I. Translated under the superintendence of Lieut.-Colonel EDWARD SABINE, R. A., For. Sec. R. S., London. Printed for Longman, Brown, Green, and Longman, Paternoster Row, and John Murray, Albemarle Street, 1846.

(FROM THE EDINBURGH REVIEW FOR JANUARY, 1848.)

KOSMOS, the adornment, the orderly arrangement, the ideal beauty, harmony, and grace, of the universe! Is there or is there not in the mind of man a conception answering to these magnificent, these magical words? Is their sound an empty clang, a hollow ringing in our ears, or does it stir up in the depths of our inward being a sentiment of something interwoven in our nature of which we cannot divest ourselves, and which thrills within us as in answer to a spell whispering more than words can interpret? Is this wondrous world of matter and of thought, of object and of subject, of blind force and of moral relation, a one indivisible and complete whole, or a mere fragmentary assemblage of parts, having to each other no inherent primordial relations? If the former, contradiction and ultimate discordance can have no place. All that is to us enigmatical *must* have its solution, however hidden for a while the word which resolves the riddle. All that shocks us as irreconcilable, *must* admit of satisfactory interpretation, could we read the character of the writing with ease and fluency. If the latter, Chaos is a reality, Polytheism a truth; since arbitrary, self-existent, and independent Powers must, on that view of the subject, agitate, without end and without hope of final prevalence, the field of Being.

It is something to have put the question in this form, uncomplicated with the idea of responsibility for its answer to any tribunal but that of the pure reason and the inborn feeling. So put, we might well leave it to be decided by the acclamation of the human race, were it not for the healthful and invigorating exercise of our faculties, and the rich enjoyment it affords to pass before us in review those grand features in the constitution of the frame of Nature which render the conclusion irresistible, and invest it with the character of a demonstrated truth rather than that of an admitted opinion.

It is true that to grasp, as by a single mental effort — to embody and realize to our conceptions the UNITY OF NATURE — to soar so high as to perceive its completeness, and enjoy the fulness of its harmony, is given neither to Man nor to Angel. The feebleness and limitation of our faculties repress such longings as presumptuous, and forbid such flights as impracticable. Yet to spring a little way aloft — to carol for a while in bright and sunny regions — to open out around us, at all events, views commensurate with our extent of vision — to rise to the level of our strength, and, if we must sink again, to sink, not exhausted but exercised — not dulled in spirit, but cheered in heart, — such may be the contented and happy lot of him who can repose with equal confidence on the bosom of earth, though for a time obscured by mists, or rise above them into empyrean day.

To some it is given to soar with steadier wing and more sustained energy; to sweep over ampler circles, and treasure up the impressions of more varied imagery. To such the ambitious but sublime idea may occur of attempting to throw off, in broad and burning outline, a picture of The Whole as it has presented itself to their aspiring conceptions. Far be it from us to reprove such aspirations. Their failures may yet be immeasurably grander than our best successes; and, as we contemplate them, a glimpse, a shadow, may impress itself which may aid us to remodel our own conceptions according to a higher ideal than any we could have formed from our more limited opportunities. Such outlines, struck with a bold hand and true to nature, though confessedly imperfect and partial, suggest in their turn, to imaginative intellects, groupings and combinations of a more recondite and deep-seated order. Transplanted onward, thus, in progressive development from observer to observer, and from mind to mind, with a constant reference to nature and experience as their prototype, it is easy to see how, while gaining in comprehensiveness, they may lose at every transfusion somewhat of their specialty, without a corresponding loss of general truth; and how, thus, a larger and more entire conception of nature in itself may by degrees arise, and come to be recognized as the common property of humanity, the permanent and ennobling inheritance of generation after generation to the end of time.

The difficulties to be encountered in such an attempt are of two opposite kinds; on the one hand, that of embracing with distinctness and truth a sufficiently extensive view, on the other that of duly suppressing detail. Such a view of nature, to be in any way successful, ought to be, in the highest possible sense of the word, *picturesque*, nothing standing in relation to itself alone, but all to the general effect. In such a picture every object is suggestive. However beautiful in itself, it is less for the sake of its intrinsic beauty than for that of the associations it calls up, and the lights which it reflects from afar, that it holds a place as an element of the work. And, as in art, intense and elaborated beauty in any particular defeats picturesqueness by binding down the thought to a sensible object, annulling association, and saturating, as it were, the whole being in its single perception; so, in throwing off such a picture of nature as the mind can take in at a view, no one portion can be suffered to appear in single completeness and ideal rotundity. Nature, indeed, offers all in her profusion, and complete in all its details; and the contemplative mind finds among them paths for all it wanderings, harmonies for all its moods. But such exuberance is neither attainable nor to be aimed at in a descriptive outline, where leading features only have to be seized, which imagination is stimulated to fill up by the grandeur of the forms, and the intelligible order of their grouping.

The origin and fount of all good writing, however, is sound and abundant knowledge. To the successful execution of such a work, a thoroughly scientific acquaintance with each component feature; a mind saturated with information, and at home in every department, is above all things requisite. The classifications of the naturalist, the surveys of the geologist, the catalogues and descriptions of the astronomer, the theories of the geometer, and the inductions of the experimentalist, must all be alike familiar, and not merely ready at a call, but present to the thought at every instant. It is, therefore, by no simply clever writer, by no mere man of vivid imagination and fluent command of language and imagery — least of all, by any ideal speculatist who may have devised a system of philosophy spun from the abstractions of his own brain, and resolving all things into some single principle, some formula embodying all possible knowledge, that such a work can be entered upon without the certainty of utter and disgraceful failure. The highest attainments in science, though necessarily inadequate to complete success in such an attempt, can alone save the adventurous mortal who shall make it from merited reproach on the score of presumption.

The author of the remarkable book before us is assuredly the person in all Europe best fitted to undertake and accomplish such a work. Science has produced no man of more rich and varied attainments, more versatile in genius, more indefatigable in application to all kinds of learning, more energetic in action, or more ardent in inquiry; and, we may add, more entirely devoted to her cause in every period of a long life. At every epoch of that life, from a comparatively early age, he has been constantly before the public, realizing the ideal conception of a perfect traveller; a character which calls for almost as great a variety of excellences as those which go to realize Cicero's idea of a perfect orator. To such an one science in all its branches must be familiar, since questions of science and its applications occur at every step, and often in their most delicate and recondite forms. The habit of close attention to passing facts, which seizes their specific features, and detects their hidden analogies, must join with the broad *coup d'œil* which generalizes all it sees, and stereotypes it in memory in its simplest and most impressive forms. To these must be added a knowledge of man and of his history in all its phases, social and political; a ready insight into human character and feelings, and a quick apprehension of local and national peculiarities. Above all things is necessary a genial and kindly temperament, which excites no enmities, but on the contrary finds or makes friends everywhere; in presence of which hearts open, information is volunteered, and aid spontaneously offered. No man in the ranks of science is more distinguished for this last characteristic than Baron Von Humboldt. We believe that he has not an enemy. His justice, candour, and moderation, have preserved him intact in all the vexatious questions of priority and precedence which agitate and harass the scientific world; and have in consequence afforded him innumerable opportunities of promoting the objects and befriending the cultivators of science, which would never have fallen in the way of a less conciliatory disposition, and of which he has not been slow to avail himself. The respect of Europe, indeed, has gone along with him to a point which has almost rendered his recommendations rules. It has sufficed that Von Humboldt has pointed out lines of useful and available inquiry, to make every one eager to enter upon them.

The idea of a physical description of the universe, as a work to be accomplished, and an object, to amass materials for which during a whole lifetime, would be a worthy and satisfactory devotion of it, had, it appears, been present to his mind from a very early epoch. For almost half a century, indeed, it had occupied his thoughts. At length, in the evening of life, he felt himself rich enough in the accumulations of thought, travel, reading, and experimental research, to reduce into form and reality the undefined vision which had so long floated before him. Not entirely, however, without some preliminary trial of strength. A course of lectures, as he informs us, had been delivered by him, both in Berlin and Paris, on the subject, about the end of 1827, previous to his departure for Northern Asia, a journey for which he had prepared himself by a course of study without example in the history of travel. On his return, after giving to the world the results of that journey, or rather the epitome of all the knowledge acquired by himself and by former travellers on the physical geography of Northern and Central Asia, in a work which would alone have sufficed to form a reputation of the highest rank; he resolved no longer to defer this realization of his early aspirations, and the result has been the work of which the volume now before us is only a commencement.

Though we cannot blame an arrangement which brings any portion of the fruits of M. de Humboldt's labours earlier before us; though aware of the hazard which passing years entail on the ultimate appearance of a work of great extent deferred already so long; and although only too glad to receive by instalments, at the convenience of the author, the payment of a self-imposed debt of such magnitude and value, yet we cannot but consider the publication of the three volumes, of which it is understood the whole will consist, separately and at long intervals, as in many respects unfortunate. Although it is now nearly four years since the work was completed, the second volume is only just on the eve of publication, and the third may possibly be yet longer delayed. Yet no work could have been undertaken, in which it would appear so needful that the impression produced be one and undivided, the unity salient and conspicuous. That the contrary course, though perhaps unavoidable, has been pursued, renders the task of duly appretiating and correctly criticizing it doubly difficult; since it is impossible to say to what extent, and in what manner, many things, which appear in the light of omissions in the first portions of such a performance, may be supplied in the sequel; or how differently the philosophy of the whole subject may come to be judged as presented by the author on a complete and on a partial view of his entire meaning. This would have been less the case, and the probability of doing injustice to the author's philosophical views greatly diminished, had the general plan of the whole work been chalked out with more precision in the introductory portion, and the nature of the contents of the subsequent volumes indicated in somewhat less vague and general terms than we find them actually to be. And the necessity for thus holding a reserve on our judgments in this respect, while considering that portion of the work which we possess, is the more imperatively pressed upon us, inasmuch as the scope of the proposed third volume, as we understand it, seems to us by far the most important in its philosophical bearings, and as that by which the character of the whole as a great philosophical work will of necessity come to be finally judged.

Such, however, we are aware, is not exactly M. de Humboldt's own impression. He must here be allowed to speak for himself: "The first volume," he says, "contains a general view of nature, from the remotest nebulæ and revolving double stars, to the terrestrial phenomena of the geographical distribution of plants, of animals, and of races of men; preceded by some preliminary considerations on the different degrees of enjoyment offered by the study of nature and the knowledge of her laws; and on the limits and method of a scientific exposition of the physical description of the universe. I regard this as the most important and essential portion of my undertaking, as manifesting the intimate connexion of the general with the special, and as exemplifying, in form and style of composition, and in the selection of results taken from the mass of our experimental knowledge, the spirit of the method in which I have proposed to myself to conduct the whole work. In the two succeeding volumes I design to consider some of the particular incitements to the study of nature,—to treat of the history of the contemplation of the physical universe, or the gradual development of the idea of the concurrent action of natural forces (Kräfte), co-operating in all that presents itself to our observation; and lastly, to notice the specialties of the several branches of science, of which the mutual connexion is indicated in the general view of nature in the present volumes."

A large portion (nearly one-fifth of the text) of the volume before us, is occupied with an introductory exposition of the various kinds or gradations of enjoyment afforded by the contemplation of nature and the investigation of her laws, and with an essay on the limitation and methodical treatment of a physical description of the universe considered as a separate and independent science—"the science of the Kosmos." The mere aspect of nature, as has been often and well observed, is a source of positive and high enjoyment; and exercises, even on rude minds, and under the sway of wild passions, if only suffered to claim attention at all, a calming and elevating influence. In all her scenes, "there is everywhere revealed to the mind an impression of the existence of comprehensive and permanent laws governing the phenomena of the universe;" before the idea of whose vastness and regularity the turbulence of human passion feels itself reproved and shrinks abashed. Whatever be the peculiar inherent or temporary character of the scene contemplated—even in her most agitated moods—this sense of the regulated and the imperturbable is never wholly effaced. We know that the storm will rage itself to rest, the angry billows subside, the earthquake roll away, and that holy calm which is her habitual mood be restored, as if it had never been broken. "That which is grave and solemn in such impressions is derived from the presentiment of order and of law, unconsciously awakened by the simple contact with external nature; it is derived from the contrast of the narrow limits of our being with that image of infinity which everywhere reveals itself—in the starry heavens, in the boundless plain, or in the indistinct horizon of the ocean."

Enjoyment of a different, and, in some respects, of a richer, because of a less overwhelming and more exciting kind, is that which depends on the peculiar physiognomy of natural scenes. Harmonizing, like music, with internal trains of thought and imagination, and with every conceivable state of mind, they awaken of themselves, as soon as presented, sentiments congenial to them, and lead the spirit, by strong associative links, through every phase of feeling. The barren monotony of one region, the varied fertility of another, the gloomy and romantic horrors of a third—the peaceful dwelling rising by the torrent's side—the misty region, where the mule seeks his track amid eternal snows—the tropical night, "when the stars, not sparkling as in our climates, but shining with a steady beam, shed on the gently heaving ocean a mild and planetary radiance,"—the deep and doubly wood-clothed valleys of the Cordilleras—the volcanic peak cleaving the clouds, from a base of vineyarded slopes and orange-groves washed by a tropical sea—the dense forest, of giant and primeval growth, swarming with every form of vegetable and animal life, now resounding to savage yells, and now to the thunder-clap, extinguishing and crushing down all other sounds,—these and a thousand other combinations find each its response in some train of human emotions and affections, which, like the lyre of Timotheus, they by turns excite and soothe.

As the poetical enjoyment of nature springs out of this its endless variety, so, on the other hand, the unity of plan, which even uncultivated minds fail not to recognize amid so much diversity, calls forth the latent germ of the philosophic spirit. When—

"—far from our native country, after a long sea voyage, we tread for the first time the lands of the tropics, we experience an impression of agreeable surprise in recognizing, in the cliffs and rocks around, the same forms and substances, similar inclined strata of schistose rocks, the same columnar basalts which we had left in Europe; this identity, in latitudes so different, reminds us that the solidification of the crust of the earth has been independent of the differences of climate. But these schists and these basalts are covered with vegetable forms of new and strange aspect. Amid the luxuriance of this exotic flora, surrounded by colossal forms of new and unfamiliar grandeur and beauty, we experience (thanks to the marvellous flexibility of our nature) how easily the mind opens to the combination of impressions connected with each other by unperceived links of secret analogy. The imagination recognizes in these strange forms nobler developments of those which surrounded our childhood; the colonist loves to give to the plants of his new home names borrowed from his native land; and these strong untaught impressions lead, however vaguely, to the same end as that laborious and extended comparison of facts, by which the philosopher arrives at an intimate persuasion of one indissoluble chain of affinity binding together all nature."

One word on this last sentence:—Is it really true that the uninstructed mind of man, thus turned loose upon nature, *does* spring, as a matter of course, to just conclusions? *Are* his homely analogies always apposite? his extempore classifications correct? his rude inductions legitimate? If so, what need of study and research? How is it, then, that we are to understand what is here intimated, and is there any sense in which it can be received as true? No doubt there is so. There are truths so large, so general, so all-pervading, that they make a part of all our experience, mix with our whole intellectual being, and imbue all our judgments, erroneous as well as correct; in this sense, at least, that we never err so far as to place ourselves in conscious opposition to them. Distorted and perverted as such truths may be in their enunciation, by their mixture with extraneous error, we find them still outstanding, redeeming by their presence, and even consecrating, that error, by being exhibited in prominent and ostentatious union with its dogmas. No absurdity would ever obtain a moment's credence, but for the presence in it of some saving particle of one of these great natural truths.

But it is to the instructed only that the contemplation of nature affords its full enjoyment, in the development of her laws, and in the unveiling of those hidden powers which work beneath the surface of things, and which, operating as physical causes, lead back the mind in the chain of causation, through the phenomena of organized life, to powers of a higher order; which, connecting themselves with the idea of Will, involve the conception of Intelligence, from which we are necessarily led to infer Design, and from Design find ourselves forced on the conclusion of Motive. It is thus, and thus only, that the contemplation of nature can be said to lead us up, by legitimate induction, to its Author,—to so much of his character, at least, as he has thought fit to reveal to us through his works. But, that it may do so we must educate our perceptions by practice and habit, till we learn to disregard specialties, whether of objects or laws, and see rather their relations and connexions, their places in a system, their fulfilment of a purpose, their adaptation to an interminable series of intersubservient ends. And this we must endeavour to do without losing sight of the objects themselves, which come at length to stand in intellectual relation to these more spiritualized conceptions, as the notion of substance does to that of quality in some of our older metaphysical theories,—as that substratum of being in which such conceptions inhere, and which serves to bind them together, give them a body, and coerce them from becoming altogether vague and imaginary. And, moreover, we must be careful to raise up no self-created phantasms of our own minds, interposing an impassable barrier to further progress, and cutting off the chain of connexion by a stern *ne plus ultra.* As the distinction drawn in the Aristotelian Philosophy between celestial and terrestrial motions operated for ages to cut off the possibility of arriving at any just views of the Planetary System, so it is perfectly conceivable that, by gratuitous assumptions of another kind, we may wilfully sever ourselves from the possible attainment of knowledge of a far higher order. Against certain notions of this description, which have obtained, or may be obtaining, currency; and others which, without being expressed in words, appear to be extensively, though tacitly, received in science, we consider it worth while to enter our protest:—

The first is, "that ancient belief, that the forces inherent in matter, and those which regulate the moral world, exert their action under the government of a primordial necessity, and in recurring courses of greater or less period. It is this necessity, this occult but permanent connexion, this periodical recurrence in the progressive development of forms, of phenomena and of events, which constitute nature, obedient to the first imparted impulse of the Creator. Physical science, as its name imports, limits itself to the explanation of the phenomena of the material world by the properties of matter. All beyond this belongs, not to the domain of the physics of the universe, but to a higher class of ideas. The discovery of Laws, and their progressive generalizations, are the objects of the experimental sciences." (*Transl.* p. 33.)

The frame of nature, moral as well as physical, according to this idea, is a piece of mechanism, which, wound up and set going, has been abandoned to itself, to evolve its changes in variously superposed periods, without choice or option, according to the combinations of an occult wheelwork. If, indeed, there were no such phenomenon as Will; if we were conscious of being thus blindly hurried along by the uncontrollable swing of the system of which we form a part, at every moment and in every action, such a system might be tenable. Periods of unknown length, superposed according to no discoverable law, lose their character of periodicity to the eye of the observer; and *periods of event*, apart from the notion of the measurement of time, similarly superposed, resolve themselves, so far as observation is concerned, into that imperfect and inadequate idea of causation which considers it as simply a determinate rule of sequence. But *Will*, admitted into any part of such a system, destroys the whole of it. The blind, unintelligent portions of the mechanism must be invested with the power, and be urged by the necessity of conforming themselves to that will, as to the original impulse which set the whole in motion; and how are we then to distinguish between those evolutions which result from a will of which we are conscious, and those which, for aught we know, may be continually resulting from a will continually in action, though concealed from our knowledge and perception?

Another notion, equally destitute, in our eyes, of positive foundation, but much more likely than the former to act prejudicially in limiting the progress even of physical knowledge, is the assumption, as old as Aristotle, that all the phenomena of nature are referable to *motions* performed in obedience to what we are in the habit of calling *mechanical* laws; that, in other words, there is no such thing as *qualitative change* unaccompanied by change of place — no causation at work other than mechanical push and pull. It is high time, we think, that this assumption should be formally called in question. We are disposed to believe that science has outgrown it. At the same time, we are quite aware into what a licentious career of wild speculation the mind is ready to rush on the removal of such a limitation; what extravagant theories we must expect to see broached, and what confusion of ideas, nay, what positive charlatanries, we must be prepared to encounter, before any clear and definite conception can emerge from the mass of images which crowd upon us on the suggestion of such a change of ground. We may indicate, however, one or two, which may perhaps carry with them some degree of distinctness, viz.: first, The intension, remission, or creation of mechanical force dependent on the presence or absence of agents, such as electricity and heat, of whose *materiality*, in the usual sense of the word, we have no proof, seeing that inertia (at least in the case of heat) forms no part of our conception of them; and secondly, the successive *quasi-undulatory* propagation of qualities—powers of affecting either the senses or material bodies by something different from mechanical impulse. It is perfectly true, that on the properties of matter only we must rely for the explanation of physical phenomena. But we conceive that those properties are only just beginning to become known to us, that we shall have to reject some which have been assumed as unquestionable, and that it is by no means improbable, that science will ere long make us familiar with others, calculated to stretch to the utmost our conception of *material* existence. Entertaining this expectation, we must here, once for all, observe, that the continual use of the word *forces* in the work before us, in such phrases as "the forces of nature"—"the concurrent action of natural forces"—grates with something approaching to a painful harshness on our ears. We should be inclined to substitute for it, wherever it occurs, the expression "physical powers," a sense which the German Kräfte might bear, we think, without violence.

A third dogma, which has of late been placed in prominence, much, as we conceive, to the detriment of sound philosophy, is that of the so-called, or rather miscalled, *positive philosophy* — an extravagant and morphological transformation of that rational empiricism, which professes to take experience for its basis; resulting from insisting on the prerogatives of experience in reference to external phenomena, and ignoring them in relation to the movements and tendencies of our intellectual nature: — a philosophy which, if it do not repudiate altogether the idea of causation, goes far, at least, to put it out of view, and with it, everything which can be called *explanation* of natural phenomena, by the undue predominance assigned to the idea of Law: — which rejects as not merely difficult, not even simply hopeless, but as utterly absurd, unphilosophical, and derogatory, all attempt to render any rational account of those abstract equation-like propositions, in which it delights to embody the results of experience, other than their inclusion in some more general proposition of the same kind. Entirely persuaded that, in physics, at least, the inquiry into causes *is* philosophy; that nothing else is so; and that the chain of causation upwards is broken by no solution of continuity, constituting a gulph absolutely impassable to human faculties, if duly prepared by familiarity with the previous links; we are far from regarding the *whole* office of experimental philosophy as satisfactorily expressed, by declaring it to consist in the discovery and generalization of laws. There are two ways of expressing every law of nature, — one which does, the other which does not, bear reference to the cause, which lies at the root of the phenomenon. It is something distinct from, and more than a mere generalization of law, which refers the planetary motions to *Force* as a Cause of motion. No acuteness would ever have sufficed to conclude the laws of perturbation from those of elliptic motion, and to detect a new planet by the mere knowledge of these latter laws, had this word, the key of the whole riddle, remained unpronounced. The craving of the philosophic mind is for *explanation*, *i. e.*, for the breaking up of complex phenomena into *familiar* sequences, or equally familiar transitional changes, or cotemporary manifestations; which, under the names of cause and effect, we are content to receive (at least temporarily) as ultimate facts, and which nothing but perfect familiarity divests of that marvellous character which they really possess, — *which are only not looked upon as miraculous because they are usual.* When we work our way up to facts of this character, physical inquiry ends, and speculation begins. Very few such ultimate facts have hitherto been arrived at in physics; and it is to the increase of their number, by future inquiry, that we must look for any prospect of erasing any one of them from the list, *i. e.*, of explaining it. No doubt explanation must ever be imperfect, if quantitative laws be wanting as a feature. But the first, at least the most necessary office of experimental philosophy, is, the detection of the *influential thing*, the *ultimate fact*, or facts, on which explanation hinges — its subsequent, and, in that sense, subordinate, though still most useful and important one; to discover the formal and quantitative laws of that influence. If, indeed, it be said that the proposition announcing these ultimate facts is *a law*, in the sense of the word intended, we protest against the abuse of language, which confounds, under one form of expression, the statement of the law itself, and the subject-matter of the law — the *quod loquimur* with the *de quo*.

With the richness of idea and command of resource which natural knowledge confers, civilization goes hand in hand. The remarks of M. de Humboldt on this part of his subject are so pointed and impressive, that we cannot refuse ourselves the pleasure of quoting them: —

"The clearer our insight into the connection of phenomena, the more easily we shall emancipate ourselves from the error of those who do not perceive that, for the intellectual cultivation and for the prosperity of nations, all branches of natural knowledge are alike important, whether the measuring and describing portion, or the examination of chemical constituents, or the investigation of the physical forces by which all matter is pervaded. . . An equal appreciation of all parts of natural knowledge is an essential requirement of the present epoch, in which the material wealth and the increasing prosperity of nations are in great measure based on the more enlightened employment of natural products and forces. . . . The most superficial glance at the present condition of European states shows that those which linger in the race cannot hope to escape the partial diminution, and, perhaps, the final annihilation of their resources. . . . The danger . . . must be averted by the earnest cultivation of natural knowledge. . . . Knowledge and thought are at once the delight and the prerogative of man; but they are also a part of the wealth of nations, and often afford to them an abundant indemnification for the more sparing bestowal of natural riches."

To all this, of course, we heartily subscribe; and we only wish that the limit M. de Humboldt has prescribed to himself would have permitted him to extend the scope of his remarks, clothed, as they are, in such animated language, to embrace a far wider range of application. The frame of Nature is not bounded by that narrow limit which is commonly understood by the term Physics. Life, thought, and moral and social relation, are all equally *natural* — equally elements of the great scheme of the Kosmos with matter and magnetism. The only imaginable reason why the sciences growing out of these ideas are not regarded and handled, or have not hitherto effectually been so, as branches of natural science and inductive inquiry, is the great difficulty of arriving at true statements of facts in some, owing to the conflict of partial interest, and the great danger and consequent heavy responsibility attending experiments in others. These obstacles can only be removed by the general enlightenment of mankind, enabling them to perceive that their true interests require truth in the statement of facts; deliberate caution in undertaking, and patience—long, calm, enduring patience—and hearty co-operation, in watching the working out of social and legislative experiments.

A great and wondrous attempt is making in civilized Europe at the present time;* neither more nor less than an attempt to stave off, *ad infinitum*, the tremendous visitation of war; and, by removing or alleviating the positive checks to the growth of population, to diminish the stringency of the preventive ones, and to subsist continually increasing masses on a continually increasing scale of comfort. May it be successful! But the only conditions on which it can be so are, that nature be laid yearly more and more under contribution to human wants; and that the masses themselves understand and go along with the exertions making in their favour in a spirit of amicable and rational conformity. To no other quarter than to the progress of science can we look for the least glimpse of a fulfilment of the first of these conditions. Neither the activity of hope, nor the energy of despair, acting by stationary means on unvarying elements, can coerce them into a geometrically increasing productiveness. Science must wave unceasingly her magic wand, and point unceasingly her divining rod. The task now laid on her, however, is not of her own seeking. She declines altogether so dread a responsibility, while yet declaring her readiness to aid, to the utmost of her powers; claiming only the privilege, essential to

* 1848.

their available exertion, of free, undisturbed, and dispassionate thought, and calling upon every class to do its duty; the higher in aiding her applications, the lower in conforming to her rules.

In that part of his work which treats of the limits and method of exposition of the physical description of the universe, M. de Humboldt takes considerable pains to represent the "Science of the Kosmos" as a separate and independent department of knowledge, distinct in scope and kind from a mere encyclopædic aggregation of physical sciences. We concern ourselves little whether in this he has succeeded in making out a useful and available distinction; admitting, as he does, that in his mode of conceiving and handling it, it is, in effect, the aggregate, by simple juxtaposition, of two separate and very unequal portions, similar in character so far as the *less* can be similar to the *more* complex. He regards it, in short, as physical geography enlarged by such a description of the heavens and their contents as shall correspond in plan and in conception (so far as our knowledge extends) to that description of the earth and its denizens which is intended by the former designation. In so far, then, as physical geography is entitled to be termed a separate and independent science, Kosmography, or the science of the Kosmos, is so also, and a more general one, including the other. A Chinese map of the globe *is* a map of the globe, and not a mere map of China, though the Flowery Land figure therein in rich detail of city, stream, and province; and though Europe, Asia, Africa, and America exist, for the most part, in mere outline, and occupying an extent of surface altogether disproportioned to their true extent and importance. This is not the fault of the Celestial Arrowsmith. Had he known more of the globe, he would have given his countrymen a better map.

Our simile, however, is faulty in one respect. What we know of the contents of space exterior to our globe we at least know truly, — at all events, we can separate our knowledge from our ignorance; and it happens, fortunately, that what escapes our view is precisely that which, if seen, would merely serve to puzzle and perplex us; while the great and obvious features which strike us are precisely those which we are best able to reduce to general laws, and to view in systematic connexion, and which reveal to us, in its grandest form, the Unity of the Kosmos. The all-pervading power of gravitation, that mysterious reality by which every material being in the universe is placed in *instant* and influential relation with every other, springs forward in a state of disengagement and prominence on the contemplation of the celestial movements which it, perhaps, might never have assumed had not the opportunity been afforded us of so contemplating it, apart from the distracting influence of corpuscular forces which, in innumerable instances, mask and overlie it in its exhibition on the surface of our planet. And again: the phenomenon of light, its uniform properties and equal velocity from whatever quarter of space it reaches us, and the certainty those properties afford of the existence of a perfectly uniform mechanism, co-extensive with space itself, continually occupied in the discharge of the most important of all offices, that of conveying at once information and vital stimulus from every region of space to every other — facts of this kind, were there no other, would suffice to force upon our minds the clear perception of a unity of plan and of action in the constitution of nature. "A connexion is maintained, by means of light and radiant heat, both with the sun of our own system, and all those remoter suns which glitter in the firmament. The very different measure of these effects must not prevent the physical philosopher, engaged in tracing a general picture of nature, from noticing the connexion and co-extensive dominion of similar forces." (*Kosmos*, p. 146., *Transl.*)

We therefore entirely agree with our author in the propriety of that arrangement of his work which gives the precedence of treatment to the celestial over the "telluric" view of nature; and prefaces the description of our own globe by that of the sidereal and planetary system. And whether such description be properly regarded as the exposition of a body of science, or (as we should rather feel disposed to look upon it) a sort of epos, a noble oratorio, or a grand *spectacle*, we are delighted to receive it at his hands, and to throw ourselves into that frame of mind for its reception which shall be best calculated to heighten the impression, and do justice to the exponent.

Taking our stand, therefore, on the extreme verge of the visible creation, let us for an instant look about us, ere we descend with him, like the angelic messenger in Milton, through stars, nebulæ, and systems, to this planetary sphere and its central sun. Where are we? *Is* there such an extreme verge? This question, which lies at the very threshold of an exposition of the Kosmos *per descensum*, is one which has so little to recommend it as a matter of discussion that we certainly should not mention it here, had it not got involved in an astronomical speculation of a very singular nature. The assumption that the extent of the starry firmament is literally infinite has been made, by one of the greatest of astronomers, the late Dr. Olbers, the basis of a conclusion that the celestial spaces are in some slight degree deficient in *transparency*; so that all beyond a certain distance is, and must for ever remain, unseen; the geometrical progression of the extinction of light far outrunning the effect of any conceivable increase in the power of our telescopes. Were it not so, it is argued, every part of the celestial concave ought to shine with the brightness of the solar disc, since no visual ray could be so directed as not, in some point or other of its infinite length, to encounter such a disc. With this peculiar form of the argument we have little concern. It appears to us, indeed, with all deference to so high an authority, invalid; since nothing is easier than to imagine modes of systematic arrangement of the stars in space (entirely in consonance with what we see around us of the principle of subordinate grouping actually followed out) which shall strike away the only foundation on which it can be made to rest, while yet fully vindicating the absolute infinity of their number. It is the conclusion only which it appears to us important to notice, as having recently been attempted to be established on grounds of direct statistical enumeration of stars of different orders of brightness by the illustrious astronomer of Pulkova, in a remarkable work, "Etudes d'Astronomie Stellaire," and even some rude approximation made to the rate of extinction. It would lead us far beyond our limits to attempt even to give a general idea of his reasonings, but one remark on the whole subject we cannot forbear. Light, it is true, is easily disposed of. Once absorbed, it is extinct for ever, and will trouble us no more. But with radiant heat the case is otherwise. This, though absorbed, remains still effective in heating the absorbing medium, which must either increase in temperature the process continuing, *ad infinitum*, or, in its turn becoming radiant, give out from every point at every instant as much heat as it receives.

Of the supposed luminiferous æther itself, as one of the material or quasi-material contents of space, M. de Humboldt says nothing. He waives, designedly, at least in the present volume, any allusion to that, and all other theoretical conceptions. The view of creation which he takes, and which we must take with him, is so purely and entirely objective, so closely confined to what Mr. Mill would call the *collocations* of the Kosmos, that even the Newtonian law of gravitation, with its noble train of mathematical consequences, is excluded from all direct and special notice. We must not, therefore, wonder, but accept it as part of the determinate plan of the work, that light itself is spoken of only incidentally, as affording a measure of sidereal distance by its velocity, and as conveying to our eyes the images of remote sidereal objects, not as they now exist, but as they existed years or ages ago; or that no account is given of the Gaussian generalizations of the theory of terrestrial magnetism—a subject, of which M. de Humboldt is so pre-eminently cognisant, that it must have required the greatest self-control, and the most entire satisfaction with his pre-conceived views of the limits of his subject, to have avoided dilating on it.

The most remote bodies which the telescopes disclose to us are, probably, the nebulæ. These, as their name imports, are dim and misty-looking objects, very few of which are visible to the unassisted sight. Powerful telescopes resolve most of them into stars, and more in proportion to the force of the instrument; while, at the same time, every increase of telescopic power brings fresh and unresolved nebulæ into view. A natural generalization would lead us to conclude, that all such objects are nothing but groups of stars, forming systems, differing in size, remoteness, and mode of aggregation. This conclusion would, indeed, be almost irresistible but for a few rare examples, where a single star of considerable brightness appears surrounded with a delicate and extensive atmosphere, offering no indication of its consisting of stars. Such objects have given rise to the conception of a self-luminous nebulous matter, of a vaporous or gaseous nature, of which these photospheres, and, perhaps, some entire nebulæ, may consist, and to the further conception of a gradual subsidence or condensation of such matter into stars and systems. It cannot be denied, however, that the weight of induction appears to be accumulating in the opposite direction, and that such "nebulous stars" *may*, after all, be only extreme cases of central condensation, such as two or three "nebulæ," usually so called, offer a near approach to. Apart, then, from these singular bodies, and leaving open the questions they go to raise, and apart from the consideration of such peculiar cases as planetary and annular nebulæ, the great majority of nebulæ may be described as globular or spheroidal aggregates of stars arranged about a centre, the interior strata more closely than the exterior, according to very various laws of progressive density, but the strata of equal density being more nearly spherical according to their proximity to the centre. Many of these groups contain hundreds, nay, thousands, of stars.

Besides these, there exist nebulæ of a totally different description; of vastly greater apparent dimension, and of very irregular and capricious forms, of which the well-known nebula in Orion is an example. They form, evidently, a class apart from the others, not only in aspect, but also as regards their situation in the heavens; for whereas the former congregate together chiefly in a great nebulous district remote from the Milky Way, or are otherwise scattered over the whole heavens (though by no means so as to form what M. de Humboldt terms a "nebulous milky way," or zone of nebulæ surrounding the sphere), these only occur in the immediate vicinity of the galaxy, and may fairly be considered, if not as integrant portions, at least as outliers of it. Their forms, therefore, may be considered as in some degree indicative of the true form of that starry stratum, could we contemplate it from a distance, so far, at least, that we may reasonably suppose it quite as irregular and complex as we observe these, its appendages, actually to be.

M. de Humboldt leans, as might be expected from one especially conversant with organic forms, to that view which represents the nebulæ as sidereal systems, in process of gradual formation by the mutual attraction of their parts, and by the absorption of the strictly nebulous element into stellar bodies. "The process of condensation," he says, "which was part of the doctrine of Anaximenes, and of the whole Ionic school, appears to be here going on before our eyes. The subject of conjoint investigation and conjecture has a peculiar charm for the imagination. Throughout the range of animated existence, and of moving forces in the physical universe, there is an especial fascination in the recognition of that which is becoming, or about to be, even greater than in that which is, though the former be indeed no more than a new condition of matter already existing; for of the act of creation itself, the original calling forth of existence out of non-existence, we have no experience, nor can we form any conception of it."

That the whole firmament of *stars* visible to us, even with the help of telescopes, belongs to that vast sidereal stratum which we call the Galaxy, seems hardly to admit of doubt. The actual form of this stratum, further than that it is not improperly characterised as such, can hardly be said to be known with any approach to certainty; but that its extent in a direct line outwards is enormously greater in some directions than in others, and that in one portion of its extent it is, as it were, cleft, and contorted, in others lengthened into processes stretching far into space, seems to rank among the positive conclusions of astronomy. In certain directions its extent would seem to be unfathomable to our best telescopes; in others, there is reason to believe we see through and beyond it, even in its own plane.

Of the distance of the stars of which this vast stratum consists, at least of some of the nearest of them, we are beginning, at length, to possess some certain knowledge. The bright star α Centauri has a measured parallax (as the observations of Henderson and Maclear teach us) of nearly a whole second (0″·9128), which places it at a distance from us equal to 226,000 radii of the earth's orbit. That of 61 Cygni has been ascertained by Bessel to be no less than 592,200 such radii, while the observations of Struve place α Lyræ at 789,600 of similar units from our system. Such is the scale of the system to which we belong, such the magnitudes we are led to regard as small, in comparison with its actual extent! The number of stars whose distance is imperfectly known to us at present is about thirty-five, seven of which may be considered as determined, with some approach to certainty, by the recent researches of Mr. Peters.

Among the countless swarm of what are commonly called fixed stars, there is not one, probably, which really merits the name. In by far the great majority, a minute, but regularly progressive, change of place is observed to take place; and from a careful examination of these movements, as observed in stars visible in Europe, it has been concluded, that a portion at least of them is only apparent, and arises from a real motion of our own sun, carrying with it the whole planetary system, towards a point in the constellation Hercules, in R. A. 259° 35′ decl. 34° 34′ north. This extraordinary conclusion, resting as it does on the independent and remarkably agreeing calculations of five different and eminent astronomers, from data afforded by northern stars, has, within the last few months, received a striking confirmation by the researches of Mr. Galloway, who has arrived at the very same conclusion, from calculations founded on the proper motions of stars in the southern hemisphere, not included among those used by his predecessors. In this path the sun moves with the prodigious velocity of 400,000 miles, or nearly its own semi-diameter, *per diem*.

Independent of the movements of translation not accounted for by this cause, several of the stars have a rotary motion, forming pairs or binary systems, called double stars, revolving about each other in regular elliptic orbits, governed by the Newtonian law of gravitation. This sort of connexion, suggested as theoretically probable by Mitchell, and demonstrated as a matter of observation by Herschel, has now been distinctly traced in fifty or sixty instances (M. de Humboldt, anticipating what will doubtless one day prove to be a fact, says 2800), among which occur examples of periodic revolutions of 200, 182, 117, 61, 44, and even 17 years, and of orbits, in some cases so excentric as to be quite cometary, in others nearly circular. Some again are concluded, with much probability, to revolve on their axes, from the observation of regular periodic changes in their lustre; while others vary in no regular and certain periods, undergoing great and abrupt changes, for which no probable cause has yet been assigned. In one remarkable instance a change of colour would appear to have taken place. Sirius, which is now one of the whitest of the stars, is characterised by Ptolemy as red, or at least ruddy. Ὁ δὲ Σείριος, ὑπόκιρρος, is his expression, speaking pointedly of its colour, and not of its scintillations.

Not the least surprising, is the actual and positive knowledge we have obtained of the *weight* or quantity of matter contained in at least one of the binary stars, 61 Cygni; from whose orbital motion, compared with its distance, Bessel has concluded that the conjoint mass of its two individuals is "neither much more nor much less than half the mass of our sun." It appears as a star of the sixth magnitude. From the photometric experiments of Wollaston on α Lyræ, compared with what we know of its distance, its actual emission of light may be gathered to be not less than 5½ times that of the sun. Sirius, which is nine times as bright as α Lyræ, and whose parallax is insensible, cannot, therefore, be estimated at less than 100 suns.

Non-luminous stars have been conjectured to exist, and Bessel even considered that some irregularities, supposed to subsist in the proper motions of Procyon and Sirius, could no other way be accounted for than by supposing them to be revolving about invisible central bodies. The illustrious astronomer of Pulkova, in the work we have already had occasion to cite, has, however, by destroying the evidence of irregularity by a careful revision of all the recorded observations, rendered it unnecessary to resort to such an hypothesis.

Neither have attempts been wanting to deduce from the proper motions of the stars, the situation in space of the "Central Sun," about which the whole firmament revolves. Lambert placed it in the nebula of Orion; Maedler, very recently, in the Pleiades, on grounds which, however, appear to us anything but conclusive.

The vast interval which separates our system from its nearest neighbours among the fixed stars is a blank which even the imaginations of astronomers have been unable to people with denizens of any definite character, other than a few lost comets slowly groping out their benighted way to other systems, or torpidly lingering *in aphelio*, expecting their recall to the source of light and warmth. In the utter insulation of this huge intervening gulph, it is impossible not to perceive a guarantee against extraneous perturbation and foreign interference, or to avoid tracing an extension of the very same principle of subordinate grouping which secures the satellites of our planets from too violent a perturbative action on the part of the central body. It thus assumes the character and importance of a cosmical law; and, while it affords another and most striking indication of the unity of plan which pervades the universe, may lead us to believe that, if other systems yet exist in the immensity of space, they may be separated from our own by intervals so immense as to appear only as dim and nebulous specks, or utterly, and for ever, to elude our sight.

Descending, now, with our guide through this *vacuum inane* to our own system, we shall for a moment depart from his arrangement to strike at once upon its central body—our own sun. This, indeed, can hardly be called a departure, since, by an extraordinary omission, we find no special notice taken by M. de Humboldt of this magnificent globe. Yet, surely, there is matter of sufficient interest in what is known and seen of its physical constitution and important peculiarities, to have justified, indeed to have required, their not being passed *sub silentio* in a physical description of the universe. If there be much, as yet mysterious, in its inexhaustible emission of light and heat, there is also much in the mechanism by which that emission is produced which is matter of ocular inspection. We know, for instance, that the sun is not simply an incandescent mass; that the luminous process, whatever its nature, is superficial only, being confined to two strata of phosphorescent clouds, floating in an atmosphere of considerable but imperfect transparency, extending to a vast distance beyond them: that these clouds are often driven asunder by tumultuary movements of astonishing energy and extent, disclosing to our eyes the dark surface below; that the region in which these movements take place is confined to an equatorial belt of about sixty degrees in breadth; being, however, comparatively much less frequent in the immediate vicinity of the equator itself. We know, moreover, that the time of its rotation ($25\frac{1}{2}$ days) stands in decided and pointed dissonance with the Keplerian law of the planetary revolutions, and that therefore the sun has *most certainly not* been formed by the simple subsidence of regularly rotating planetary matter gradually contracting in dimension by cooling; a fact which the advocates of the nebulous hypothesis must, therefore, render some other account of.*

The primary planets known to us at the present moment are sixteen† in number, including no less than five which have been added to the list since the publication of the Kosmos in 1845. The discovery of one of these, Neptune, by the theory of gravitation, as delivered by Newton, and matured by the French geometers, on the mere consideration of the recorded perturbations of the remotest planet previously known, will ever be regarded as the most glorious intellectual triumph of the present age. If anything could enhance its claim to be so considered, it is the assurance given us of the exceedingly firm grasp by which theory has seized on this most complicated subject: by

* It offers no real difficulty to the advocates of that hypothesis. In their view the sun must be regarded as the centre of subsidence of all the matter whose elastic movements have contradicted each other and terminated in collision. (H. 1857.)

† Thirty-four at present. (H. 1857.)

the fact of the discovery having been made almost simultaneously by two geometers of different nations, pursuing different courses of investigation, each in entire ignorance of the other's proceedings, and arriving at what may fairly be termed the same identical place of the yet unseen planet. It is not a little remarkable that astronomy, the oldest, and as it might be considered, the maturest among the sciences, is perhaps at this moment the most rapidly progressive of any, such is the novelty as well as the magnitude of the facts which every year brings forth.

M. de Humboldt, in this division of his subject, presents us with a rapid, but an extremely striking and well-digested view of the "collocations" of our system; that is to say, of the actual arrangement and distribution of its masses in respect of their magnitudes, densities, and distances from the sun, their times of rotation on their axes, and the extent of their provision with satellites. We have never met with a better *exposé* of these particulars, grouped as they are under a variety of aspects, with the object of bringing into view the general relations, if any, which exist between them.

"It has been proposed to consider the telescopic planets," now eight in number, between Mars and Jupiter, "with their more excentric, intersecting, and greatly-inclined orbits, as forming a middle zone, or group, in our planetary system; and if we follow out this view, we shall find that the comparison of the inner group of planets, comprising Mercury, Venus, the Earth, and Mars, with the outer group, consisting of Jupiter, Saturn, Uranus," (and Neptune), "presents several striking contrasts. The planets of the inner group, which are nearer the sun, are of more moderate size, are denser, rotate round their respective axes more slowly, in nearly equal periods, differing little from twenty-four hours, are less compressed at the poles, and, with one exception, without satellites. The external planets. . . . are of much greater magnitude, five times less dense, more than twice as rapid in their rotation round their axes, more compressed at the poles, and richer in moons in the proportion of seventeen" (eighteen) "to one."

So soon as we descend to particulars, however, we find these general relations broken in upon by continual exceptions. The history of the discovery of Neptune has afforded a signal instance how little reliance could be placed on a *law of collocation*, which had begun to be considered as a fundamental relation pervading the whole system. Still, as such laws, partially carried out, they possess a peculiar interest, especially when we consider the exactness of numerical relation which holds good in several instances, and which leads irresistibly to speculate upon causes, as is the case with all close numerical coincidences, which nothing can persuade us to believe purely accidental when they take place in matters of fact. *Why*, we are tempted to ask, do the diurnal rotations of Mercury, the Earth, and Mars, agree to an hour ?* *Why* are the densities of the Sun, Jupiter, Uranus, (and ? Neptune) exactly alike, and just one fourth of the Earth's ? Again, among the satellites, *why* are the

* In the number of the Edinburgh Review, clxxv. p. 194, line nineteen, *for* minute *read* hour. This erratum conveys, as the passage there stands, a very exaggerated impression.

periodic times of Saturn's third and fourth satellites respectively, *precisely* double those of the first and second? And *why* are the rotations of the satellites, generally, on their axes performed in *precisely* the same times as their revolutions about their respective primaries? Of this last-mentioned coincidence, indeed, a mechanical explanation is given (Kosmos, p. 155. *Trans.*) which we are aware rests on high authority. It pre-supposes, however (which our author does not appear to have recollected), an original, *very near* adjustment to exact coincidence; and even with this admission we remain by no means satisfied of its validity. It appears to us that the very smallest deviation from *perfect* coincidence, originally subsisting, would destroy all tendency to that accumulation of matter on one diameter of the satellite, and consequent permanent elongation of its figure, which the further steps of the so-called explanation require.

By far the most wonderful and mysterious bodies of our system are the comets. Their number is immense, their variety of aspect infinite, their magnitude astounding. Apart from the magnificence of their appearance, and the interest attaching to their excentric orbits, and utter contempt of the ordinary planetary conventions in their excursions into space, they have become to us instruments of physical inquiry; and the study of their motions has disclosed to us features in the constitution of our system of which we should otherwise have had no idea, and afforded opportunities, which, but for them, had been altogether wanting of completing our knowledge of the masses of the planets themselves. Their almost spiritual tenuity enables them to feel as it were, and to manifest by a sensible retardation the resistance of a medium pervading the planetary spaces, while the direction of their tails always turned from the sun, and the enormous velocity with which these singular appendages have appeared on some occasions to be projected in the opposite direction to the solar gravity, has afforded more than a presumption of the existence of repulsive as well as attractive forces in our system. It would be endless to recount the singularities presented by these bodies. Some have had two tails, one (1744) six, and some none at all, though otherwise large and conspicuous. Many have been seen in bright sunshine and at noon-day, as was the case with the recent magnificent one of 1843. The tails of some have equalled, and even surpassed in length, the radius of the earth's orbit; and through those of the comets of 1819 and 1823 the earth itself is supposed to have passed. The famous comet of Lexell passed twice (1767 and 1779) among the satellites of Jupiter, and approached the earth in 1770 within six times the distance of the moon. Several of them return in known periods; the celebreted comet of Halley in 76·871 years; that of Encke in 3·316; that of Biela in 6·599, and that of Faye in 7·29 years. The climax to the bizarreries of these singular bodies was afforded in 1846 by one of these last-mentioned comets (that of Biela), which was actually seen to separate itself into two; which, after thus parting connexion, continued amicably journeying along side by side without further mutual disturbance.

The fall of masses of stone, of iron, and of ashes and other substances from the heavens, is a fact now so thoroughly well attested, that every doubt as to its reality has long since vanished. The latter phenomenon may not unreasonably be attributed to volcanic eruptions, or to matter swept from the surface of the earth by tempests and whirlwinds, carried to a vast height, and deposited at great distances from its origin; and such, indeed, appears to have been the case in many well authenticated instances. We have before us a portion of a sheet of 200 square feet, of a substance exactly similar to cotton felt, and of which clothing might be made, which fell at Carolath, in Silesia, in 1839. On microscopic examination it is found to consist of delicate matted and bleached confervæ containing infusoria; and was, therefore, doubtless, raised from its natural site, the dried bed of some lake or marsh, and wafted to the place of its fall by a storm.

But no such explanation will apply to the astounding phenomenon of the sudden fall of blocks of stone or iron of several pounds, nay, tons in weight.

"A presumptuous scepticism," says M. de Humboldt, "which rejects facts without examination of their truth, is, in some respects, even more injurious than an unquestioning credulity. It is the tendency of both to impede accurate investigation. Although for upwards of

2000 years the annals of different nations had told of falls of stones, which, in many instances, had been placed beyond doubt by the testimony of irreproachable witnesses; although the Bætylia formed an important part of the meteor worship of the ancients, and the companions of Cortes saw, at Cholula, the aërolite which had fallen on the neighbouring pyramid; although caliphs and Mongolian princes had had swords forged of fresh-fallen meteoric iron; and even although human beings had been killed by the falling stones (viz., a friar at Crema on the 4th of September, 1511, a monk at Milan, 1650, and two Swedish sailors on board a ship in 1674); yet, until the time of Chladni, who had already earned for himself imperishable renown in physics by the discovery of his figure-representations of sound, this great cosmical phenomenon remained almost unheeded, and its intimate connection with the planetary system remained unknown."

We can pardon some degree of scepticism, on a subject apparently so marvellous, before the assemblage of recorded facts had brought a mass of independent and agreeing evidence to bear upon the general mind, nauseated as it had become by tales of monkish miracle and travellers' wonders. Chladni wrote in 1794, and his work had effectually shaken this scepticism, and excited general attention, when, on the 26th of April, 1803, a shower of stones, thousands in number, and several of them weighing many pounds, was hurled over a district of between twenty and thirty square miles in extent, by the explosion of a globe of fire in mid-day and in a clear sky, vertically over the town of l'Aigle, in Normandy. This was precisely the opportunity to inquire minutely into all the circumstances of the event, and to place them on official record. Accordingly, at the instance of the French Academy of Sciences, the government commissioned M. Biot to proceed to the spot, examine witnesses, and collect every particular. His report on this event, which forms part of the memoirs of the Institute for 1806, leaves no room for doubt as to its reality. Trees were broken, houses struck, the ground ploughed up, the actual stones picked up or dug out in vast abundance. Many persons had narrow escapes, and one was slightly wounded. A list published by Chladni (Ann. du Bureau des Longitudes, 1825) enumerates upwards of 200 instances of similar occurrences, collected from the annals of all nations, China included; among which we observe no less than sixteen recorded in the British Isles subsequent to 1620, one of which (May 18. 1680) took place in London. Subsequent research has added largely to this list, and new occurrences of the kind are continually happening. Many of the masses which have so fallen have been of great magnitude. To say nothing of the enormous weight of some of the blocks of iron supposed to be of meteoric origin; the stone which fell at Ægospotamos was as large as two mill-stones; and that which fell at Narni, A. D. 921, formed a rock projecting four feet above the surface of the river. A mass of this magnitude, so distinct in its nature from the materials of the surrounding rocks, and in a locality so very definite, might surely yet be found by persevering search. Facts of this kind preclude all idea of their being formed in the air from floating vapours, while their difference from all known volcanic products or minerals excludes their reference to a terrestrial origin. Volcanoes in the moon were for a time resorted to, and M. de Humboldt (note 69.) is at some pains to prove this opinion untenable. We believe it to be now entertained by no one. Their planetary nature is the only remaining account which can be given of their origin; and this opinion he of course adopts, classing them with the other admitted members of our system. The phenomena of their explosion, and the violent, though transient and merely superficial heat which they undergo at the moment of their fall, may perhaps be considered as militating against such an origin. But we perceive nothing in these circumstances incompatible with the necessary consequences of such a rencontre. Arriving with planetary velocity at the confines of our atmosphere, where the air is many thousand, perhaps million times rarer than at the surface of the earth, such a body would carry before it the air on which it immediately impinged, compressing it to an enormous *relative* extent against its own surface, before the *absolute* compression could reach such a point as to determine its lateral escape. Now, it has been shown by Poisson (Ann. de Chim. xxiii. 341.) that the latent heat of a given weight of air is greater, the lower the pressure under which it exists. A given quantity (by weight) of air, therefore, at those elevations contains more latent heat than the same quantity at the earth's surface. When condensed, therefore, it will give out *more* heat than would be elicited by the same extent of *relative* condensation from air of ordinary density, which we know to be capable of producing ignition, even under very moderate degrees of sudden compression. A source of sudden and transient heat of almost any conceivable intensity, is thus provided in immediate contact with the surface of the stone, which it would fuse and partly vaporize, while the sudden and violent expansion of the parts immediately beneath the fused film must necessarily cause decrepitation and disruption of fragments. In short, there is no part of the phenomenon which this explanation does not reach. Mere friction against the atmosphere, as suggested by Poisson, seems quite insufficient to produce incandescence.

That a resemblance should be conceived to exist between those globes of fire which throw down stones and those which only gleam and are extinct, or which terminate with a harmless, though often very terrific explosion, is not to be wondered at. Yet the analogy founded on mere optical resemblance would hardly suffice to prove a community of nature or origin. Accordingly, little or no attempt was made to connect these formidable visitors with the innocuous spectacle afforded by shooting stars or train-accompanied meteors, till 1833, when a brilliant display of the November meteors, on the 12th and 13th of that month, repeated on the same days of the following year, brought to recollection a similar display witnessed by M. de Humboldt in 1799, in America. On comparison of dates, it was perceived, with astonishment, that they precisely coincided. The extraordinary fact has since been established by observation, and by the assemblage of ancient and modern records, that meteoric showers occur *periodically* on certain given days of the year, though not of every year, and especially on the 12th—14th November, and the 9th—11th of August; the latter epoch being the most uniform in respect of the intensity of the phenomenon. Another fact, not less striking, has emerged in respect of the directions effected by the meteors in their flight. They diverge, apparently, from fixed points in the heavens, whose longitudes are 90° in advance of the actual places of the earth in the ecliptic at the epochs in question. Such apparent divergence, by the rules of perspective, is the criterion of a real parallelism; and we are thus carried onwards to the inevitable conclusion of a cosmical origin and common direction of motion, in groups or flights of these bodies, which the earth encounters in its annual path, and which are presumed to form rings or planes more or less interrupted about the sun, revolving according to planetary laws. We agree with M. de Humboldt in considering the general conclusion as perfectly well established, and as justifying his admission of them into the rights of recognised membership of the planetary system.

The zodiacal light is another of those luminous

phenomena to which a cosmical origin has always been ascribed: —

"The earliest distinct description" of it "is contained in Childrey's 'Britannia Baconica' (1661). Its first observation may have been two or three years earlier. Dominic Cassini has, however, incontestably the merit of having been the first (in 1683) to investigate its relations in space. . . . It may be conjectured with much probability that the remarkable light, rising pyramidically from the earth, which, in 1509, was seen in the eastern part of the sky for forty nights in succession from the high table land of Mexico (and which I find mentioned in an ancient Aztec manuscript in the Codex Tellerio-Remensis, in the Royal Library at Paris), was the zodiacal light." (*Transl.* p. 189.)

This light, as M. de Humboldt justly reasons, cannot be the solar atmosphere in the ordinary sense of the words. But we cannot so readily admit the conclusion he draws, that it is an extremely oblate *ring* of lucid vapours revolving in space between the orbits of Venus and Mars. An extent much beyond the earth's orbit, at all events, seems incompatible with its pointed or pyramidal form and termination at a certain apparent distance from the sun, instead of being continued all around the heavens. Nor can we perceive any good reason for ascribing to it an annular form, wholly exterior to the orbit of Venus. The passage which he cites from Cassini (note 96.) in support of this opinion appears to us by no means susceptible of this interpretation; nor are we aware of any observations which necessitate such a conclusion, contrary as it is to the opinion generally received on the subject.

Descend we now to our own globe, "from the region of celestial forms to the more restricted sphere of terrestrial forces; from the children of Uranus to those of Gea;" from the contemplation of matter obedient to comparatively few and simple impulses and laws, offering no indications of qualitative diversity — to matter under the influence of molecular forces of excessive complication, and laws very imperfectly understood, exhibiting fundamental diversities of quality, affording endless scope to agencies which scarcely appear to resolve themselves into the simple conception of mechanical effort, and whose active principles, electricity and heat, present themselves to us under aspects now reminding us of the ordinary forms of matter by their quantitative relations to tangible bodies, and now eluding our grasp by a subtilty which seems to transcend our notions of corporeal existence. Here, too, we become conversant with organic life in all its infinite diversities and stages of manifestation, and in all its adaptations to external conditions; as a something superposed upon and subsequent to matter. Here, too, we encounter voluntary motion as something again superposed upon mere organic development; and here, too, the life of instinct and the life of thought, rising higher and higher by successive but gradual steps, till at length one vast bound lands us in HUMANITY, with all its hopes and visions of something yet beyond. Such is the field we have now to enter upon —

"The wide, th' unbounded prospect lies before us;"

but its richness, no less than its extent, forbids our lingering on its outskirts in idle contemplation of its glories.

The path followed by M. de Humboldt in threading the labyrinth of this vast mass of knowledge is, perhaps, on the whole, the best which could have been adopted to preserve a continuity of course, and to bring the phenomena to bear on each other with due regard to causal sequence.

He first, under the general head of "Terrestrial Phenomena," gives us an outline of those broad features which have relation to the mass of the earth as a whole; and in which the acting forces and powers are considered in their mean or average intensity, or as acting on the largest scale, unaffected by local causes. The features which admit of being so presented, are those which refer to the dimensions and figure of the earth, its mean density and temperature; and the evidences, such as we possess them, of an increase in both these respects, in descending from its surface to its centre. Terrestrial magnetism too, and the disturbances, whatever be their origin, which the magnetic power of the earth undergoes upon the great scale, during "magnetic storms" and auroral displays, as well as those secular variations which modify all its local manifestations, according to laws yet unknown, but whose influence extends to the whole globe, find a natural place in this division of the entire subject.

Under the general notion of the "reaction of

the interior of the earth on its exterior," which affords, as it were, the canvas on which to depict the phenomena of earthquakes, volcanoes, hot springs, &c., we recognise the impress of that theory of geological dynamics which represents the external solid crust of the globe as in a continual though exceedingly slow process of contraction, by refrigeration, on its internal liquid contents, by which it becomes placed in a state of strain which from time to time, and according to local circumstances affording facilities for disruption, relieves itself by fracture and by the ejection of a portion of the liquid matter. Such, at least, seems to be the conception implied in the word *reaction*, which presupposes *action*. The want of an original *primum mobile* competent to the production of the volcano and the earthquake as general, and not as local phenomena, is imperatively felt in geology.

As consequences of this reaction, appearing indifferently on every part of the earth's surface, we have the ejection of *erupted* or "*endogenous*," and the production of *metamorphic* rocks, together with upheavings and subsidences of portions of the earth's crust of greater or less extent, which in the course of ages modify the distribution of sea and land over the surface of our planet. Simultaneous with these changes, but referring themselves to a totally different order of causes the seat of which is wholly exterior to our globe, and which depend entirely on the action of the sun and moon as the ultimate causes — the *prima mobilia* — of all those oceanic and atmospheric movements to which continents owe their destruction and reproduction, we have the continual formation of new strata at the bottom of the ocean; their gradual condensation by increase of pressure as more and more of their materials become accumulated; and their ultimate consolidation by the invasion of heat from beneath, in virtue of those general laws which regulate the movement of heat from point to point of bodies, the surface of which is maintained at a temperature, which, for this purpose, may be regarded as invariable. From the combination of the two orders of events arising from the continued action of these two classes of causes, each proceeding in perfect original independence of the other, but each in its progress continually modifying the conditions under which the other acts; and so producing a compound cycle, or rather interminable series, of excessive intricacy; depend all geological phenomena, properly so called. Meanwhile, on this interwoven tissue, as if not yet sufficiently complex, is superposed another cycle of causation in the electro-magnetic relations of the globe, which, though uninfluential as respects the movement of masses, is no doubt powerfully so in the mineralogical arrangement of their particles, in the production of planes of false cleavage in the strata, and in the filling up, by metalliferous and other mineral veins, of the fissures which intersect them. To this class of mineralogical causes (on whose action the researches of Becquerel, Fox, and

Hunt have thrown some light, but which stands in need of much more extensive and assiduous inquiry), we are somewhat surprised to find no allusion made in the work before us.

Among the materials of subverted and reconstructed continents, occur the buried remains of their former inhabitants. Palæontology, therefore, and the evidence it affords, in conjunction with other circumstances attending the materials and position of strata, leads us naturally to the consideration of the state of the surface of our globe in former epochs, in relation to its habitability by various orders of organic beings, and more especially to its distribution into sea and land.

"We here indicate a connecting link between the history of the revolutions our globe has undergone, and the description of its present surface, — between geology and physical geography — which are thus combined in the general consideration of the form and extent of continents. The boundaries which separate the dry land from the liquid element, and the relative areas of each, have varied greatly during the long series of geological epochs: they have been very different, for example, when the strata of the coal formation were deposited horizontally upon the inclined strata of the mountain limestone and the old red sandstone; when the lias and the oolite were deposited on the keuper and the muschelkalk; and when the chalk was precipitated on the slopes of the green sand and the oolitic limestone. . . . . . Maps have been drawn representing the state of the globe in respect of the distribution of land and water at these periods. They rest on a more sure basis than the maps of the wanderings of Io, or even than those of Ulysses, which at best represent but legendary tales, whilst the geological maps are the graphic representations of positive phenomena."

We find ourselves thus introduced to the domain of physical geography, or the description of the actual state of the earth's surface in its three great divisions,—those of land, sea, and air,—as prepared for the habitation of organic beings, and as exhibiting the play of all those complex agencies on which depend the distribution of temperature and moisture, aerial and oceanic currents, and those conditions which, under the general title of climate, determine the abundance and limits of vegetable and animal forms. A general view of organic life and the distribution of plants and animals, infinitely less copious in detail than we should have expected from the exceeding richness of M. de Humboldt's information on this subject, and a short chapter on Man, close the text; which is followed by a series of notes, indicating the authorities from which the statements throughout are derived, and full of a vast mass of other information, so interesting, so recondite, so various as to leave us lost in admiration, both of the reading which could amass, and the discrimination which could select it.

The dimensions and figure of the earth constitute a branch of inquiry on which, perhaps, more pains, labour, and refinement have been lavished than on any other subject of human research. "The history of science," says M. de Humboldt, "presents no problem in which the object obtained, the knowledge of the mean compression of the earth, and the certainty that its figure is not a regular one, is so far surpassed in importance by the incidental gain which, in the course of its long and arduous pursuit, has accrued in the general cultivation and advancement of mathematical and astronomical knowledge." In fact, however, the benefit conferred has not been confined to these. The continual heaping on of refinement upon refinement, in respect both of instruments and methods, has been far from a mere barren and ostentatious accumulation. On the contrary, it has overflowed on all sides, and fertilized every other field of physical research, by the example it has set, and the necessity it has imposed of exactness of numerical determination, mathematical precision of statement, and rigorous account taken of every influential circumstance; as well as by the numerous physical elements whose exact measures and laws it has incidentally required to be known as data. By the improvement of our knowledge of these, the aspect of all science has been changed, and the apparently disproportionate application of talent and cost which have been brought to bear upon the subject, repaid with interest. The fixation of national standards of weight and measure, which has become indissolubly interwoven with it, has ever marked, and will ever continue to mark, the highest point to which human skill and refinement in the application of science to practical objects are capable of attaining.

In stating the result of these inquiries, M. de

Humboldt follows the determination of Bessel in 1841. A better authority he could not have selected, and it is worth while to notice (since he has omitted to do so) the precise coincidence of this determination with that of Mr. Airey in 1831, from the assemblage of all the geodesical measurements *then* procured,—a coincidence amounting in fact to identity, the difference between the two statements of the earth's equatorial diameter being but 234 *feet*, between those of the polar only 296, and of the compression 38. Neither can we omit to mention here the only considerable accession to our knowledge on this head since the publication of "The Kosmos," viz., the rectification of Lacaille's erroneous arc at the Cape, by the admirable and indefatigable Maclear (performed at the hazard and almost at the sacrifice of his life), which has removed for ever one of the great stumbling-blocks in the way of general and exact conclusions on this subject.

The ellipticity of the earth, as Playfair has shown, can by no means be taken as affording even the slightest evidence of the entire primitive fluidity of its whole mass. Even when that of the internal strata is taken into the account, if there be any degree of mobility, from whatever cause arising short of entire and simultaneous fluidity, among its materials, this would ultimately conform its internal arrangement, as the sea does its external form, to the elliptic model. We do not mean to deny the strong presumption, however, that such fluidity does prevail at a certain depth: "Tolerably accordant experience has shown that in Artesian wells the average increase of temperature, in the strata passed through, is 1° of the Centigrade thermometer for 92 Parisian feet of vertical depth (54·5 English feet for 1° Fahr.). . . . If we suppose this increase to continue in an arithmetical ratio, a stratum of granite would be in a state of fusion at a depth of nearly 21 geographical miles." The phenomena of hot springs in countries where volcanic eruptions have long since ceased; "direct observation of the temperature of rocks in mines; and, above all, the volcanic activity of the earth, ejecting molten masses from opened clefts or fissures, bear unquestionable evidence of this increase for very considerable depths in the upper terrestrial strata." Still we can determine nothing with certainty respecting the depth at which the materials of our rocks exist, "either in a softened and still tenacious state, or in complete fusion; respecting cavities filled with elastic vapours; *the condition of fluids heated under enormous pressure;* or the law of the increase of density, from the surface to the centre." One thing only is certain, that the density *does* so increase, since the wonderfully agreeing conclusions arrived at by Cavendish, Reich, and Baily (for such they ought assuredly to be considered, the difference between Baily and Reich amounting to no more than one twenty-eighth part), abundantly demonstrate a mean density for the whole mass of five and a half, which is double that of basalt, and more than double that of granite; substances which undoubtedly emanate from very great depths beneath the surface.

The mean temperature of the globe is supposed to have attained so nearly an invariable state, that since the time of Hipparchus, and in an interval of 2000 years, it has not diminished by one three-hundredth of a degree of Fahrenheit's thermometer.

This conclusion rests on the records of ancient eclipses, which having taken place in conformity with the theory of gravitation, implies the invariability of our unit of time or of the length of the day, during the interval. Hence Laplace has concluded, and the conclusion may be regarded as certain, that the length of the day, or the time of rotation of the earth on its axis, has not diminished by one hundredth part of a second. Hence also we are entitled to conclude that its mean radius has not diminished by a single yard in that interval. So far we are on sure ground: and if we consent to disregard as merely superficial, the transfer of matter from a higher to a lower level by oceanic and atmospheric abrasion, and the counteracting effect of volcanic ejections, — if, moreover, we set as in a balance one against the other, the upheavings of mountain chains, such as our own times have witnessed in the Andes, and the subsidences of extensive districts, such as are going on in Scandinavia, the conclusion, as relates to temperature, must be admitted as valid, however it may be supposed to militate against the refrigeratory theory above alluded to.

The mean temperature at which the *surface* of the earth is maintained, if we consider the average of the whole globe, depends solely on external causes, the only one of which, worth considering as really influential, is the sun's radiation. Of the constancy or variability of this from year to year, or from century to century, we know nothing, though from the analogy of periodical or changeable stars we may surmise anything. But it by no means follows that this ignorance, on a point of such immense importance, is to continue. It is to the temperature of the ocean, continually and carefully observed in those parts of its surface where its changes are least (in the equatorial region, from 10° N. to 10° S.) that we must look, with the greatest probability of ultimate success, for the solution of this difficult but interesting problem. In these regions, the observations and researches of M. de Humboldt himself have established the fact of "a wonderful uniformity and constancy of temperature over spaces of many thousand square miles." It is here, therefore, that observations directed to this object can be made to the greatest advantage, and least exposed to the influence of casual and temporary disturbance. We know of no class of observations deserving more the attention of voyagers: and the more so, as the recent results of Mr. Caldecott respecting the temperature of the *soil* at considerable depths in India, have brought into evidence *enormous* differences, amounting to 6° between the *mean temperatures* of the earth and air at the same spot. Such might indeed have been expected on a careful consideration as to the different agencies of wind and rain on the one hand, and solar and nocturnal radiation on the other, in determining the respective averages, but they stand in striking contradiction to the generally received opinion of the necessary equality between the two means in question. It ought to be remarked, that M. de Humboldt, when stating this opinion (p. 165. *Tr.*), and the practical application of it recommended by Boussingault, expresses himself with hesitation, if not with doubt on its subject.

The power of magnetism, and the polarity of the magnetic needle, appear to have been known to the Chinese from the most remote antiquity. Extracted from the annals of See-ma-thsian, a Chinese historian cotemporary with the destruction of the Bactrian empire by Mithridates I., we find the following extraordinary relation. "The emperor Tching-wang (1110 years before our era) presented to the ambassadors of Tong-king and Cochin China, who dreaded the loss of their way back to their own country, five magnetic cars, which pointed out the south by means of the moving arm of a little figure covered with a vest of feathers." To each of these cars, too, a hodometer, marking the distances traversed by strokes on a bell, was attached, so as to establish a complete dead reckoning. (Humboldt, Asie Centrale, xli.; Kosmos, 171.) Such inventions, we cannot but observe, are not the creation of a few years, or a few generations. They presuppose long centuries of previous civilization, and that too "at an epoch cotemporary with Codrus and the return of the Heraclides to the Peloponnesus"—the obscure dawn of European history! Even the declination of the needle, or its deviation from the true meridian, was known to this extraordinary people at the epoch in question.

Two views of terrestrial magnetism may be taken. The one is that which makes the earth itself, or a large portion of the substance of it, intrinsically magnetic in that sense in which a loadstone is so. This view (which is at all events general, and but for the secular variations of the magnetic curves, would be even now perhaps the best which could be taken) is vindicated by M. de Humboldt to our admirable countryman Gilbert, whose ideas were, in all physical matters, far in advance of his age (note 142.). It was the knowledge of these variations which led Halley to the formation of his wild as well as inadequate theory of an internal globe revolving within the external shell of the earth. If the mass of the globe be magnetic in the sense of the loadstone, it is scarcely conceivable that the local distribution of magnetic power on its surface should be otherwise than permanent. That it is not so—that the magnetic curves, one and all, are in a continual state of slow but regular change, *sweeping round upon the two hemispheres in contrary directions* (by which very act their forms are undergoing continual modification), we cannot help receiving as an indication that the seat of the earth's magnetism, if not entirely atmospheric, is at least so far superficial as to be subject to a large amount of external influence: seeing that they bear relation neither to any fixed lines in the globe itself on the one hand, nor to any determinate directions in external space on the other. The explanation of these secular variations is perhaps the obscurest problem which the "Physique du Globe" has yet offered for solution; and its solution, when known, cannot fail to carry with it the explanation of every other part of the phenomena.

Meanwhile it is certain that the phenomena of the magnetic needle, and its direction at each point of the surface, may, to a certain extent, be imitated on an artificial globe, by passing round it at the surface a due system of electro-magnetic currents. This was actually done by the late Professor Barlow. To a slowly and *secularly variable* system of electric currents, therefore, whether atmospheric or terrestrial, all probability refers us as the cause of the earth's magnetism. And here we are brought to a stand, not only by the very imperfect state of our knowledge in respect of atmospheric electricity, of all the branches of meteorology the least advanced; but also by our ignorance of the actual forms of the magnetic curves over many and extensive regions of the earth, to say nothing of their secular changes. This blank area, however, is happily diminishing rapidly under the pressure of surveys set on foot in pursuance of that noble plan of co-operative magnetic research which (thanks in the first instance to M. de Humboldt's powerful recommendation) has been adopted and acted on by our own and other Governments upon a scale and with a sequence and energy to which no age has furnished a parallel. Within the interval, short of ten years, since the adoption of this system, the whole area of the Antarctic Ocean has been added to the domain of exact magnetic knowledge by the expedition under Sir James C. Ross, and by the subsequent survey of Lieutenants Moore and Clerk. British North America has become in like manner known ground by the survey of Lieutenant Lefroy, to which has been, or is in the course of being, added that of the United States by Locke, Loomis, Bache, and other able and indefatigable observers. The expedition of Sir John Franklin, speedily, we trust, to return crowned with merited success, taken in conjunction with the survey of Hudson Bay, accomplished in the course of last summer by Lieutenant Moore, will complete our knowledge of the northern coast, and give to the continent of North America its due significance on the magnetic chart of the globe. Nor are these the whole, or anything like the whole, of the acquisitions recently made and still making in this direction, which, however, our limits will not permit us further to dilate on, or to give their merited tribute of applause to the indefatigable exertions of the able editor of the work before us, in deducing from the vast mass of observations thus continually pouring in, the true forms of the magnetic curves, and in particular of the isodynamic lines and ovals which, although the last to be received into the list of magnetic elements, have proved the most interesting and important of any. The service thus rendered to magnetic science, it is in fact impossible to overappreciate.

Whatever idea we may form of the greater and more regular magnetic system of our globe, there can hardly remain a doubt as to the reference of the diurnal and annual periodic fluctuations of the magnetic elements to electric currents in the earth or atmosphere caused by solar excitement. Nor can there be any hesitation in referring to sudden and violent disturbances of electrical equilibrium, from whatever cause arising, those mysterious phenomena to which M. de Humboldt (the first to observe, or at least strongly to draw attention to them) has given the expressive name of magnetic storms, and in which the needle is agitated simultaneously over vast regions, whole continents, nay, even in some cases, *over the whole surface of the globe.* Of these the most remarkable on record is that of the 24th and 25th of September, 1841, which was observed at Toronto, in Canada, at Prague, at the Cape of Good Hope, at Van Diemen's

Land, and at Macao. And here we cannot omit to notice the very remarkable coincidence of date between this and a great and extraordinary disturbance, which has quite recently been observed at Toronto, and of which the account by Lieutenant Lefroy is before us. The range of the needle, in respect of horizontal direction, on this occasion exceeded 4°, and the fluctuation in respect of horizontal intensity surpassed *a twentieth part of its total amount.* Now this disturbance (which was observed at Greenwich, though to not quite so great an extent) also took place on the 24th of September! A coincidence of this kind, should it be repeated, like that of the meteoric showers, would lead us irresistibly, and as an *instantia lucifera*, to look outwards, into the planetary spaces, for the cause of these singular phenomena.

Intimately connected with these irregular magnetic disturbances, and characterized by M. de Humboldt as the final discharge which restores the magneto-electric equilibrium, wrought to a climax of tension during their continuance, is the Aurora or polar light. Of one variety of this superb phenomenon, that which consists in luminous beams and dancing streamers, terminating in a corona round the place of the elevated magnetic pole, he gives a most picturesque and beautiful description. The other, rarer, and less vivid in its phases, but perhaps in some respects even more interesting; that which consists in quiet luminous masses, either insulated or forming more or less regular arches transverse to the magnetic meridian, and *drifting constantly with a slow and steady movement* southward, he passes in silence. In both we recognize, by many indications, the presence of matter in the higher regions of the atmosphere, rendered luminous by the passage of electricity, but differing in the two cases in the mode of its arrangement, and perhaps, too, in elevation; the arrangement in the former being in lines parallel to the dipping-needle; in the other sometimes in amorphous masses, at others with a strong tendency to a transverse position. Is it possible that the distinction between the magnetic and diamagnetic forms of matter, brought to light by Faraday's late researches, may play a part in these arrangements?

The height of the auroral phenomena has been a subject of very varying estimation, and if we allow that, as M. de Humboldt expresses it, "every observer sees his own aurora as certainly as he sees his own rainbow," it must be evident that no parallactic mode of determining its height is practicable. This, however, applies only to the first of the above-mentioned species of Aurora, where, from the number and rapid coruscations of the streamers, no one can be individualized and definitely fixed. The luminous masses and transverse arcs of the other variety have assuredly an optical *reality*—are *objects*, and capable of being seen in their true geometrical places by any number of spectators at once. It is impossible, in short, that a body of light, steady enough to be definitely referred by one observer to one given direction in space, and by another to another at the same instant, should not have an objective locality. The arcs of October 17. 1819, and March 29. 1826, whose heights, as calculated by Dalton from very positive data, appear to have been nearly equal (100 —110 miles), were certainly in this predicament; nor do we consider his conclusions as at all shaken by the objections advanced against them by Dr. Farquharson. On the other hand, M. de Humboldt appears disposed to doubt the reality of auroral *streamers* having been seen below the clouds; but on this head the observations of the last-named excellent observer on the Aurora of February 24. 1842, are so positive and circumstantial, as to leave no room for doubt. The crackling or hissing sound, reported to accompany their displays in high latitudes, he considers as altogether apocryphal. It is not among the least puzzling features of auroral phenomena, that although so intensely *magneto-electric as actually to interfere with the free transmission of messages along the electric telegraph*, experiments made during their continuance with very sensitive electrometers have hitherto given only negative results, since, during the finest Auroras, no change in the electric tension of the atmosphere has been detected. ("Kosmos," 186., *Tr.*)

On the subject of earthquakes and volcanoes, those great manifestations of internal telluric activity, there is probably no geologist now living who can speak so largely from personal knowledge as M. de Humboldt—who has had such opportunities of studying their phenomena in that region of the globe where they are habitually developed on the grandest and most terrific scale, as an eye-witness, or by diligent and immediate inquiry on spots the recent scenes of some of the greatest catastrophes on record. The tremendous convulsions which, in 1797, destroyed Riobamba, with the loss of between 30,000 and 40,000 lives in a few minutes, with "a sudden and mine-like explosion, a vertical action from below upwards," which hurled the corpses of many of the unfortunate sufferers several hundred feet in height on a neighbouring mountain, and across a river, took place only three years before his arrival in Quito, the city lying still in ruins, and every particular, of course, vividly fresh in the recollection of the survivors. The catastrophe which destroyed Cumana took place in the same year. The personal narrative of his travels has made us familiar with the volcanoes of Quito, Mexico, and Chili, and given to the names of Cotopaxi, Pichincha, Tunguragua, and Jorullo, a terrible, yet fascinating, celebrity. With his extraordinary account of the last-named volcano, with its Malpais and Hornitos, there are probably few of our readers unacquainted.

We shall not enter here into any of the speculations current among geologists which have for their object to render an account of the ultimate origin of earthquakes, and the immediate seat of their first impulse. It is to their propagation along the superficial strata, and especially with the mode in which that propagation is dynamically effected, that inquiry can be most usefully, because most effectively, directed. Every one, indeed, is agreed that it is in some sense undulatory; but probably no two geologists have hitherto exactly agreed as to the sense in which that term is to be taken: whether, for instance, the undulation be analogous to that of a fluid surface, or of a stretched sheet, or, lastly, to that by which waves are propagated through elastic media in the conveyance of sound and light, viz., not by lateral tension or by gravity, but by the direct elastic action of the particles on each other. It is here that experience furnishes us with an unequivocal indication in the recorded velocity of their propagation, estimated by M. de Humboldt at twenty-eight geographical miles per minute, which, however, is probably underrated, and which, at any rate, exceeds double that of sound; a velocity, as Mr. Mallet has justly remarked in a paper read before the Royal Irish Academy in 1846, incompatible with any imaginable mode of propagation but that last alluded to. This is, accordingly, the view of the subject which Mr. Mallet adopts, and which, on the whole, appears to render a clear and intelligible account of many of the apparently bizarre and capricious phenomena with which the records of these events abound; such, for example, as the reversal of the stones of a pavement, and the twisted obelisks of Stephano del Bosco by the Calabrian earthquake; the confusion of fields and boundaries; and the strangely irregular intermixture of lines of violent action with others of comparative repose, resulting from nodal intersections and interferences of shocks arriving at the same point from different origins or by routes of different lengths. Such interferences, we must observe, are expressly indicated by M. de Humboldt (p. 192.) as resulting from intersecting earthquake waves, "as in intersecting waves of sound;" adding, moreover,—

"The magnitude of the waves propagated in the crust of the earth will be increased at the surface, according to the general law of mechanics by which vibrations transmitted in elastic bodies have a tendency to detach the superficial strata."

What may be the mechanical law here alluded to we know not. Probably the scaling off of brittle coatings from hard bodies by a blow. But we cannot help supposing the true mode of earthquake propagation (by waves of elastic compression) to have been apprehended with very considerable distinctness in penning this passage, though not seized and worked out, as it might have been, into a regular theory. We will only notice, in further illustration of the explanatory power of this mode of conceiving the matter, the facility with which the singular effect of vorticose motion is accounted for by the crossing of two waves of horizontal vibration, which, as in the theory of the circular polarization of light, compound, at their point of intersection, a rotary movement.

That a theory so simple, and, we may add, so obvious, has not been earlier propounded and received, can only be accounted for by the vast scale of the phenomena and the amplitude of the earthquake wave, which causes the wave itself, as "an advancing form," to escape notice, and the molecular motions only by which it is propagated to be perceived. For in this theory we are to bear in mind that man and his works, in respect of these gigantic movements, are but what the sand spread by Chladni on one of his vibrating plates is to the sonorous vibration it furnishes the means of examining.

What the auroral discharge is to the "magnetic storm," in M. de Humboldt's view of that phenomenon, and, as appears to us, with far more correctness, the volcano in eruption is to the earthquake — the relief of tension and the restoration of equilibrium. Innumerable instances of this connexion might be adduced, but the subject is rather trite, and our limits begin to warn us that we have yet a wide extent of ground to travel over, and we must therefore pass over, not without regret, the evidences of diminishing volcanic action afforded by the phenomena of Solfaterras and hot springs, as well as those of interior heat generally, as manifested in the continued ejection of carburetted hydrogen, of which See-tchuan, in China, and Fredonia, in New York, offer the most striking examples; as well as those of carbonic acid which, in many parts of Germany and on the Rhine, "indicate the last remains of volcanic activity in and near its ancient foci in an earlier state of the globe."

In the "Geological Description of the Earth's Crust," two distinct classifications or arrangements are followed, which, perhaps, we can hardly better characterize in contrast with each other than as *genetic* and *historical*. The former is in consonance with that view of superposed causalities which we have taken of geological phenomena in general. It refers itself to the presumed origin, and not to the historical order, of the matters classified. This would naturally divide the rocks of which the earth's crust is composed into two orders: *endogenous*, having their origin from the internal activity of the earth; and *exogenous*, arising from the degradation of continents by external force, and their reconstruction in new localities by aqueous deposition. But these causes being in perpetual and simultaneous action, it becomes necessary to admit two other members into this general classification, in whose formation as they exist at present both orders of genetic cause have had a share; those namely, first, in which deposited rocks have been altered in texture, density, and mineralogical characters by subterraneous heat either slowly invading them by conduction from below, or suddenly applied by eruptive energy forcing melted matter into contact with them, and introducing new materials into their composition by sublimation (as in the view taken by Von Buch of the Dolomitic limestone of the Tyrol). The second member of the series resulting from this complex action comprises rocks constructed by re-cementation of fragments and pulverized matter, whether produced by the violence of eruptive agency, or by the slower process of water-washing and the action of torrents or debacles. Thus we have at length a fourfold division of the materials of the earth's exterior, into erupted, sedimentary, metamorphic, and conglomerate rocks.

In subdividing the eruptive rocks little importance would attach to oryctognostic character, except in so far as it can be connected with indications of the depth from which they may have been erupted, the scale upon which their expulsion from the bosom of the earth may have been effected, and the state of fluidity at which they may have arrived at the surface. These give rise to a system of characters partly mineralogical and partly geological, in which granite and syenite stand at the lower end of the scale, and basalt and superficial lavas at the upper, while porphyries, greenstones, serpentine, hypersthene rock, and trachyte, fill up the intermediate stages. Some particulars, given by M. de Humboldt, respecting the superposition of granite, will be found interesting, when we recollect at how comparatively late a period the idea of overlying granite was considered almost to amount to a contradiction in terms:—

"In the valley of the Irtysch, between Buchtarminsk and Ustkamenogorsk, granite covers transition slate for a space of four miles, and penetrates it *from above downwards* in narrow branching veins, having wedge-shaped terminations. . . . As granite covers argillaceous schists in Siberia and in the Departement de Finisterre (Ile de Mihau), so does it cover oolitic limestone in the mountains of Oisons (Fermonts), and syenite and chalk in Saxony near Weinböhla."

To these instances we may add the valley of Lavis, in the Tyrol, near Predazzo*, where it overlies dolomite. The true reason for the rarity of these granite superpositions is doubtless to be sought in the very slight degree of fluidity of the upper portions of the upheaved masses, and their vast thickness, which permits but rare opportunities for escape of the more liquid matter from below. A beautiful granite dyke is seen intersecting granite perfectly similar, and no doubt nearly cotemporaneous, on the summit of the Paarl Rock† near Stellenbosch, in South Africa, as if the fissured rock had been re-cemented in the very act of rising by an upward injection, which in cooling has arranged itself in parallel layers, nearly at right-angles to the general direction of the vein.

Sedimentary rocks are necessarily classified according to their geological order of superposition, and are made to consist of—1. Argillaceous schists of the transition series, including the Silurian and Devonian formations; 2. Carboniferous deposits; 3. Limestones; 4. Travertin; 5. Infusorial masses. From this series M. de Humboldt excludes all

* Visited September 1. 1824. The spot is called Canzocoli. (H. 1857.)

† Visited January 16. 1836. (H. 1857.)

purely mechanical deposits of sand and detritus, regarding them as in strictness belonging to the conglomerate division. The abundance of limestones in the latter portions of this series he considers as a result of the decreasing heat of the superficial waters allowing of their absorbing carbonic acid from an atmosphere overcharged with that element.*

The process of metamorphism (a term first introduced into geology, we believe, by Lyell) is very obscure. That electrical action is often concerned in it, we can hardly doubt. The portion of M. de Humboldt's work which treats of it is full of interest, but we cannot afford room for remark or extract, further than to notice the singular difficultiés which beset any geological account of the vast beds of pure quartz, from *seven to eight thousand feet in thickness*, characteristic of the Andes of South America. In the older Plutonic theories, indeed, these would be easily dealt with. Modern speculation, however, is scarcely hardy enough to draw so largely on internal heat as would be necessary to fuse and erupt such masses of so intractable a substance. Their consolidation from sandy deposits by partial fusion under the transforming influence of adjacent rocks (as Murchison

* The exceeding readiness with which newly precipitated carbonate of lime subsides in warm water, compared with what takes place in cold, especially when certain saline substances are present, is a chemical fact which may have some bearing on this point.

proposes to account for the phenomena of the Caradoc sandstones) is subject to hardly less difficulties. The chemistry of long-continued heat under pressure, the production of artificial simple minerals, and the imitation of metamorphic changes on rocky substances, by contact with heated matter, open a field of inquiry deserving of more cultivation than it has hitherto obtained.

The same reason which renders it necessary to limit our remarks on this portion of the subject of geology, compels us to pass over entirely the view which M. de Humboldt takes of the historical department of that science, and the order of succession of the forms of animal and vegetable life which modern geological research has revealed to us as the denizens of our planet in the previous stages of its existence. We should do so with extreme regret (since the sketch which is given, though in the utmost degree condensed, is arranged in a very luminous and masterly manner,) were it not that, although ranking high as a geologist, his own personal contributions to that science belong rather to the lithological than to its palæontological department; and were it not too that an extensive knowledge of the main features of these grand disclosures is very generally diffused in this country. We shall prefer, therefore, to devote what room remains to us to those subsequent portions of his work, where the light which he directs upon them is mingled with many and bright rays emanating immediately from himself.

Among the leading features of that part of the general contemplation of nature which relates to the PHYSICAL GEOGRAPHY of our globe in its actual state, we must regard, first, the quantity of land raised above the water; next, the configuration of each great continental mass in horizontal extension and vertical elevation. That all, or nearly all, the existing land has been so raised, M. de Humboldt regards as an established truth, and considers a considerable part of the height of all the present continents to be due to "the eruption of the quartzose porphyry, which overthrew with violence the first great terrestrial Flora, the material of our coal beds." Previous to this, the portion supporting land vegetation was exclusively insular; nor was it until the epoch of the older tertiary formations that the great continents approached to their present form and extent.

The ratio of sea to dry land is stated at 270 or 280 to 100, or in round numbers as about 3 to 1, the islands amounting to one twenty-third of the continental masses. As regards the general distribution of sea and land, M. de Humboldt confines himself to observing that the northern hemisphere contains nearly three times as much land as the southern, and the eastern (from the meridian of Teneriffe) far more than the western. This mode of statement, however, conveys a much less lively and distinct impression of the law of distribution than the division (suggested by Colson, *Phil. Trans.* vol. xxxix. p. 210.) of the globe into two hemispheres, a terrene and an aqueous one, the former having Great Britain, the latter her antipodes, for its vertex.* In fact, if we endeavour to include the maximum of land in one hemisphere, and that of water in the other, according to our present knowledge of the globe, we shall find as the centre of the terrene hemisphere a point in the south of England somewhat eastward of Falmouth. With exception of the tapering termination of South America, the land in the other is wholly insular, and were it not for New Holland, its amount would be quite insignificant. As protuberance above the sea level indicates comparative levity, are we not thence entitled to conclude the non-coincidence of the centre of gravity of our globe with its centre of figure, the denser portion being situate beneath the South Pacific ?

On the general form of the land we find some striking remarks. The southern terminations of the great continental masses affect the pyramidal form, which is repeated on a smaller scale in the peninsulas of India and Arabia, &c., while generally, prolonged appendages, both to the northward and southward, affect a meridional direction. Eastern and western coasts, we may add, are for the most part rounded, though the eastern occasionally present instances of angular forms (as Brazil and Labrador in America, Azania (Adel) in Africa, Oman in southern and Tschutschki in

* See a chart of the two hemispheres on the horizon of London. Hughes. London, 1839.

northern Asia. The major axis of the Asiatic continent (to which Europe is a peninsula) is at right angles to that of the American ; though perhaps South America is rather to be considered as analogous to Africa, not only from its remarkable similarity of general form, but also from the singular thread-like adhesion of each to its neighbouring northern mass. Were these threads broken, every commercial relation, and almost every climate of the civilized world, would undergo the most remarkable changes.

"The general direction of the land of Europe is from south-west to north-east, and is at right angles to the direction of the great fissures, which is from north-west to south-east, extending from the mouths of the Rhine and the Elbe, through the Adriatic and Red Sea, and the mountain system of Puschti-koh in Luristan, and terminating in the Indian Ocean. This rectangular intersection of the Continent in the direction of its principal extent, has powerfully influenced the commercial relations of Europe with Asia and the north of Africa, as well as the progress of civilization on the formerly more flourishing shores of the Mediterranean."

M. de Humboldt has been at great pains to arrive at a knowledge of the mean elevations of the chief continental masses above the sea level, which (in English feet) he states as follows. For Europe 671 feet, North America 748, Asia 1132, South America 1151. For Africa we have no sufficient data. "Laplace's estimation of 3078 feet (French) as the mean height of continents, is at least three times too great. The illustrious geometer was conducted to this erroneous result by hypothesis as to the mean depth of the sea" (note 360). The chain of the Pyrenees, if equably spread over France, would raise its surface, according to his estimate, 115, and the Alps over Europe 21·3 English feet. The former of these estimates certainly gives us a greater idea of the magnitude of the natural barrier between France and Spain than any ordinary exaggeration of language or poetical description would do. M. de Humboldt closes this part of his subject with the following comfortable reflection: —

"Since Mont Blanc and Monte Rosa, Sorata, Illimani, and Chimborazo, the colossal summits of the Alps and the Andes, are considered to be among the most recent elevations, we are by no means at liberty to assume that the upheaving forces have been subject to progressive diminution. On the contrary, all geological phenomena indicate alternate periods of activity and repose. The quiet which we now enjoy is only apparent; the tremblings which still shake the surface, in every latitude and in every species of rock, — the progressive elevation of Sweden, and the appearance of new islands of eruption, — are far from giving us reason to suppose that our planet has reached a period of final repose."

The phenomena of the ocean may be considered with reference to its depth, temperature, density, and to its motions as agitated by waves, tides, and currents. With respect to its depth, except near shores and in frequented tracks, we know almost nothing. Theoretical considerations indicate a *mean* depth of "a small fraction of the ellipticity of the earth," which can hardly be interpreted at more than four or five miles. Ross sounded (in 15° 3′ south, 23° 14′ west) without finding bottom at 27,600 feet (about five miles and a quarter), which is the greatest depth yet attained.

As regards the temperature of the ocean, the observations of Kotzebue in his voyage round the world appears first to have indicated, those of Beechey in his voyage to the Pacific to have (so far as they go) supported, and those of Sir James C. Ross in his recent Antarctic voyage to have established almost beyond a doubt, the extraordinary fact that the deep sea water, below a certain level determined by the latitude, is of one invariable temperature throughout the globe, *and that temperature a very low one*, the calculations of Lenz, founded on Kotzebue's results, giving 36° Fahr., and those of Ross 39°·5. The depth at which this temperature is attained, according to the latter authority, is 7200 feet at the equator, diminishing to 56° 26′ south latitude where it attains the surface, and the sea is of equal temperature at all depths. Thence again the upper surface of this uniform substratum descends as the latitude increases, and at 70° has already attained a depth of 4500 feet. Similar phenomena would appear to occur in proceeding from the equator northward, the circle of constant temperature being repeated nearly in the same latitude. Thus the ocean is divided into three great regions, two polar basins in which the surface temperature is below, and one medial zone in which it is above 39°·5, being 80° at the equator, and at the poles of course the freezing point of sea water. It will be very readily understood that in this statement there is nothing repugnant to hydrostatical laws, the compressibility of water insuring an increase of density in descending within much wider limits of temperature than here contemplated.

The physical consequences of this great law, should it be found completely verified by further research, are in the last degree important. One of them, noticed by Ross, is, "that the internal heat of the earth exercises no influence upon the mean temperature of the ocean," *a conclusion not very easy to reconcile with the theory of central heat itself*, or at least with its regular distribution. Another is the complete destruction of the notion of submarine currents setting from the poles towards the equator, caused by the subsidence of cold water in high latitudes. On the contrary, the actual disposition of things would necessitate a constant *superficial* flow of cold water from the poles towards the equator, and of warm from the equator towards the poles, in abatement of the polar and equatorial excesses of level; a mingling of these overflows on, or about, the parallels of latitude where the mean temperature is found; and their descent there in maintenance of a continual, but merely superficial triple system of circulation. If any deep-sea currents could arise at all from such a state of temperature, it must be in consequence of the descent of water rendered salter by evaporation at the tropics, unless indeed (as is conceivable) the circulation of salt as well as of heat should be also confined to the superficial strata. Enough, however, of these considerations, which are leading us astray from our guide.

M. de Humboldt passes very cursorily over the vast and complex subject of the tides, into the somewhat flagging interest of which a fresh vitality has been of late years infused by the striking researches of Whewell into the laws of propagation of the tide wave, which he has taken up as a matter of inductive inquiry; thereby exchanging the slow and arduous struggle of the geometer with almost insuperable obstacles, for the animating pursuit of practical laws. The elaborate inquiries of Airy also into the combined theory and practice of tide observation, have added to this reviving interest, and their joint labours have made this part of the Newtonian doctrine once more an English subject, which it had long well nigh ceased to be. On the other hand, the great ocean currents resulting from the general set of the trade winds and the friction of the tide wave on the bed of the ocean (adopting Weber's view of undulatory motion), are described with much spirit. The great current of the gulf stream, to which we are indebted for the genial warmth of our south-western coast, is one result of this movement, and is too well known by the descriptions of all voyagers, and the elaborate researches of Rennell, to require notice here. Not so the counterpart of this current in the South Pacific, first brought into notice by M. de Humboldt in 1802. This current drifts the cold water of the South Seas along the western coast of South America, as far as the extreme north-westerly projection of that coast, where it is suddenly deflected outwards in a due west direction into the open ocean, and there ultimately lost. At this point its waters are nearly 24° Fahr. colder than those of the general surrounding ocean, and so sharply marked is its course, that a ship sailing northwards passes quite suddenly from cold into hot water.

As the scene of a wonderfully diversified and exuberant life, both vegetable and animal, but especially the latter, the ocean also claims our attention. To say nothing of those colossal forms which divested, by the buoyancy of the medium in which they subsist, of the incumbrance of *weight*, are left free to exert the whole of their giant power to overcome its resistance, we find in the minuter forms of animal existence an unbounded field of admiring contemplation.

"The application of the microscope increases still farther our impression of the profusion of organic life which pervades the recesses of the ocean, since throughout its mass we find animal existence, and at depths exceeding the height of our loftiest mountains the strata of water are alive with polygastric worms, cyclidæ, and ophrydinæ. Here swarm countless hosts of minute luminiferous animals, mammaria, crustacea, peridinea, and ciliated nereides, which, when attracted to the surface by peculiar conditions of weather, convert every wave into a crest of light.

The abundance of these minute creatures and of the animal matter supplied by their rapid decomposition is such that the sea-water itself becomes a nutritious fluid to many of the larger inhabitants of the ocean. If all this richness and variety of life, —

M. de Humboldt goes on to add, in that vein of thoughtful poesy in which he indulges in several parts of this work, and to which, in truth, it owes much of its charm,

"—containing some highly organised and beautiful forms, is well fitted to afford, not only an interesting study, but also a pleasing excitement to the fancy; the imagination is yet more deeply, I might say, more solemnly, moved by the impression of the boundless and immeasurable which every sea voyage affords. He who, awakened to the inward exercise of thought, delights to build up an inner world in his own spirit, fills the wide horizon of the open sea with the sublime idea of the infinite; his eye dwells especially on the distant line where air and water join, and where stars arise and set in ever renewed alternation. In such contemplations there mingles, as in all human joy, a breath of sadness and longing."

As the sea, no doubt, holds in solution some small proportion of every soluble body in nature, so, besides the two great chemical elements of which dry air consists, and its variable constituent of aqueous vapour, there is probably no vaporizable body of which the atmosphere does not contain some trace. And from what we know of the influential part played in the economy of nature by one or two of these subordinate constituents, we can hardly doubt that others, whose presence has not hitherto been actually detected by analysis, have functions of high importance assigned to them in that economy. On the carbonic acid, which constitutes less than the two thousandth part of the atmosphere, all vegetation depends for its supply of carbon; and Liebig has shown that to the presence of ammonia, in far less proportion, the rain water owes its fertilising power. To the occasional production of ozone, the most powerfully bleaching and oxidating substance in nature, by electric discharges, though in proportion inconceivably minute, we probably owe the disinfection of the air from a variety of noxious miasmata, thus verifying, by one of the most delicate results of scientific inquiry, the vulgar notion of the purifying agency of thunder-storms.

Meteorology, however, has no concern with these minute chemical admixtures — the only distinction it recognises is that of air and vapour, and this only because these form, in fact, two distinct, and to a great extent independent, atmospheres, subject each to its own peculiar laws (and those laws widely different), and each reacting on the other solely by mechanical impulse and resistance. In the movements and affections of these two atmospheres by the sun's heat, the one permanent in material and constant in quantity, the other in a continual state of renovation and destruction; we recognize, as in geology, the simultaneous agency of two distinct systems of causation, superposed and modifying each other's effects — but with this advantage on the side of meteorology, that their agency is limited to definite annual and diurnal cycles, corresponding to those of the supply of solar heat, rendering their study, so far, easier. Here also we have to deal with electricity as a third element, but we strongly incline to the opinion, that its agency as a meteorological *cause* is exceedingly limited, indeed that it may be altogether left out of the account as productive of any meteorological effect of importance on the great scale.

It is by no means, however, in its general connexion as a science, that M. de Humboldt considers this vast and complex subject. The view which he takes of it regards only its final and practical bearings on climate as a part of physical geography, and that under very general heads, viz., the variation of atmospheric pressure, the climatic distribution of heat, the humidity of the atmosphere, and its electric tension. Each of these heads will afford us room for a few remarks.

All those meteorological phenomena whose period is diurnal may be studied, as he very justly observes, in their greatest simplicity, and therefore to the greatest advantage, between the tropics and especially under the equator. For this there are two reasons: first, that the sun's meridian altitude varies but little throughout the year; and secondly, that the equatorial zone is symmetrically related to the two hemispheres. In particular the diurnal fluctuation of barometric pressure pursues a march so regular that we may infer the hour of the day from the height of the mercurial column, without an error, on the average, exceeding fifteen or seventeen minutes. "In the torrid zone of the new continent," he says, "I have found the regularity of this ebb and flow of the aërial ocean undisturbed either by storm, tempest, rain, or earthquake, both on the coasts and at elevations of nearly 13,000 feet above the sea." The total diurnal oscillation amounts, under the equator, to 0·117 in., diminishing gradually as the latitude increases. This fluctuation has usually been compared to the tides of the ocean, but has, in fact, no theoretical connexion with it. It is a compound phenomenon arising from the superposition of two perfectly distinct diurnal oscillations, [each going through its complete period in twenty-four hours;] * the one taking place in the aërial atmosphere, and arising from its alternate heating and cooling, which produce a flux and reflux over the point of observation; the other arising in the aqueous atmosphere by the alternate production and destruction of vapour by the heat of day and cold of night. The resolution of the hitherto puzzling part of this phenomenon, viz., its double diurnal wave into two single ones, following different laws, and noncoincident in their phases, does honour to the sagacity of Dove, fol-

* *Sic* in the original. But the thing is mathematically impossible. To produce a double maximum in twenty-four hours, *i. e.* a double diurnal wave, a term going through its period in twelve hours *must* be introduced. Such a term may arise from alternating day and night winds in those localities where the double maximum of pressure really exists. (H. 1857.)

lowed up as it has since been by the laborious researches of Colonel Sabine, to whose discussion of this point (note 382.) we particularly direct our readers' attention.

The gradual depression of the barometer in proceeding from tropical latitudes either way to the equator, was first noticed by M. de Humboldt himself. Its explanation is easy, viz., the continual efflux of heated air upwards from the equator towards the poles. Hence, by the effect of the earth's rotation on the currents setting in below to supply the void, arise the trade winds, and in the amount of this depression, which does not exceed two tenths of an inch, we have a measure of the motive power which originates these great currents. The connexion of the trades with the monsoons, and the varying winds of higher latitudes, is beautifully placed in evidence by the law of rotation of the wind lately discovered by Dove, a conclusion following so simply and naturally from the very same principle on which Hadley originally explained the constant easterly direction of the trades (the difference of rotatory velocity on different terrestrial parallels), that it is only astonishing it should so long have escaped notice. As regards the local distribution of barometric pressure, the most extraordinary fact which has yet appeared in meteorology is, perhaps, the general depression of the mercury to the enormous amount of an entire inch over the whole Antarctic Ocean, established by the late observations of Ross.

The chief elements of climate are heat and moisture; but it is neither on the extremes of heat or cold, moisture or dryness, experienced on rare occasions, that the character of a climate depends. Climatology is throughout a matter of averages, and is best studied and best understood by the graphical depiction of such averages, obtained by many years of careful observation according to a method proposed and carried out by M. de Humboldt himself, in 1807. In this system, all those points on the earth's surface which have equal mean annual temperatures are connected by a system of curves called *Isothermal;* those, again, in which the mean temperatures of the hottest summer months are alike, by another system of *Isothēral* curves; and those in which the mean winter temperatures agree, by a third, or *Isocheimōnal* system.

The law of distribution of heat over the surface of the globe, is best apprehended by the study of the first of these systems of curves, respecting which researches subsequent to those of M. de Humboldt have led to general and very remarkable conclusions. In the northern hemisphere only, are the forms of the Isothermal curves known with any degree of exactness. In this Sir D. Brewster places two points, or *poles of maximum cold*, on the 80th parallel of latitude, and in nearly opposite longitudes (95° W. and 100° E.), of which the mean temperature is 3½° Fahr., and about which as foci the Isothermal lines form a system of spherical lemniscates, imitating in general form those beautiful curves exhibited by polarized light in biaxal crystals. The meridians of these poles pass almost diametrally through the main bodies of the American and Asiatic continents, while two other meridians nearly at right angles to them traverse the Polar Sea, running out along the north Atlantic down the west coasts of Europe on the one hand, and nearly through Behring's Straits into the Pacific on the other. These then are the meridians respectively of greatest cold and warmth, and it is impossible not to recognize in them the effect of extensive tracts of land in high latitudes in increasing, and of sea in diminishing the intensity of cold as we approach the pole. Kämtz's projections confirm this result, so far as the general form of the isothermic ovals is concerned, but place their foci in rather lower latitudes, the one near Chatankoi in the Samoiede country, the other nearly upon Barrow's Strait. The succession of these lines followed along their intersections with the east coast of America, as compared with the west coast of that continent and of Europe, places the mean climate of the whole of the former coast in striking and disadvantageous contrast with that of both the latter, and abundantly explains the early prevalent, though mistaken impression, of a general deficiency of genial warmth in the New World as compared with the Old.

The influence of great tracts of land remote from sea coasts, owing, doubtless, to the greater clearness of sky arising from the defect of moisture, tends to exaggerate both the summer heat and the winter cold, but the latter in a higher degree than the former. Accordingly we find the Isothēral curves in the interior of the great continents of the northern hemisphere affecting a greater convexity towards the north, and the Isocheimōnal less so as compared with the lines of mean temperature. The effect of this is to produce in those regions *extreme* or excessive climates in which violent summer heat is succeeded by intense winter cold. Of such, M. de Humboldt gives instances in Tobolsk, Barnaoul, and Irkutzk, in whose summers, for weeks together, the thermometer remains at 86° or 87° Fahr. while their winters exhibit the severe *mean* temperature of $-0°{\cdot}4$ to $+4°{\cdot}0$ of the same scale, or 40° lower than the mean winter temperature of London.

On the other hand, the proximity of the sea for many and obvious reasons tends to mitigate and equalize the fluctuations of temperature, and where this tendency, as on the west coast of Ireland and the south-west coast of England, conspires with a generally favourable position as regards the Isothermic curves, an approach to perpetual spring prevails. "In the north-western part of Ireland, in lat. 54° 46′, under the same parallel with Konigsberg" (where even our holly cannot survive), "the myrtle flourishes as luxuriantly as in Portugal." The winter mean temperature of Dublin is actually $3°{\cdot}6$ higher than that of Milan.

The effect of such local peculiarities is, of course,

strongly marked in vegetation, which M. de Humboldt exemplifies in the growth of the grape, and the production of *drinkable* wine. This condition, he observes, necessitates a mean summer temperature of at least 64°·4 Fahr., a mean annual temperature not below 49°·2, and a mean winter one above 32°·8. These conditions are all amply satisfied and exceeded along our southern coasts: so that it is clear that not merely drinkable, but respectable, wine might be grown there: and if, at very early periods of our history, we find that such was the practice, we may observe that, owing to the diminution of the obliquity of the ecliptic, we are placed, so far as summer temperature is concerned, in a *somewhat* less favourable situation than at the epoch of the Roman occupation. The difference amounts to 13′, by which the summer sun comes less northward than at the epoch alluded to.

"I have, in no part of the earth, not even in the Canary Islands, in Spain, or in the south of France, seen more magnificent fruit, especially grapes, than at Astrachan. With a mean annual temperature of 48°, the mean summer temperature rises to 70°·2, which is that of Bordeaux; while not only there, but still more to the south, at Kislar (in the latitude of Avignon and Rimini), the thermometer sometimes falls, in winter, to −13° or −22° Fahr."

Ascent into a higher region of the atmosphere has the same depressing effect on temperature with increase of latitude. The fact is universally known —the cause, perhaps, less familiarly so. Were there no atmosphere, a thermometer freely exposed (at sunset) to the heating influence of the earth's radiation, and the cooling power of its own into space, would indicate (if the dip of the horizon be neglected) a medium temperature between that of the celestial spaces (−132°) and that of the earth's surface below it (82° at the equator, −3½° in the Polar Sea). Under the equator, then, it would stand, on the average, at −25°, and in the Polar Sea at −68°. The presence of the atmosphere tends to prevent the thermometer so exposed from attaining these extreme low temperatures; first, by imparting heat by conduction; secondly, by impeding radiation outwards. Both these causes are more effective in proportion to the density of the air in contact with the thermometer, which is, therefore, always maintained at a degree higher than those named, and approaching more nearly to the temperature of the soil, the lower the level of the station.

The habitual dryness of the upper regions of the atmosphere is another general fact, the causes of which are not usually neatly conceived. It is partly apparent, partly real. In proportion to the rarity of the air about any moist surface, evaporation is freer, the drying process goes on more rapidly, and superfluous moisture is more speedily exhaled. Mere facility of exhalation, however, is not to be construed as any proof of extreme deficiency of moisture in the air. On the other hand, however, such deficiency really and necessarily exists. If there were never any rain, snow, or dew, the aqueous atmosphere would be co-extensive with the aërial one, and each stratum of the latter in a state of exact saturation. Every act of precipitation (no matter how produced) unsettles this state of things, and withdraws from the total mass of the air some portion of its entire amount of vapour. As such precipitations, therefore, are constantly going on in some place or other, the atmosphere, as a mass, though incumbent on a wet and evaporating surface, is necessarily always deficient in moisture. And for the very same reason, every superior stratum is relatively deficient in comparison with that immediately beneath it, from which its supply is derived. In point of ultimate causation, there is a constant drain upon the aqueous contents of the atmosphere, arising from changes of temperature. This drain extends to all its strata; but while the lower renew their losses from a surface hygrometrically *wet*, the upper draw their supply intermediately from sources more and more deficient in moisture.

In intimate connexion with these general relations stands the striking and picturesque phenomenon of perpetual snow on mountain summits, and the causes which determine the altitude of its inferior limit in different regions. The snow-line necessarily descends to the level of the sea, in latitudes where the mean temperature is beneath the freezing point, and rises, generally speaking, as we approach the equator, where, in South America, or Cotopaxi and Chimborazo, it attains a level not inferior to that of the *summit* of Mont Blanc. On the southern declivity of the Himalayas, in latitude 31°, its level may be stated at 13,000 feet, while yet, on their northern slopes, under the influence of radiation from the high lands of Thibet (11,500 feet in mean elevation) it attains a height of 16,600 feet. Such, indeed, is the influence of local circumstances, and especially of the extreme dryness which prevails aloft in the southern prolongation of the chain of the Andes, that in the western or maritime part of that chain, in lat. 18° S., the snow-line is found nearly 2700 feet higher than under the equator; and even so far as 32½° south, the volcano of Aconcagua, 1400 feet higher than Chimborazo, has, on one occasion, been seen *entirely free from snow, by the mere effect of evaporation*, being not at the time in a state of eruption. (*Kosm. Tr.*, p. 329.)

According to the alternation of the seasons, the lower line of *actual* snow oscillates between limits more or less extensive, according to the difference of the summer and winter temperatures at the place; but besides this annual oscillation, successions, which appear to us casual, of cold, warm, dry, and wet seasons, winds, &c., give rise to fluctuations in the amount of accumulated snow, which manifest themselves in the slow alternate prolongation and recess of glaciers, a subject which M. de Humboldt passes over with slighter notice than we should have expected. The arduous and indefatigable researches of Professor Forbes, one of the greatest, if not the very greatest, of Alpine travellers since Saussure, and his ingenious theory of glacier motion, have heightened to an extraordinary degree the interest of this branch of terrestrial physics, and might, we think, have secured his name a place beside those of Vernetz, Charpentier, and Agassiz, in the briefest possible mention of the subject.

The electricity of the atmosphere is a subject too inconsecutively studied, and too little understood, to admit of any distinct, general, and positive conclusions being drawn respecting it. We have ventured to hazard an opinion that the part it plays, in phenomena properly called meteorological, is rather that of an effect than a cause, whatever influence its development may have on organic life in stimulating the nerves and promoting the circulation of the juices (both, we apprehend, much overrated). Our limits, however, forbid us to assign the grounds for this opinion, and the mention of organic life reminds us that we have yet another field to traverse in M. de Humboldt's guidance. But here too we shall imitate his own brevity, confining himself as he does to the *general* influence of temperature and climate on the distribution of organic forms, to the physiognomy of different countries imparted by the greater or less predominance of those families of plants which are called "social," and to the similar influence of elevation above the sea and increase of latitude, and waiving, as it would seem designedly and of purpose, all mention of a subject the most prominent and the most interesting in natural history. We allude to the local distribution of genera and species, not as affected simply by diversity of climate and soil, but by locality *as such*, according to laws which almost seem to have had reference not so much to the mere fitness of this or that climate, &c., for this or that species, as to some more general object, such as that of superinducing the utmost possible diversity of organism and assemblages of organized beings on the face of material creation. This forbearance is the more disappointing, because it is precisely from M. de Humboldt himself that the first impulse of philosophical speculation and inquiry in this direction was given, and that there is, therefore, no one to whom we should more naturally look up for large and general views on the subject, or for satisfactory impressions as to the aspect in which the facts actually present themselves to those who alone are fully competent to judge of them. In stating these great facts, it is by no means necessary to go into questions of origin (which he very properly declines to do). There may or there may not have been local centres of creation, whence, in all geological epochs, species have spread themselves. But the matter of fact, the observed *laws of collocation*, strongly marked as they are, appear of paramount importance, and constitute the most salient features of the geography of plants and animals. "Each hemisphere," says M. de Humboldt in his Personal Narrative, "produces plants of different species; and it is not by the diversity of climates that we can attempt to explain why equinoctial Africa has no Laurineæ, and the New World no heaths; why the Calceolariæ are found only in the southern hemisphere; why the birds of the continent of India glow with less splendid colours than those in the hot parts of America; finally, why the Tiger is peculiar to Asia, and the Ornithorhyncus to New Holland."

The total diversity of all the plants and animals of New Holland from those of all other countries; the complete separation of the Old from the New World in their representation of natural families not only in their living, but in many of their fossil productions, is part only of a general system of regional repartition which pervades the whole scheme of organic life; a *fact* of the first magnitude, whatever be the speculative aspect in which it may be regarded.

Man, "subject in a less degree than plants or animals to the circumstances of soil and to meteorological conditions, and escaping from the control of natural influences by the activity of mind and the progressive advancement of intelligence," forms everywhere an essential part of the life which animates the globe. In considering the great questions which ethnology presents, M. de Humboldt avows his conviction of the superior weight attributable to those arguments which support, over those which combat a community of origin and a gradual branching forth into established varieties or races. He observes, however, that,

"As in the vegetable kingdom and in the natural history of birds and fishes, an arrangement into many small families proceeds on surer grounds than one which unites them into a few sections embracing large masses; so also, in the determination of races, it appears preferable to establish smaller families of nations. In the opposite mode of proceeding, whether we adopt the old classification of Blumenbach into *five* races, . . . or that of Prichard into *seven* . . . it is impossible to recognize in the groups thus formed any true typical distinction — any general and consistent natural principle. The extremes of form and colour are separated indeed, but without regard to nations which cannot be made to arrange themselves under any of the above-named classes."

Language is the main clue we have to guide us through the labyrinths of ethnology; but it is one which must be followed with caution, and with all the light which history can throw upon its application.

"Subjection to a foreign yoke, long association, the influence of a foreign religion, a mixture of races, even when comprising only a small number of the more powerful and the more civilized immigrating race, have produced in both continents similarly recurring phenomena: viz., in one and the same race two or more entirely different families of languages; and in nations differing widely in origin, idioms belonging to the same linguistic stock."

Where history fails, however, as is the case with the barbarous nations of the New World, and those which in other regions are fast disappearing before

European encroachments, language, physical resemblance, and similarity of customs (when not traceable to general principles of human nature) are all the guides which are left to us in tracing the affiliation of races. That aiding and warning light withdrawn, it behoves us to be all the more scrupulously careful in collecting and preserving unimpaired and undistorted whatever vestiges of human language still subsist. And here we must enter our protest, we fear an unavailing one, against the supineness which suffers those invaluable monuments, the unwritten languages of the earth, to perish with a rapidity yearly increasing, without one rational and well-concerted effort to save them in the only mode in which it can be done effectually, viz., by reducing them to writing *according to their exact native pronunciation* through the medium of a thoroughly well considered and digested Phonetic alphabet. About sixty well-chosen, easily written, and *unequivocal* characters, completely exemplified in their use by passages from good writers in the principal European and Eastern languages, would satisfy every want, without going into impracticable niceties; and we earnestly recommend the construction and promulgation of a manual of this kind for the use of travellers, voyagers, and colonists, as a matter of pressing urgency, to the consideration of philologists, ethnologists, and geographers, in their respective societies assembled.*

* Many attempts at the construction of such alphabets have been made, but none at all satisfactory. That of Young

We have been so intent on the subject-matter of the work before us, as to have left little space for comment on the mode of its presentation to the English reader. The author has been especially fortunate in his translator (translatress we should rather say, since, in the style of its execution, we have no difficulty in recognizing the same admirable hand which gave an English garb to Baron Wrangell's Expedition to the Polar Sea). So perfect a transfusion of the spirit and force of a very difficult original into another language, with so little the air of a translation, it has rarely been our fortune to meet with. To the editor it is indebted for several very interesting and instructive notes (to some of which we have had occasion specifically to draw the reader's attention) relating

(Lectures, ii. 276) is perhaps the most complete in its analysis of speech, though still defective, and in some points erroneous —his system of characters wretched. Gilchrist's is perhaps the best known, and in profession nothing short of absolute universality; but its author (a Scotman) was altogether defective in *ear*, and his examples in consequenee self-contradictory—his system of writing confusion itself. The *Fonotipik kariktur*, devised by the ingenious Mr. PitmUn and his associates for the speedy and effectual abrogation of the English language, would have considerable merit were it not founded on an essentially English instead of a cosmopolitan view of the vowel sounds as represented by European letters, and therefore sure to be rejected by every foreign philologist. Yet even this, enlarged to suit the exigencies of the case, would be preferable for temporary use to the present no-system in which each traveller in his diary, and each missionary, in formal grammar and dictionary, confounds and for ever mars, as seems good in his own eyes, the pronunciation he pretends to fix. (Note of 1848). See Appendix.

to a variety of subjects, on which, either from personal observation on the most extended scale, or from laborious and systematic discussion of the observations of others, he is entitled to every attention.

While the preceding pages were in progress, we have been favoured with the perusal, in proof sheets, of a portion of the second volume of the "Kosmos" (translated and edited as above), containing, under the title of "Incitements to the Study of Nature," a series of beautiful and brilliant essays, of the highest literary merit, and full of scholarship, classical research, and artistic feeling, on the reflex action of the imaginative faculty when excited by the contemplation of the external world, as exemplified in the production of poetic descriptions of nature (especially of wild and landscape scenery), and in landscape painting. For examples of the former kind, M. de Humboldt lays under contribution the literature of all ages and nations, from ancient India to modern Europe, entering largely into the influence exercised by the peculiar aspect of society in each on the development of this form of the poetic sentiment, which he regards, and justly, as the first expansion of the heart towards a recognition of the unity and grandeur of the Kosmos. In like manner the art of landscape painting is traced from its first origin as the mere background of historical composition or scenic decoration, to its grand developments in the seventeenth century — to "Claude Lorraine, the idyllic painter of light and aërial distance, Ruysdael's dark forest masses and threatening clouds, Gaspar and Nicholas Poussin's heroic forms of trees, and the faithful and simply natural representations of Everdingen, Hobbima, and Cuyp." The gradual emancipation of the art from its trammels, as a subordinate auxiliary, and its assumption of an ideal of its own embodying, are shown to be ever found in connexion with increasing knowledge and observation of nature, consequent on advancing cultivation. To such poetic descriptions and depicted scenery, as well as to the view of exotic products assembled in collections, hot-houses, and museums, he traces much of that lively impulse which stimulates young and excitable minds to foreign travel for the sake of knowledge, and to the prosecution of physical study at home. These essays form a graceful and elegant episode, interposed between the more massive and austere divisions of the general subject, the "Physical Description of the Universe," which we have passed in review, and the "History of the Contemplation of Nature;" and will be read with equal enjoyment by the poet, the artist, and the philosopher.

Of the "History of the Contemplation of Nature" one section only has reached our hands: sufficient, however, to convey a notion, and to correct an impression we had formed, as to our author's intended mode of handling this part of his matter. The history with which he proposes to present us would appear to be not so much a history of Physical Science in the gradual development of its theories, as *a history of objective discovery*, a review of those steps in the progress of human cultivation which have prepared the way and furnished the materials for science such as we now possess it. With every successive expansion of society the views of mankind have become enlarged as to the extent and construction of the globe we inhabit, the objects it offers to contemplation, the elaborate structure of its parts, and its relation to the rest of the universe. Great events in the world's history have from time to time especially facilitated and promoted this enlargement of the horizon of observation; such as the migrations of nations, remarkable voyages, and military expeditions, bringing into view new countries, new products, new relations of climate. Great epochs too, in the history of the knowledge of nature, are those in which accident or thought has furnished artificial aids, new organs of sense and perception, by which man has been enabled to penetrate more and more deeply either into the profundity of space, or into the intimate constitution of the animate and inanimate objects which surround him. In tracing these epochs and following out the course of these events so far as they bear upon the object in view, availing himself of all the light which modern research has thrown on the early history of civilization, whether from the study of ancient monuments, or the critical comparison of written records, M. de Humboldt has opened out for himself a field nearly co-extensive with literature itself, and one peculiarly fitted to his own powers and habits of thought, which, as our readers need not to be informed, have made its higher walks — Æsthetics, History, and Antiquarian and Monumental Lore—quite as familiar to him as those of Science. We should do injustice, however, both to him, and to those whose office it may be to render an account of the further progress of this work, by further anticipation, and shall, therefore, content ourselves with adding that, should the conclusion correspond (as we doubt not) with these beginnings, a work will have been accomplished every way worthy of its author's fame, and a crowning laurel added to that wreath with which Europe will always delight to surround the name of Alexander von Humboldt.

QUETELET ON PROBABILITIES.

1. *Lettres à S. A. R. le Duc régnant de Saxe-Cobourg et Gotha sur la Théorie des Probabilités appliquée aux Sciences Morales et Politiques.* Par M. A. QUETELET, Astron. Royal de la Belgique, &c. &c. 1 vol. in 8vo. 1846. Chez M. Hayez, à Bruxelles.
2. *Letters addressed to H. R. H. the Grand Duke of Saxe-Cobourg and Gotha on the Theory of Probabilities as applied to the Moral and Political Sciences.* By M. A. QUETELET, Astronomer Royal of Belgium, Corresponding Member of the Institute of France, &c. &c. Translated from the French by OLINTHUS GREGORY DOWNES, of the Œconomic Life Assurance Society. London. Charles and Edwin Layton, 150. Fleet Street: 1849.

(FROM THE EDINBURGH REVIEW, JULY, 1850.)

EXPERIENCE has been declared, with equal truth and poetry, to adopt occasionally the tone, and attain to something like the certainty, of Prophecy. In the contemplating mind the past and the future are linked by a bond as indissoluble as that which connects them in their actual sequence. Metaphysicians may dispute concerning the nature of causation; and it will always, no doubt, be difficult to explain and demonstrate the objective reality of that relation; but the reality, as an internal feeling, of the expectation that what has happened under given circumstances will happen again under precisely similar circumstances, is independent of metaphysical dispute, and above it. It is an axiom drawn from the inward consciousness of our nature, by involuntary generalization. We acknowledge it expressly or impliedly in every instant of life. It is the practical ground of every sane transaction. Instinctive in childhood—or if not instinctive, the direct result of the earliest, simplest, and most powerful associations,—it becomes, however, entangled with conditions and modifications; as reason enlarges her sphere of vision, and we learn to question the absolute similarity of circumstances in any two assigned cases. But though puzzled for a while, and baffled as by a verbal quibble, the impression itself is not destroyed or weakened. We begin early to distinguish between relevant and irrelevant circumstances; to attend only to the former, and to disregard the latter. Upon this ground Inductive Science takes her stand and erects her axioms, making it her business to ascertain, in each case, what are the really relevant circumstances on which events depend, and to analyze the complicated web of phenomena into a system of elementary and superposed uniformities, to which we assign the name of inductive theorems, or laws of nature.

One of the greatest steps which have yet been made in the philosophy of Logic—a step which may almost be termed a discovery when we consider the inveteracy of the habits and prejudices which it has cast to the winds—is that recently taken by Mr. Mill*, in showing that all reasoning (meaning thereby the investigation of truth as distinguished from the mere interpretation of a formula) is from particulars to particulars, and in thence assigning to general propositions their true character, and to the syllogism its true office. But while a vast accumulation of rubbish, which obscured the basis of all sound philosophy, has thus been swept away, a condition of affairs is disclosed which, at first sight, seems to annul our prospect of attaining to any general knowledge whatever,—at least in those of its departments in which analogies are not at once perceived to be identities. No one has ever yet contended that our knowledge of special facts is intuitive. The questions, therefore, at once arise, 1st. What sort of security we have for the truth of any assertion concerning any external thing or fact which has not been made a matter of direct observation; and, 2ndly. What measure have we of the degree or amount of that security, supposing we possess it in some degree, and supposing

* System of Logic, 2nd ed. chap. iii., on the functions and logical value of the Syllogism. Perhaps Mr. Mill may be considered as only following out more emphatically the views originally taken by Berkeley on this subject, but which seem to have dropped so far out of notice as to give their revival all the force of novelty.

absolute and mathematical certainty to be unattainable?

Now, with regard to the first of these questions, it must at once be admitted that no conclusion from inductive reasoning, *i. e.* from the observed to the unobserved, can enjoy more than a provisional security. If the unbroken experience of all observers, in innumerable instances, be really no ground for extending the conclusion to one unobserved instance admittedly parallel, then and in that case inductive argument should have no influence on human belief. But if, on the other hand, such large and uniform experience of the past is irresistibly felt to warrant a conclusion as to the future, we should then confidently adopt that conclusion, though with a distinct perception and admission of a risk of error more or less infinitesimal, which we make up our minds to disregard. And it is thus that we come to rest in *practical*, as distinct from *mathematical*, certainty, in all physical inquiry, and in all the transactions of life.

It is to express the perception, and enable us to speak consistently, and at the same time definitely, concerning the amount, of this risk, that the term PROBABILITY has been invented — a term having reference to our ignorance of the analysis of events, and of the efficient causes which really *necessitate* the successive steps by which they arise; and that not generally, but with special and personal reference to the party using that term; so that the same physical relation — the same historical statement—the same future event—may have very different degrees of probability in the eyes of parties differently informed of the circumstances, the causes in action, the reputation for veracity of the testifying authors, or their opportunities of knowing the facts related.

The scale of probability, as viewed in its greatest latitude, obviously extends from the assured *impossibility* of the event contemplated to the *certainty* that it will happen. The total interval between these extremes, either of which is complete knowledge, is occupied by higher or lower degrees of expectation or belief, determined by the partial knowledge we happen to possess, and may be regarded as a natural unit susceptible of numerical subdivision into fractional parts—much as the interval, from the freezing to the boiling point on the thermometric scale may be subdivided into aliquot parts or degrees. Properly speaking there is no natural numerical *measure* of a mental impression, any more than of a corporeal sensation; but in both cases we are sure that higher degrees in the numerical scale may well represent greater intensities of the impression, and in both there is proof that equal increments of a certain element purely ideal in the one, though possibly substantial in the other,—answer to equal numerical differences on the scale; and that the greater or less abundance of this element, in some way or other, *determines* the degree of intensity of the impresssion in question.

But the scale of probability plainly admits of a much more precise graduation than that which would merely mark a general increment or decrease, inasmuch as it is obviously capable of an exact bisection, marked by a definite state of mind,—that, namely, where the mind is completely balanced between the expectation of the event happening and not happening; and this state is therefore indicated by assigning $\frac{1}{2}$ as the measure of probability in its case. In fact the *non-happening* of an event is in itself an event; and in the case of a balanced state of mind this event is held to be as probable as the other; so that the unit of certainty must be taken as equally divided between them. In reference to this state of neutrality, then, the words "probable" and "improbable" present a meaning. An event is "probable" when its probability numerically estimated exceeds $\frac{1}{2}$,—"improbable" when it falls short of that fraction.

The certainty of an event is not usually spoken of in common parlance as a probability,—as 0 is not commonly called "a *number*," nor "the whole" "*a part*." Continuity of mathematical language, however, obliges us to identify a probability having 1 for its measure with certainty. Yet there seems to be some psychological cause, some involuntary mental action in the sort of leap which most men make from a high probability to absolute *assurance*,— bearing no remote analogy to the sudden consilience or springing into one (with an immediate sentiment of tangible reality) of the two images seen by binocular vision, when gradually brought within a certain proximity; or as some eminent authorities in the higher logic seem to have become impressed with a conviction of the *necessary* truth of certain physical axioms, which others continue to regard only as inductive propositions of very great generality. There is no doubt that minds differ materially in their readiness to make this spring, and to acquiesce in probable propositions as if certain.

Into the delicate and refined system of mathematical reasoning, now generally known as the "Calculus of Probabilities," the metaphysical idea of Causation does not enter. The term *Cause* is used in these investigations without reference to any assumed power to effect a given result by inherent activity. It simply expresses *the occasion* for a more or less frequent occurrence of that result, and may consist quite as well in the removal of an impediment as in any direct agency. The distinction is that taken by metaphysicians between the *efficient* and *formal* cause. The result itself, too, is regarded not as a magnitude or phenomenon susceptible of varieties of degree according to the intensity of causation, but merely as an event which *must either happen or not happen*; and which will happen more or less frequently, according to the facilities so afforded for its happening under the action of its proper but unknown physical or moral causes, be they what they may, or the impediments interposed to defeat them. Moreover, the sort of events contemplated in establishing the fundamental principles of this calculus are such as in their simplicity, absolutely exclude one another without the possibility of compromise, or passing into each other by insensible gradation. Hence the frequency in its reasonings, of illustration by the drawing of balls of different colours, or otherwise differently marked, from urns; the distinction between the colours or marks in such cases being obviously absolute, and mutually exclusive. Such events are commonly said by writers on the subject to be *contrary* to each other. We should prefer the word *complementary*, as we should "hypothesis" or "occasion" for "cause,"—and we think the subject would acquire an accession of clearness by this change in its nomenclature. The distinction itself is most important, and requires to be steadily borne in mind in all applications of this calculus, the chief delicacies in which depend on duly resolving any contemplated event into a determinate succession, or simultaneous combination, of other elementary events mutually exclusive and yet presenting equal facilities for their occurrence.

It requires also to be dwelt on with some emphasis in another point of view, as establishing a chain of relation between the province of this branch of science and that of Physics which concerns itself with *efficient* causes, on the one hand; and with Natural Theology, which refers phenomena to *final* ones, on the other. So considered, it lies at the root of all philosophical inquiry. Chance, indeed, is admitted into its reasonings as the expression of our ignorance of agents, arrangements, and motives, but with the express view to its exclusion from their results. We speak of it as opposed to human certainty, not as opposed to Providential design.* And, as the first step towards narrowing its domain, we endeavour to form a correct estimate of its extent. Among all the applications of this calculus by far the most important are those which come directly in aid of physical, social, and moral inquiry, by enabling us to measure either the degree of rational reliance we may place on numerical data (the fundamental elements of Physical science), or the decisiveness with which we are justified in pronouncing the existence of a formal cause or determining condition, from the records of a succession of phenomena. Such conditions once placed in evidence and rendered matter of practical certainty, we hand them over to reasoners of another kind, to discover by appropriate inquiries or experiments in what they consist, and what other offices they may fulfil in the great arrangements of creation.

It is matter of familiar observation and experience that a *single* occurrence of an event, accompanied by any circumstance then for the first time noticed, is enough to raise a considerable amount

* De Morgan, Encyc. Metropol., art. *Probabilities.*

of expectation that a recurrence of the same circumstance will issue in the reproduction of the same event. The one becomes indissolubly associated with the other, and is *connoted* with it; that is to say, set down in memory as one of its distinctive marks. A man with a black crape over his face presented a pistol at me yesterday, alone and at nightfall, and demanded my purse. I shall never see a craped face in future (especially if alone and at dusk) without expecting also to see the pistol and hear the unwelcome demand. The unusual event and the unusual circumstance become associated in imagination, never after to be disunited; and even when further experience may have shown that they often occur disjoined, the occurrence of what has been once set down as a mark or sign of a highly painful or pleasurable incident continues to agitate us with a feeling we cannot shake off, however condemned by reason. In infancy or early youth, when all phenomena are new and striking, and all pains and pleasures vivid, these earliest *connotations* make a deep and indelible impression, and become either the germs of knowledge or the roots of prejudice. Now it may be worth while to inquire what account the theory of Probabilities gives of this impression, apart from all metaphysical considerations. What is the numerical measure of the expectation (derived from a simple consideration of equipossible combinations) that, of two well-characterized events, each of which has been once, and once only, observed, and then in connexion with the other, the next appearance of the one will be accompanied with that of the other? The happening of one event (A) (no matter which) may be considered as equivalent to inserting the hand into an urn containing no other than black and white balls, at least one of each, but without any further restriction as to their numbers, absolute or relative; while the coincident happening of the other event (B) may be assimilated to the drawing thence a ball of the one or the other colour, the opposite colour being held *thenceforward* to denote its not happening. The second happening of the event (A) will therefore come to be assimilated to a second insertion of the hand into the same urn, *the ball first drawn not being replaced*,* and a second happening of (B) will be expressed by the drawing thence of a ball of the same colour as the first: its not happening by the contrary colour. Under these circumstances, an exact analysis of all the possible combinations assigns $\frac{2}{3}$ for the probability *antecedent to the first drawing* that the second drawing will produce the same colour as the first; or, as commonly expressed, there are two chances

* This is essentially involved in the conditions. Though we may presume, or guess, that a combination which has once happened may happen a second time, we are not *sure* that it can. There may be an impossibility in the very nature of the events that it should If we replace the ball first drawn, we leave no room for the contingency that the supposed dependent event may be unique in its kind, and having once happened can never happen again.

to one in favour of such a result. It is never without its instruction to trace this sort of parallel between mental impressions and abstract numerical relations. As in the theory of sound, we are led to perceive that the uninstructed ear, in a manner unknown to us, feels out the exact coincidence of numerical ratios, and the sense is delighted with such coincidence; so here we find that a sentiment arises in the uninstructed mind—we know not how, yet irresistibly—to which exact science enables us to trace a parallel, if not to see a reason, in the numerical preponderance of favourable over unfavourable cases, in an indefinite and absolutely unknown multitude of combinations.

As Probability is the numerical measure of our expectation that an event will happen, so it is also that of our belief that one *has* happened, or that any proposed proposition *is* true. Expectation is merely a belief in the future*, and differs in no way, so far as the measure of its degree is concerned, from that in the past. It may be more difficult to weigh the credibility of human testimony than to reason on contingencies in passing events; but the difficulty exists only in making the estimation, not in the mode of calculating on it when made. Numerically speaking, a certain percentage of every man's assertions is incorrect; and the way in which overwhelming probabilities may arise from the accumulation of such imperfect statements

* Brother Jonathan applies the word "expect" indiscriminately to past, present, and future.

on the one hand, or in which all reasonable reliance may be destroyed by successive hearsay transmission on the other, is not among the least interesting subjects of consideration in this calculus.

The theory of Probabilities has been characterized by Laplace, one of those who have contributed most largely to its advance,— as "good sense reduced to a system of calculation;" and such, no doubt, it is. But it must be especially noticed that there is hardly any subject to which thought can be applied, which calls for so continuous an application of that excellent quality, or in which it is easier to make mistakes from simple want of circumspection. And, moreover, that its reduction to calculation is attended with difficulties of a very peculiar nature, such as occur in no other application of mathematical analysis to practical subjects, arising out of the great magnitudes of the numbers concerned, which defeat the ordinary processes of arithmetical and logarithmic calculation, by exhausting the patience of the computer, and require special methods of *approximate* evaluation to bring them within the compass of human industry. These methods form a conspicuous feature of the general subject, and have furnished scope for very extraordinary displays of mathematical talent and invention. That very large numbers will inevitably be concerned in questions where numerous and independent contingencies may take place, and in any order or mode of combination, will be apparent to any one who considers the astonishing *fecundity* of such combinations numerically estimated, when the combining elements are many. For example, the number of possible "hands" at whist (regard being had to the trump) is 1,270,027,119,200.

The calculus of Probabilities, under the less creditable name of the doctrine of Chances, originated at the gaming table; and was for a long time confined to estimating the chances of success and failure in throws of dice, combinations of cards, and drawings of lotteries. It has since effectually obliterated the stain of its cradle, as there is no monitor more severe, no lecture which can be delivered on the certain ruin which attends habitual gambling more emphatic than may be found in its demonstrations. Questions of this kind, it is true, are still retained in treatises on the subject; nor indeed can they be conveniently dispensed with, since they furnish the simplest and readiest illustrations of the combination of independent events, and the superposition of contingencies arising out of them, which belong essentially to its principles. They, however, form a very insignificant part of its applications, in comparison with the problems which its scope at present takes in, and which its modern developments have enabled it to handle.

Its first advances towards the dignity of a distinct branch of Mathematics are attributable to the celebrated Blaise Pascal, and his no less celebrated contemporary and correspondent Fermat, — both reasoners of extraordinary acuteness, and who seem to have been specially attracted (like many of their followers) by the close reasoning and careful analysis its problems demand for their successful issue. Subsequent to these, but still among its earlier contributors, we find the distinguished names of Huyghens (to whom we owe the first treatise on the subject), those of the Grand Pensionary De Witt, Hudde, and Halley (with whom originated its application to the probabilities of life and the construction of tables of mortality), and that of James Bernouilli, who may be considered the first philosophical writer on the subject. To him we owe the demonstration of two great fundamental theorems or laws of Probability, as applied to the results of very numerous trials of any proposed species of contingency: viz., 1st, that in any vast number of trials there is a demonstrably greater probability that the events will happen in numbers proportional to *their respective chances in a single trial*, than in any other *specified* proportion; and, 2dly, that a number of trials may always be assigned so great, as to make the probability of the events happening in numbers falling within any assigned limits of deviation from that proportion, however narrow, approach to certainty as nearly as we please. The first of these propositions has the air of a truism, when the meaning of its terms is not nicely weighed. But the second is obviously of paramount importance; since it goes to take the totality of results obtained in any sufficiently extensive series of trials, almost out of the domain of chance, and to place in evidence the influence of any "cause" or circumstantial condition common to the whole series, which may give even a trifling preponderance of facility to any one of the classes of events contemplated over the rest.

Common sense, it may perhaps be said, would tell us as much as this. No doubt it might suggest some such propositions as likely enough to be true; and the usual course of inductive reasoning up to causes tacitly assumes their truth. But when we come to demand what number of trials may reasonably be expected to bring out into prominence a very small given preponderance of facility? or to declare within what limits of accuracy such preponderance may reasonably be expected to be represented on the upshot or final average of a given number of trials?—or, lastly, what is the probability that on a given number of trials such an average will represent the preponderant facility in question within given limits of exactness?—all of them, and especially the last, evidently practical questions of much interest; we find ourselves forced to appeal from the unaided judgment of simple good sense, to strict numerical calculation,—taking for its basis not a mere *aperçu* but a rigorous demonstration of the truth of the propositions above stated. This is very much the case with all the more important conclusions of this theory; when generally enunciated, they are almost universally seen to be pretty plainly conformable to ordinary clear-judging apprehension of their relations. Even the apparently paradoxical conclusions by which we are occasionally startled, lose that aspect when their exact wording is duly attended to, and all the conditions implied in it clearly apprehended. It is their applicability to exact computation, and the handle they afford thereby for precise determinations useful in practice, which give them all their value.

Problems of the class above mentioned were first successfully treated by De Moivre, to whom also we owe the happy idea of applying Stirling's theorem to approximate to the ratio of the high numbers which enter into such calculations, without which they would be impracticable. From these it would appear but a small step to pass to what may be deemed the *inverse calculus* of Probabilities, which applies the knowledge gained by the observation of past events to the prediction of future, by concluding from the succession of facts observed the respective degrees of probability of the existence of each out of several equipossible determining conditions, and thence starting as it were anew, and ascertaining from the knowledge thus acquired the probability of an event or events similarly determined *in futuro*. It was reserved, however, for another member of the gifted family of Bernouilli to make this step, which has in some respects changed the whole aspect of the subject, and given to it that degree of importance it possesses as an auxiliary of the inductive philosophy.

It may perhaps be doubted whether subsequent

writers have added very materially to the intrinsic philosophy of the subject, though there can be no hesitation as to the value of the improvements they have made in its methods of procedure, whether in point of elegance or power; the extension given to its formulæ; or the numerous and important applications made of its principles, especially in those cases (which comprise almost all the really interesting ones) where the transition has to be made from the finite to the infinite, from the limited though often large number of possible combinations which its simple and more elementary problems offer, to the *literally infinite* multitude which the gradation of natural causes and influences obliges us to consider, and which calls for the perpetual employment of the most refined theories, and the most delicate and abstruse applications of the integral calculus. In all these respects the great work of Laplace ("Théorie Analytique des Probabilités") stands deservedly pre-eminent; occupying in this department of science the same rank and position which the "Mécanique Analytique" of his illustrious rival Lagrange holds in that of force and motion, and marking (we had almost said) the *ne plus ultra* of mathematical skill and power. So completely has this sublime work been held to embody the subject in its utmost extent, and to satisfy every want of the theorist, that an interval of a quarter of a century elapsed from the date of its appearance (1812) before any further original contribution of moment was made to the theory.

The valuable memoir of Poisson, published in 1837, on the probability of judicial decisions * (which contains a *résumé* of the whole theory of Probabilities), though admirable for its clear exposition of principles and elegant analysis, can hardly be said to have carried the general subject much beyond the point where Laplace left it.

It may easily be imagined that a work like this of Laplace, followed at a short interval by an admirable *exposé* of its contents by himself ("Essai Philosophique sur les Prob."), could not fail to make a lively impression and to excite general attention. Laplace possessed in an eminent degree the talent of stating the most profound results of his own geometry in a style at once philosophical, luminous, and pleasing. Few works have been more extensively read or more generally appreciated than this Essay and that on the "Système du Monde" by the same author. There is in both a breadth and simple dignity corresponding to the greatness of the subjects treated of, a loftiness of style, the direct result of generality of conception, and which is felt as adding to rather than detracting from clearness of statement, and a masterly treatment which fascinates the attention of every reader. Nowhere can be found so great a body of important discoveries, so consecutively linked together, and so distinctly and impressively

* Recherches sur la Probabilité des Jugemens en Matière Criminelle et en Matière Civile ; précédées des Règles Générales du Calcul des Probabilités. Paris, 1837.

announced. It is not, perhaps, too much to say, that were all the literature of Europe, these two Essays excepted, to perish, they would suffice to convey to the latest posterity an impression of the intellectual greatness of the age which could produce them, surpassing that afforded by all the monuments antiquity has left us.

Previous to the publication of the "Essai Philosophique," few except professed mathematicians, or persons conversant with insurances and similar commercial risks, possessed any knowledge of the principles of this calculus, or troubled themselves about its conclusions,—regarding them as merely curious, and perhaps not altogether harmless speculations. Thenceforward, however, apathy was speedily exchanged for a lively and increasing desire to know something of a system of reasoning which for the first time seemed to afford a handle for some kind of exact inquiry into matters no one had ever expected to see reduced to calculation and bearing on the most important concerns of life. Men began to hear with surprise, not unmingled with some vague hope of ultimate benefit, that not only births, deaths, and marriages, but the decisions of tribunals, the results of popular elections, the influence of punishments in checking crime—the comparative value of medical remedies, and different modes of treatment of diseases—the probable limits of error in numerical results in every department of physical inquiry—the detection of causes physical, social, and moral,—nay, even the weight of evidence, and the validity of logical argument—might come to be surveyed with that lynx-eyed scrutiny of a dispassionate analysis, which, if not at once leading to the discovery of positive truth, would at least secure the detection and proscription of many mischievous and besetting fallacies. Hence a demand for elementary treatises and popular exposition of principles, which has been liberally answered.

Among the valuable works of this kind in the French and English languages which have appeared since the epoch in question, we may notice more especially Lacroix's "Traité Élémentaire du Calcul des Probabilités; Paris, 1822," and the several encyclopædic essays and articles on the subject by Sir John Lubbock and Mr. Drinkwater (Bethune), in the Library of Useful Knowledge, by Mr. Galloway in the Encyclopædia Britannica (since published separately in a small and compendious form—a work of great merit and utility), and by Mr. De Morgan in the Encyclopædia Metropolitana. To the last-mentioned treatise, as well as to two admirable chapters on the subject in the recent elaborate work by the same author on the Formal Logic, we may refer as containing, *par excellence*, the clearest views of the *métaphysique* of the subject, and the most satisfactory analysis of the state of the mind as to belief or disbelief, and the degree of assurance afforded by the conclusions of the calculus in cases where the data themselves are vague and uncertain, which can any where be found. All or any of these works will afford the English student a perfect insight into the mathematical treatment and reasonings of the subject, and consequently serve as an abundant preparation for the study and mastery of Laplace's great work; but we would caution all who desire to enter upon the more general and intricate parts of the theory, never for an instant to lose sight of special examples and numerical particulars, since nothing can exceed the bewilderment of ideas experienced by the tyro in this department of mathematics, who trusts himself *with both feet off the ground* to the whirl of symbols and notations in which those who are accustomed to ride these storms know how to guide their course, and even seem to feel a wild and fierce delight in the turmoil.

There is, however, a very large portion of those who desire to know something of the results at which thinking men have arrived in this as in all other departments of knowledge, to whom a book full of mere algebraic formulæ and calculations must remain for ever sealed. These are not necessarily or generally persons of despicable acquirements or intellect; nor is this their curiosity to be slighted as devoid of a reasonable object or motive. They desire to understand with a view to apply. Mathematicians, in common with men of high science in all departments, have long since begun to perceive that they have to address a mixed audience of a highly important and respectable character—an audience by no means disposed to treat them with derision or distrust, but, on the contrary, to regard them as their fitting instuctors in matters within the scope of their legitimate pretensions, if only they will condescend to make themselves intelligible. Learned jargon such an audience will not endure. Charlatanerie of every description it can detect and chastise. Common-sense statement driven home by pointed illustration, and an earnest endeavour to inform, are what it eagerly desires, and in such a spirit is assuredly entitled to receive at the hands of those able to afford it.

The work now before us is conceived on these principles, and on this view of the duty devolving on those who have advanced beyond the ordinary limits of knowledge, to pause occasionally in their onward career, and inform the world, in plain terms and without exaggeration, whither they have got, and what they see beyond, which may make it worth while either for themselves to continue in the track, or for others to follow in it; as well as to render easy and intelligible to all whom it may concern the practical application of the information acquired. Its author is a teacher well worth listening to, and may claim attention on the excellent ground that he has himself approached his subject in a practical manner, through a long and severe apprenticeship, to the actual collection of data in a great variety of departments, and to the deduction from them of definite results of unmistakeable value and import, by the rules and principles he professes to teach.

M. Quetelet has been long and advantageously known as an ardent and successful cultivator of science. No one has exerted himself to better effect in the collection and scientific combination of physical data in those departments which depend for their progress on the accumulation of such data in vast and voluminous masses, spreading over many succeeding years, and gathered from extensive geographical districts,—such as Terrestrial Magnetism, Meteorology, the influence of climates on the periodical phenomena of animal and vegetable life, and Statistics, in all the branches of that multifarious science, political, moral, and social. Peculiar facilities (or rather opportunities which he has improved into facilities) have been afforded him for such researches by his position in his own country, where he has filled the leading and responsible office (there at least, as in France, so considered) of perpetual Secretary to the Belgian Academy of Sciences, as well as that of Director of the Royal Observatory of Brussels, an institution which owes its establishment mainly to his solicitations, and its remarkable efficiency as a "physical observatory" entirely to his activity and perseverance. Placed as a member of the Central Statistical Commission in direct communication with his government, whose confidence he deservedly possesses, he has been enabled to suggest and carry out a variety of useful and important improvements, both in the forms and objects of statistical registry. The centre of an immense correspondence, he has moreover succeeded in inspiring numerous and able coadjutors, not only in Belgium but in other countries, with a similar zeal, impressing them with his views and securing their aid in carrying out a system of definite and simultaneous observation. No one threw himself with more entire devotion into the system of combined magnetic and meteorological observation set on foot by the British and other governments, and which has been productive of, and continues to produce, such useful and valuable results to science. And it will not cease to be remembered that while in one special branch of combined inquiry (that directed to tracing the progress of atmospheric waves across Europe and the Atlantic), France stood aloof and furnished not one solitary instance of co-operative observation (thus interposing herself as a desert might have done between England and the rest of the Continent), Belgium, influenced by M. Quetelet's instances and example, supplied corresponding and very valuable observations from five stations.* By many who may be little able to estimate such claims on our attention, our author will yet be regarded with interest as the preceptor of a prince whose conduct and virtues have endeared him to every Englishman.

In considering the manner in which he has

* Subsequently increased to seventy stations over all parts of Europe, held in correspondence with Brussels. (Rapport adressé au Ministre de l'Intérieur sur l'état et les travaux de l'Obs. R. de Bruxelles, 1845.)

executed his task, we beg to protest *in limine* against the form of letters to an illustrious personage in which it appears, and of which (though as he informs us, so originally written) a very moderate amount of subsequent alteration would have divested it. Although nothing which can be considered adulatory occurs in any of them, yet every reader must feel that a certain portion of each letter is out of place as regards his own information, and, so far, an interruption to the consecutiveness of his thoughts. It is a certain quantity of non-luminous matter, interposed between the author and his meaning, which serves no good purpose, since it neither pleases nor relieves. The objection is general against all such artifices of communication as letters—dialogues—catechisms, &c., if the subject be a scientific one and the object of the work didactic. They are like pebbles in the bed of a stream, which may make it sparkle and please the eye and ear when the thought is but loosely engaged. But the welling waters of scientific lore should be clear, glassy, and unrippled, offering their inmost depths to a quiet and contemplative gaze, and neither distracting by murmurs nor dazzling by irregular reflections.

A comparatively small portion of the work, the first and least extensive only of four divisions into which it is broken, and an appendix in the form of notes containing tables and formulæ, are devoted to the theory of Probabilities in the abstract, and to the illustration of its fundamental axioms and propositions; all which have been so repeatedly and so well laid down and elucidated in the various treatises we already possess, that it is hardly possible to place them in any very new and more than usually striking light. The distinction between mathematical and moral expectation belongs to this part of the subject, and can hardly be put more pointedly than it was originally done by Buffon, who first called attention to it.

"If two men were to determine to play for their whole property [supposed equal, and with equal risks], what would be the effect of the agreement? The one would only double his fortune, and the other would reduce his to nought. What proportion is there between the loss and the gain? The same that there is between all and nothing. The gain of the one is but a moderate sum; the loss of the other is numerically infinite, and morally so great that the labour of his whole life may not suffice to restore his property."

It was on such considerations that Daniel Bernouilli was led to propose, as a rule for estimating the value of a very small pecuniary or other material advantage, its *relative* value as compared with the total fortune of the party benefited, and for the moral as distinguished from the mathematical expectation of such advantage, that relative value multiplied by the probability of its accruing. On this or some equivalent mode of estimation is founded the principle of the subdivision of risks, which, rightly understood, so as to preserve their absolute independence while multiplying their number, is the best guarantee of commercial security. It is by such subdivision carried to an extreme point, that insurance and annuity offices thrive, and that benefit societies might do so, were it not for the single great risk which the dishonesty of entrusted agents throws in their way as a fearful stumbling-block.

In the case of savings' banks, this is, in fact, the only risk; and, as experience has too recently* and abundantly shown, a most imminent and fatal one. To annihilate this risk by a perpetual and searching superintendence, carried even to the utmost stretch of suspicious vigilance, obnoxious as it may appear, is the paramount duty of all who connect themselves with them as managers or trustees. Of the general benefit of such institutions, which, by guaranteeing the security of the produce of successful exertions, tend to cherish habits of industry, prudence, and frugality, no one can entertain a doubt. It is in this point of view that a certain considerable amount of national indebtedness, so far from meriting denunciation as an evil, ought to be regarded as an indispensable element and engine of civilization. In its practical working it resolves itself into the establishment of a savings' bank on a vast scale, administered with what may be considered a perfect exemption from the consequences of dishonesty in its officials, and subject only to the inconvenience (no doubt a considerable one), of its deposits being withdrawable only at a market value, — but that market the fairest, readiest, and openest which can anywhere exist. Yet it is too commonly forgotten by those who

* 1850.

deprecate taxation, while insisting on the objects for which taxation is instituted, and which alone it can secure, that the interest on savings' bank deposits is derivable only from that source, and that every depositor is as truly (and in some respects even more emphatically) a tax-holder—as the proprietor of consols.

To render the consequences of our actions certain and calculable as far as the conditions of humanity will allow, and narrow the domain of chance, as well in practice as in knowledge, is so thoroughly involved in the very conception of law and order as to make it a primary object in every attempt at the improvement of social arrangements. Extensive and unexpected fluctuation of every description, as it is opposed to the principle of divided and independent risks, so it also, by consequence, stands opposed to the most immediate objects of social institutions, and forms the element in which the violent and rapacious find their opportunities. Nothing, therefore, can be more contrary to sound legislative principle than to throw direct obstacles in the way of provident proceedings on the part of individuals (as, for instance, by the exorbitant taxation of insurances), or to encourage a spirit of general and reckless speculation, by riding unreservedly over established laws of property, for the avowed purpose of affording a clear area for the development of such a spirit on a scale of vast and simultaneous action. The sobering influence of an upper legislative assembly, refusing its sanction to the measures demanded, or spreading it over time, can alone repress or moderate these epidemic outbreaks of human cupidity: and its mission is abandoned, and its functions *pro tanto* abdicated, if it retreat from the performance of this duty.

The first and most important application of the calculus of Probabilities (since it applies to all departments of science, and affords a measure of the degree of precision attained in all numerical determinations) is that which relates to means and limits, and forms the second division of M. Quetelet's work. A general idea of the sort of questions contemplated in this department of the theory, and the kind of relations they involve, may be conveyed by the following simple case. Suppose a man to throw stones at random, and without any aim. From the marks left by any given number of them, however great, on a wall, we could obtain no impression, or a fallacious one, of his intention. All that we could conclude from their evidence would be, that, if he aimed at anything, it was not a point in the surface of the wall, and that only stray missiles had struck it. But, suppose he had been practising with a rifle at a wafer on the wall; which, being subsequently removed, we were required to indicate at once the situation it had occupied, and his skill as a marksman. It is obvious enough that, from the evidence of a great number of shot marks, both might be determined, at least with a certain degree of approximation, and with a probability of error less in proportion to their number. The theory of Probabilities affords a ready and precise rule, applicable not only to this, but to far more intricate cases: it is this: that the most probable determination of one or more invariable elements from observation is that in which the sum of the squares of the individual errors or aberrations from exactness which the observations imply, shall be the least possible. In the case before us the "errors" are the distances of the shot-marks from the point where the centre of the wafer was fixed; to ascertain which we have, therefore, to resolve the geometrical problem (a very elementary one)—"to find a point such that the sum of the squares of its distances from a certain number of given points shall be a minimum,"—a problem which is, in effect, identical with that of finding their centre of gravity. As to the skill of the marksman, it may be estimated in two different ways:—1st, by ascertaining what is the probability that he will place a single shot within a given distance: this may be done by counting the number of marks within that distance of the point ascertained as above, and dividing it by the total number: or, 2ndly, by ascertaining within what distance of the mark he would probably (*i. e.* more probably than the contrary, or with a probability exceeding one half) place it: this may be done by describing circles about the wafer's place (found as above) for a centre, and measuring the radius of that which just includes half the total number of marks. For it is obvious that, so far as the evidence before us goes, and judging only from the numbers of instances favourable or unfavourable, there is just as great a presumption that he will shoot within as without that circle; and, if it be ever so little enlarged, the scale will turn in his favour.

Suppose the rifle replaced by a telescope duly mounted; the wafer by a star on the concave surface of the heavens, always observed for a succession of days at the same sidereal time; the marks on the wall by the degrees, minutes, and seconds, read off on divided circles; and the marksman by an observer; and we have the case of all direct astronomical observation where the place of a heavenly body is the thing to be determined. Or we may substitute for the wall the floor of a lofty building or deep mine, and for the marksman an experimenter dropping, with all possible care, smooth and perfectly spherical leaden balls from a fixed point at the summit of the building or the mouth of the mine, with intent to determine, by the means of a great number of trials, the true point of incidence of a falling body,—a physical experiment of great interest. We might, if we pleased, instance more complicated cases, in which the elements to be determined are numerous and not *directly* given by observation, but with such we shall not trouble our readers: suffice it to say that the rule above stated, or, as it is technically called, the "Principle of Least Squares," furnishes, in all cases, a system of geometrical relations characteristic of the *most probable* values of the magnitudes sought, and which, duly handled, suffice for their numerical determination.

This important principle was first promulgated, rather as a convenient and impartial mode of procedure than as a demonstrable theorem, by Legendre. Its demonstration was first attempted by Gauss,—but his proof is in fact no proof at all, since it takes for granted that in the case of a single element, variously determined by *any finite number of observations however small*, the arithmetical mean is the most probable value,—a thing to be demonstrated, not assumed, not to mention other objections. Laplace has given a rigorous demonstration, resting on the comparison of equipossible combinations, infinite in number. His analysis is, however, exceedingly complicated, and, although presented more neatly by Poisson; and in this work, by M. Quetelet stripped of all superfluous difficulties, and reduced to the most simple and elementary form we have yet seen, yet must of necessity be incomprehensible to all whose knowledge of the higher analysis has not perfectly familiarized them with those delicate considerations involved in the transition from finite differences to ordinary differentials. Perhaps, therefore, our non-mathematical readers will pardon us if we devote a single page to what appears to us a simple, general, and perfectly elementary proof of the principle in question, requiring no further acquaintance with the transcendental analysis than suffices for understanding the nature of logarithms.

We set out from three postulates. 1st, that the probability of a compound event, or of the concurrence of two or more independent simple events, is the product of the probabilities of its constituents considered singly; 2dly, that there exists a relation or numerical law of connexion (at present unknown) between the amount of error committed in any numerical determination and the probability of committing it, such that the greater the error the less its probability, according to some regular LAW of progression, *which must necessarily be general and apply alike to all cases, since the causes of error are supposed alike unknown in all;* and it is on this ignorance, and not upon any peculiarity in cases, that the idea of probability in the abstract is founded; 3dly, that the errors are equally probable if equal in numerical amount, whether in excess, or in defect of, or in any way beside the truth. This latter postulate necessitates our assuming the function of probability to be what is called in mathematical language *an even function*, or a function of the square of the error, so as to be alike for positive and negative values; and the postulate itself is nothing more than the expression of our state of *complete* ignorance of the causes of error, and their mode of action. To determine the form of this function, we will consider a case in which the relations of space are concerned.

Suppose a ball dropped from a given height, with the intention that it shall fall on a given mark. Fall as it may, its deviation from the mark is *error*, and the probability of that error is the unknown function of its square, *i. e.* of the sum of the squares of its deviations in any two rectangular directions. Now, the probability of any deviation depending solely on its magnitude, and not on its direction, it follows that the probability of each of these rectangular deviations must be the same function of *its* square. And since the observed oblique deviation is equivalent to the two rectangular ones, supposed concurrent, and which are essentially independent of one another *, and is, therefore, a compound event of which they are the simple independent constituents, therefore its probability will be the product of their separate probabilities. Thus the form of our unknown function comes to be determined from this condition, viz., that the product of such functions of two independent elements is equal to the same function of their sum. But it is shown in every work on algebra that this property is the peculiar characteristic of, and belongs only to, the exponential or antilogarithmic function. This, then, is the function of the square of the error, which expresses the probability of committing that error. That probability decreases, therefore, in geometrical progression, as the square of the error increases in arithmetical. And hence it further follows, that the probability of successively committing any

* That is, *the increase or diminution in one of which may take place without increasing or diminishing the other.* On this, the whole force of the proof turns. (H. 1857.)

given system of errors on repetition of the trial, being, by postulate I., the product of their separate probabilities, must be expressed by the same exponential function of the sum of their squares however numerous, and is, therefore, a maximum when that sum is a minimum.

Probabilities become certainties when the number of trials is infinite, and approach to practical certainty when very numerous. Hence this remarkable conclusion, viz., that if an exceedingly large number of measures, weights, or other numerical determinations of any constant magnitude, be taken,—supposing no bias, or any cause of error acting preferably in one direction, to exist—not only will the number of small errors vastly exceed that of large ones*, but the results will be found to group themselves about the mean of the whole, always according to one invariable law of numbers (that just announced), and *that* the more precisely the greater the total number of determinations.

Such being the case, and the law of distribution of errors over the whole range of possible error being known, it becomes practicable to assign the relative numbers of cases in which the errors will fall respectively within and beyond any proposed limit on the average of an infinite number of trials, and thence to assign, *à priori*, the probability of committing in any single future trial — not a given specific amount of error, but an error *not*

* See note at the end of this Essay.

exceeding that limit, provided only the probable error of a single trial be known; which, as we have seen, can always be ascertained on the evidence of foregone experience, if very extensive. To illustrate this, we may recur to the case of a marksman aiming at a target. Suppose, that on counting the marks left by his practice, it has been found, on the result of a great number of (say 1000) trials, that half his shots had struck within 10 inches of the centre. About this point let circles be described, the first at 2 inches distance, and others at distances progressively greater by 2 inches at a time. Then it will be found, on counting the marks within the areas of these several circles, that their numbers, up to the tenth circle or to 20 inches distance, will run as follows: — viz., 107, 213, 314, 411, 500, 582, 655, 719, 775, 823. Within the 15th circle, or 30 inches, already 957 shots will be found to have struck; and within 40 inches, 993. Only one out of the whole thousand will be found beyond the 25th circle, or have erred so far as 50 inches from the point aimed at; and not one in 20,000 (were the practice prolonged so far) would stray beyond the 30th or err 60 inches. Computations of this sort are rendered exceeding easy by a table, originally calculated by Kramp, with a widely different object, which is given in the notes to M. Quetelet's book, and more *in extenso*, with differences, at the end of Mr. Galloway's work above noticed.

What is yet more remarkable is, that the skill with which the trials are performed is absolutely of no importance so far as the *law* of distribution of the errors over their total range is concerned. Were our marksman, for instance, only half as skilful, or to have 20 instead of 10 inches as the expression of his probable error, we have only to double the diameters of all the circles, and his shots will be found distributed among them according to the same succession of numbers. An important consequence follows from this: viz., that rude and unskilful measurements of any kind, if accumulated in very great numbers, are competent to afford precise mean results. The only conditions are the continual *animus mensurandi*, the absence of bias, the correctness of the scale with which the measures are compared, and the assurance that we have the *entire range of error* at least in one direction within the record.

In a matter so abstract, and on which, at first sight, human reason would appear to have so little hold, it is assuredly satisfactory to find the same conclusion, and *that* one so positive and definite, reached by different roads and from different starting points. It is not easy to imagine two principles of demonstration having less in common than that given above with that of Laplace, Poisson, and Quetelet. Yet the conclusions are identical, and the verifications afforded by experience in all cases where the trials have been sufficiently numerous, and care taken to guard against bias, have been of the most unequivocal character.

Some of these verifications, adduced by M. Quetelet as instances of the practical application of his rules of calculation in the theory of means and limits, have an interest independent of their value as such. They form part of a series of researches in which he has engaged extensively on the normal condition, physical and moral, of the human species, and, *inter alia*, as regards its physical development, in respect of stature, weight, strength, &c. By the assemblage of data collected from the experience of others, as well as his own, he has arrived at a variety of interesting conclusions as to the law of progressive increase and decay in all these respects, of the *typical* individual, of either sex, during the period of life, which are given at large in his work "Essai de Physique Sociale."* We shall offer no apology for placing one or two of these before our readers.

From the 13th volume of the "Edinburgh Medical Journal," M. Quetelet extracts a record of the measurement of the circumference of the chests of 5738 Scotch soldiers of different regiments. The measures are given in inches, and are grouped in order of magnitude, proceeding by differences of 1 inch, each group containing of course (we presume) all that differ by less than half an inch in excess or defect from its nominal

* Sur l'Homme et sur la Développement de ses Facultés ; ou Essai de Physique Sociale. Paris, Bachelier, 1835.

value. The extreme groups are those of 33 and 48 inches, and the respective numbers in the several groups stand arranged as in the table below.* Supposing each measure exactly performed, these, therefore, may be taken as the results of nature's own measurements of her own model; and the question whether she recognizes such a model ? is at once decided by inspection of the groups, in which the *animus mensurandi* is broadly apparent. It is equally so that such model would fall within the group of 40 inches. An exact calculation of the mean, allowing to each group a weight in proportion to the number it contains, assigns 39·830 inches as the circumference of the chest of this model.

Now this result, be it observed, is a *mean* as distinguished from an *average.* The distinction is one of much importance, and is very properly insisted on by M. Quetelet, who proposes to use the word mean only for the former, and to speak of the latter as the "arithmetical" mean. We prefer the term average, not only because both are truly arithmetical means, but because the term *average* carries already with it that vitiated and vulgar association which renders it less fit for exact and philosophical use. An average may exist of

* Inches -	33	34	35	36	37	38	39	40	41	42	43	44	45	46	47	48	Totals.
Groups as per Observations -	3	18	81	185	420	749	1073	1079	934	658	370	92	50	21	4	1	5738
M. Quetelet -	4	17	63	185	419	765	1054	1140	961	629	321	125	40	9	2	1	5738
Our calculation -	6	21	72	200	433	746	1024	1103	943	639	341	145	50	12	2	1	5738

the most different objects, as of the heights of houses in a town, or the sizes of books in a library. It may be convenient, to convey a general notion of the things averaged; but involves no conception of a natural and recognizable central magnitude, all differences from which ought to be regarded as deviations from a standard. The notion of a mean, on the other hand, does imply such a conception, standing distinguished from an average by this very feature, viz., the regular march of the groups, increasing to a maximum, and thence again diminishing. An average gives us no assurance that the future will be like the past. A mean may be reckoned on with the most implicit confidence. All the philosophical value of statistical results depends on a due apprehension of this distinction, and acceptance of its consequences.

The recognition of a *mean*, as thus distinguished from a mere average, among a series of results so grouped in order, depends on the observance of a conformity between the law of progression in the magnitude of the groups, and the abstract law of probability above stated, from which every consideration has been excluded, but the reality of *some* central truth, and an intention of arriving at it, liable to be baffled by none but purely casual causes of error. And the test to be applied, in this and all similar cases, is this. Is it possible to assign such a mean value, and such a probable error as shall alone, by the simple application of the table of probabilities, reproduce the numbers under the several groups in order, with no greater deviations than shall be fairly attributable to a want of observations numerous enough to bring out the truth? In the instance before us, the answer to this inquiry is contained in the results of calculation as compared with fact in the table above referred to. The mean we have used is 39·830 inches, and our probable error 1·381 inches. Those of M. Quetelet differ somewhat from these values, which accounts for the trifling discrepancy of the results.

The coincidence admits of being placed in even a more striking light. In the complete expression, by theory, of all the groups in a statement of this kind, three elements are involved — the mean value — the maximum group *having that mean for its centre* — and the probable error. And to determine these, it ought to suffice to have before us three terms of the series. Suppose then we take for our data the numbers corresponding to 35, 39, and 43 inches, viz., 81, 1073, and 370, given by observation. Then, by a computation of no great difficulty, there will result for the mean value, 39·834 inches, and for the probable error 1·413 inches, both agreeing almost precisely with those already stated. For the greatest possible group of an inch in amplitude the same calculation gives 1161, which is in obvious accord with observation. No doubt, then, can remain, as to the reality of a typical form, from which all deviations are to be regarded as irregularities. On this M. Quetelet observes,—

"I now ask if it would be exaggerating to make an even wager that a person little practised in measuring the human body would make a mistake of an inch in measuring a chest of more than 40 inches in circumference. Well! admitting this probable error, 5738 measurements made on the same individual, would certainly not group themselves with more regularity as to the order of magnitude than these 5738 measurements made on the Scotch soldiers; and if the two series were given us without their being particularly designated, we should be much embarrassed to state which series was taken from 5738 different soldiers, and which was obtained from one individual with less skill and ruder means of appreciation." (P. 92., *Transl.*)

This is assuredly an over-statement. So far from less skill being supposed in the measurements of the individual, the probable error of nature is nearly half as much more than that assumed here for the term of comparison (1 inch); and it is clearly beyond the bounds of any supposable negligence or rudeness of practice, to commit such errors as the extreme registered deviations (7 inches one way, and 9 the other), in a series of such measurements however multiplied, or even half those amounts.

We are thus led to the important and somewhat delicate question,—What we are to consider as reasonable limits, in such determinations—beyond which, if deviations from the central type be recorded, they are either to be referred to exaggeration, or regarded as monstrosities.

The answer to this question must evidently depend, first, on the "probable" deviation from the mean or typical value; secondly, on the number of cases experience has offered, or within which we agree to limit our range of speculation. We have already seen that 20,000 might be betted against 1, that an observed deviation, one way or other from the type, will not exceed sixfold its "probable" value; and therefore we shall have double that amount of chances against such a deviation in either direction separately. Among 40,000 individuals, therefore, we are entitled to expect to find one so far deviating from the mean type in excess, and one in defect. Beyond this the probabilities decrease with extreme rapidity. Thus, for a 7-fold deviation, we must seek our specimen among 263,000; and, for an 8, 9, 10-fold, among 4,760,000, 250,000,000, and 25,000,000,000 respectively.

We might apply these numbers to the case of giants and dwarfs, if we had any dependable data from which the mean human stature and its probable deviation could be ascertained. From an interesting discussion of the measurements of 100,000 French conscripts, taken at the age of 20 years, and arranged in groups, inch by inch, M. Quetelet concludes a mean height of 63·947 inches (English measure), with a probable deviation of 1·928 inches. The numbers in the respective groups (with certain exceptions at the lower limit) run in satisfactory accordance with the law of abstract probability, and afford complete evidence of the existence of a central type, uniform, or nearly so, for the French nation. Allowing (according to the tables given by M. Quetelet in his "Essai de Physique Sociale") 0·43 inches for the growth from the 20th year to adult stature, we may take 5 ft. 4½ in. (English) for the adult height of a typical Frenchman, with a probable deviation of almost exactly 2 in. Calculating on these data, we should expect to find in the existing population of France (taken at 12,000,000 adult males), one individual of 6 ft. 9 in. in height; in that of the whole world only one of 6 ft. 11 in.; and, in the whole records of the human race, not more than one of 7 ft. 1 in. The corresponding dwarfs would be respectively 4 ft. 1 in., 3 ft. 11 in., and 3 ft. 9 in.

The actual limits, both in excess and defect, we need hardly observe, are very much wider. M. Quetelet, on the authority of Birch, assigns 17 in. as the minimum of human stature authentically recorded. "The celebrated dwarf Bebé, king of Poland, was taller." The most celebrated dwarf of recent times, C. Stratton (*alias* Tom Thumb), exceeds this limit by 10 in.* Taking 17 for the

* Cardan saw a man in Italy of full age not above a cubit (21·9 in.) high. He was carried about in a parrot cage.—*Wern. Club. Transl. Pliny's Hist. of Nature,* ii. 200. note. Suetonius mentions a Roman knight exhibited by Augustus in the theatre "tantum ut ostenderet" (we quote from memory) "quod erat *bipedali* minor, librarum xxvii. et vocis immensæ."

minimum, and allowing an equal deviation in excess from the *conscript* type, our author fixes his gigantic limit hypothetically at 9 ft. 3 in. Even this we are disposed to extend. Disregarding such pigmies as the Swedish body guard of Frederick the Great (8 ft. 3 in.); Byrne, the celebrated "Irish giant" (8 ft. 4 in.) whose skeleton adorns the Hunterian collection; the Dutch giant of Schoonhaven (8 ft. 6 in.), attested by Diemerbroeck and Ray; and the Emperor Maximin — we have the testimony of Pliny to an Arab, named Gabbara (9 ft. 9 in.), "the tallest man that hath been seen in our age" (*Nat. Hist.* book vii. Wern. Cl. Transl. ii. 200.), and to the preservation, as curiosities, in a vault in the Sallustian Gardens, of the bodies of "two others, named Pusio and Secundilla," higher than Gabbara by half a foot (10 ft. 3 in.). The mummies might be counterfeited: the living Arab, exhibited by Claudius, would hardly escape some scrutiny. But, even in modern times, we have testimony, to which we cannot refuse at least the epithet of respectable, to the existence of giants who might well claim companionship with him of Basan, whose bedstead measured nine cubits "after the cubit of a man," or the Philistine, whose stature is expressly stated at six cubits and a span (11 ft. 5 in.). Thus, Dr. Thomas Molyneux, "an excellent scholar and physician," and a Fellow of the Royal Society, describes a well-formed human *os frontis* preserved in the School of Medicine at Leyden, from whose dimensions, carefully measured by himself, he concludes it to have belonged to an individual between 11 and 12 ft. in height: "a goodly stature," he remarks, "and such as may well deserve to be called gigantick." * Molyneux accompanies this description with notices of several other cases, which it may perhaps be worth while to recall attention to: as, for example, that of a skeleton seen and measured by Andreas Thevet, cosmographer to Henry III., king of France and Portugal, which belonged to a man 11 ft. 5 in. in height, who died in 1559.† And again, the cases of a man nearly, and a woman quite, 10 ft. in height, are attested by Beccanus, in his "Origines Antwerpianæ," 1569, as an eye-witness, the man living within ten miles of his own residence. We find mention by Dr. Degg (*Phil. Trans.* xxxv. p. 363.) of the exhumation, in 1686, at Repton, of a human skeleton 9 ft. long.

As regards men of seven feet in stature, so many

* Dr. Molyneux was brother to the celebrated astronomer of that name. (See *Phil. Trans.* vol. xv. p. 880. and vol. xxii. 487.) He gives engravings of this extraordinary bone, accompanied by one of a similar bone of ordinary size for comparison. Its dimensions are stated as follows. Coronal suture, measures along its course from orbit to orbit, 21 in. Breadth from back to front to junction of the bones of the nose, 9·1 in.; from side to side following the convexity of the skull, 12·2 in. Just as our last revise is going to press, we are informed, on the highest authority, that this *might* have been a case of hydrocephalus, though it *could not* have been known as such to Molyneux.

† The *head* was 37 in. in circumference. The leg bones measured fully 3 ft. 4 in.

cases are recorded that they can hardly be termed gigantic ; and, whatever we may think of such extreme cases as 11 or 12 feet, it seems impossible to hesitate at admitting 9 ft. 6 in. as a stature which may be exceeded, and perhaps even 10 ft. attained, without *monstrosity* in the proper sense of the word. We must, therefore, conclude that the "probable" deviation of nature's workmanship from her universal human type cannot possibly be less than the double of that resulting from the French measurements: a conclusion which ought to excite no surprise; since it is impossible to reason from a single nation (and that decidedly undersized, and of remarkable uniformity as to habits of life) to the whole species.*

Practically speaking, nothing can be simpler or more easily stated than the rules for handling any given series of determinations of a single *quæsitum* supposed to be arranged to our hands in regular progressive groups, with a view to derive from it numerically the only things which it is really im-

* M. Quetelet, as above stated, makes the mean French conscript of 20 years 63·947 in.; but he elsewhere (*Essai de Phys. Soc.* ii. 14.) states it at 1·615 met. or 63·583 in., which, with a growth of 0·433 in. to the adult age, gives only 5 ft. 4·02 in. for the typical French stature in 1817. The Belgian type (*Essai de Ph. Soc.* ii. 42.) is 5 ft. 7·8 in. That of the English non-manufacturing labourers in the neighbourhood of Manchester and Stockport, he states at 1·775 met. or 5 ft. 9·88 in. at the age of 18, which gives 5 ft. 10·75 in. for the adult type in Lancashire. The mean between extremes of 17 and 120 in. is 5 ft. 8·5 in., which is by no means improbable as a *general* standard.

portant to know, viz., the *most probable value*, the *probable error* of a single determination, and the *weight* of the result as compared with that similarly derived from a different and independent series. But when the data are otherwise grouped, which is a case by no means of unfrequent occurrence, or when a portion only is regularly arranged in groups, and all above or below certain limits massed together in the gross without regard to grouping, much delicacy subsists in deciding, according to just principles, on the exact amount of all these elements; and it would have added much to the practical utility and value of M. Quetelet's work had he given some examples of this nature, with plain and brief rules or formulæ for their working. This is the more to be regretted, because we are actually left at a loss to decide by what numerical process his mean results, where stated, have been arrived at in some of the examples set down. For instance, in that of the Scotch soldiers, where all the groups are regular and all stated, we find it merely mentioned incidentally that the mean is "a little more than 40 inches," whereas the really most probable mean is 39·830, while that which the course of the figures in the tabulated working of the example would appear to indicate as resulting from an equipartition of the numbers of cases in excess and defect is 39·525. Again, in the example of the conscripts, where the extreme groups are massed undistinguishably, the rule of equipartition, according to its simplest and most obvious application to the tabulated figures, would place the mean at 63·939 inches, whereas we find it indicated rather than stated, as follows: "*If it be observed* that the mean height is about 63·947 inches." The difference, it is true, is trifling in itself, but becomes of consequence when the object is from the figures set down to discover by what process they have been obtained.

We come now, however, to that highly interesting part of the work before us which treats of the study of causes; in general; and in the peculiarly complex form it assumes, in those moral and social inquiries, the data for which are gathered by statistical enumeration. A few remarks on the part which the theory of probabilities plays in these inquiries will not be out of place here.

This theory is connected with the general philosophy of causation and with inductive inquiry in two distinct ways—the one theoretical, and the other practical. When we see an event happen several times in succession in some particular manner, there arises, in the first place, a *primâ facie* probability that it will happen once more in that manner; which, if the number of repetitions be large, forms of itself a very cogent ground of expectation. But the probability that such repetition has not been merely fortuitous, but has resulted from a determining, or at least a biassing cause, increases with each repetition in a far higher ratio, than the simple probability of the once more happening of the event itself. The distinction is that between a geometrical and an arithmetical progression. Thus, for example, the expectation that the sun will rise to-morrow, grounded on the sole observation of the fact of its having risen a million times in unbroken succession, has a million to one in its favour. But to estimate the probability, drawn from that observation, of the existence of an influential cause for the phenomenon of a daily sunrise, we have to raise the number 2 to the millionth power — thus producing a number inexpressible in words and inconceivable in thought, and the ratio of this enormous number to unity, is that of the probability of the phenomenon having happened *by cause*, to that of its having happened *by chance*. The theorem on which depends this curious application of the doctrine of *probabilities* to the expulsion from philosophy of the idea of *chance*, is known to Geometers by the name of its first promulgator, Bayes. It must be observed, that as to the nature of the cause thus insisted on, the calculus says nothing. There may be opposing causes, and a daily struggle between them for the mastery. In this case we are simply forced to admit that the arrangements of Nature are highly favourable to the successful exertion of the one, and highly unfavourable to the other.

It is however as a practical auxiliary of the inductive philosophy that we have chiefly to contemplate this theory. Its use as such depends on that mutual destruction of accidental deviations from the regular results of permanent causes which always take place when very numerous instances are brought into comparison. Examples of this sort have been already adduced, and might be multiplied indefinitely in every department of practical inquiry. Indeed, every phenomenon which Nature offers on the great scale may be regarded as such. Nothing can be more irregular and uncertain than the action of the wind on the waters,—yet, in the most violent storms, the *general* surface of the ocean preserves its level. What more fortuitous than the fall of a drop of rain in a shower, or the growth of a blade of grass? Yet the soil is uniformly irrigated, and the unbroken sheet of verdure testifies to the resultant equilibrium of that and a thousand other causes of inequality. These things, it will perhaps be said, are the results of Providential arrangement. No doubt they are so; but it is an arrangement working through a complication of secondary causes and contingencies,—on which man, if he will philosophize at all, is obliged to do it by reference to the laws of probability. Still there is no one who is not astonished, in cases where what we are obliged to call contingency enters largely, to find not only that the mean results of several series of trials agree in a wonderfully exact manner with each other, but that the very errors of individual trials—precisely those portions of the special results which are purely attributable to that which is contingent in the process—group themselves around the mean with a regularity which would appear to be the effect of deliberate intention.

"This singular result" (says M. Quetelet) "always astonishes persons unfamiliar with this kind of research. How, in fact, can it be believed that errors and inaccuracies are committed with the same regularity as a series of events whose order is calculated in advance? There is something mysterious, which however ceases to surprise when we examine things more closely."

The rationale of this mystery is this. Where the number of accidental causes of deviation is great, and the maximum effect of each separately minute in comparison of the result we seek to determine,—great total deviations can only arise from the conspiring of many of these small causes in one direction,—the more that so conspire the greater the deviation. Now all combinations being equally possible *individually*, and those combinations which can alone give rise to the extremes of error being necessarily very much fewer in number than those which result in moderate amounts of deviation, we easily perceive that the opportunities for the occurrence of great errors are much rarer than for small ones. And this is in fact the reasoning, which, carried out by exact analysis (assimilating the causes of *plus* and *minus* error to black and white balls in an urn), takes the form of that demonstration of the law of probability, which we have above spoken of as devised by Laplace and simplified to the utmost by M. Quetelet.

There still remains behind, however, this inquiry,

—which we have known to occur as a difficulty to intellects of the first order,— *Why* do events, on the long run, conform to the laws of probability? What is the *cause* of this phenomenon as a matter of fact? We reply (and the reply is no mere verbal subtlety), that events do not so conform themselves,— the fact to the imagination,— the real to the ideal,—but that the laws of probability, as acknowleged by us, are framed in hypothetical accordance with events. To take the simplest case, that of a single contingency,—the drawing of one of two balls, a black and a white. We suppose the chances equal, in theory; but, in practice, what is to assure us that they are so? The perfect similarity of the balls? But they need not be similar in any one quality but such as may influence their coming to hand. And, on the other hand, the most perfect similarity in all visible tangible, or other physical qualities cognizable to our tests is not such a similarity as we contemplate in theory, if there remain inherent in them, but undiscernible by us, any such difference as shall tend to bring one more readily to hand than the other. The ultimate test, then, of their similarity in that sense is not their general resemblance, but their verification of the rule of coming equally often to hand in an immense number of trials: and the observed fact, that events *do* happen according to their calculated chances, only shows that *apparent* similarities are very often *real* ones.

The application of this calculus to the detection

of causes turns essentially upon this view of the conformity in question, and of the nature and delicacy of this *test by indefinite multiplication of trials* which we are enabled, in many cases, to apply to mixed phenomena. All experience tells us, that where efficient *causes* are known, but from the complication of circumstances cannot be followed out into their specific results, we may yet often discern plainly enough their *tendencies*, and that these tendencies *do* result, in the long run, in producing a preponderance of events in their favour. Were it asked, Why do the strong men, in a general scramble, carry off the spoil, and the weak get nothing? the reply would be, that such is not the fact in every instance; that, although we cannot go fully into the dynamics of the matter, we can clearly see the mode of action in some individual struggles, and that in the whole affair there is a visible enough *tendency* to the defeat of the weaker party. Again, when we reverse this process of reasoning, and declare our conviction that success in the long run is a proof of ability, we give this name to some personal quality or assemblage of qualities which, acting as an efficient cause through a complication of events we do not pretend to penetrate, has a tendency in that direction which issues in success. Here the tendency becomes known by observation, and the nature of the cause is concluded from the nature of the tendency, by appeal to experience, which, in some instances, has shown us the cause in action, and informed us of its direct effect. But it may happen that observation may plainly enough indicate the direction of a tendency which yet experience has not enabled us to connect with any known cause. And it may further happen that this tendency, which we are driven to substitute in our language for its efficient cause, may be so feeble—whether owing to the feebleness of the unknown cause, its counteraction by others, or the few and disadvantageous opportunities afforded for its efficacious action (general words, framed to convey the indistinctness of our view of the matter)—as not to become known to us but by long and careful observation, and by noting a preponderance of results in one direction rather than another.

And thus we are led to perceive the true, and, we may add, the only office of this theory in the research of causes. Properly speaking, it discloses, not causes, but tendencies, working through opportunities,—which it is the business of an ulterior philosophy to connect with efficient or formal causes; and having disclosed them, it enables us to pronounce with decision, on the evidence of the numbers adduced, respecting the reliance to be placed on such indications,—the degree of assurance they afford us that we have come upon the traces of some deeply-seated cause,—and the precision with which the intensity of the tendency itself may be appretiated.

Such tendencies are often apparent enough, without any refined considerations, or reference to any calculus. Thus, on the consideration of thirteen instances of coincidence between the direction of circular polarization in rock crystal, with that of certain oblique faces in its crystalline form,—it was asserted that the phenomena were connected in that invariable manner which is one of the characters of efficient causation. The chances against such a coincidence happening thirteen times in succession by mere accident are more than 8000 to 1; and this, therefore, was the probability that some law of nature, some cause, was concerned. Subsequent observation has brought forward no exception; but, on the contrary, other cases of a similar character have arisen, which go to place the observed tendency in *uncounteracted* connexion with the efficient cause —which, however, still remains concealed.*

It is, however, the extreme delicacy of the test above spoken of—that property it possesses of bringing out into salience and placing in indisputable evidence, by sufficient multiplication of

* So again, an examination of the elements of all known cometary orbits has disclosed a tendency to direct or eastward motion, increasing in the degree of its prominence with the approach to coincidence of the orbit with the plane of the ecliptic,—and especially marked in the cases where calculation has assigned elliptic elements to the orbit Here we have a tendency pointing to a cause, still unknown, but with whose effects we are so far familiar that we can trace its action throughout the planetary system, with only two known exceptions among its most remote and insignificant constituents, and those of a very undecided character.

observations, any preponderance, however small, among the efficient causes in action — that it becomes applicable to those complicated cases in which we find it resorted to. As an instance of this nature, we shall take a phenomenon which has engaged the attention of all who have written on probabilities, from Laplace downwards; one which has been much insisted on by M. Quetelet, and on whose acknowleged obscurity his inquiries have at length thrown a ray of light; viz., the excess of the number of births of male over that of female infants. As a matter of observation, the phenomenon is indisputable; but it requires the assemblage of a great number of instances to bring it out into evidence. In individual experience, or in the birth registers of a parish or small town, the tendency to excess on the male side is quite overlaid and concealed by accidental irregularities. It is otherwise when those of great cities or whole nations are consulted. The irregularities then disappear by mutual destruction, and the result exhibits the tendency in question in its full prominence. If we extract from the population returns of England and Wales the total numbers of registered births in the seven years, from 1839 to 1845 inclusive, we find 1,863,892 males and 1,772,491 females, the excess being 91,401 on the male side, or 105·157 males to 100 females. Suppose it were urged that this may, after all, be a purely accidental excess. It might be said, not without apparent plausibility, that as it would be the height of improbability to expect in so vast a number an exact equality, so, on the other hand, an excess of 91,401, which, though a large number in itself, is yet but $2\frac{1}{2}$ per cent. on the total number of cases, *does not seem so very improbable.* To this theory replies that, where such high numbers are concerned, it *is* so:—that the case assumed in the objection is identical with that of drawing 3,636,383 balls out of an urn containing black and white balls in equal proportion and infinite in number, and that the expectation of drawing such an excess of one colour in such a number, so far from a mere moderate unlikelihood, is, in fact, equivalent, supposing the chances equal, to the expectation of throwing an ace 643 times successively, with a single fair die.* Even on a total of 20,000 births we might bet many thousand millions to one that the same relative preponderance would not be found, were the chances even.

It is abundantly evident, therefore, that we have here arrived at a proof of a tendency which must be taken as a law of human nature under the circumstances in which it exists, at least in this country; and the constancy with which the proportion is maintained in successive years, and even in different nations, is not less striking than the fact itself, and shows it to be a result of deep-seated causes, acting with almost absolute uniformity on great masses of mankind. Thus in the seven

* The chances against throwing an ace only nine times in succession, are ten millions to one.

years from which the above ratio has been concluded, taking them *seriatim*, we find 104·8, 104·7, 105·3, 105·2, 105·4, 105·4, 105·2, on totals averaging about half a million each; while in France a similar comparison gives 105·9, 105·7, 106·1, 106·2, 105·8, 105·9, 105·9, on nearly double the total numbers. As to the causes of this most striking phenomenon, much speculation has, of course, prevailed; but the inquiries of M. Quetelet into the statistics of marriage have rendered it extremely probable* that the relative ages of the parents very materially influence the sex of the offspring, and that the effect is therefore a resultant one, due to this physiological cause, acting through the medium of all those prudential and moral considerations which in civilized states determine the relative ages of parties contracting marriage. This view of the subject is strongly corroborated by a separate examination of the registers of illegitimate birth, which indicate an excess of only 3 instead of 5 per cent.

The causes, or tendencies indicative of causes, which may be disclosed by the assemblage and comparison of numerous recorded instances, are classed by M. Quetelet under three heads: constant, variable, and accidental. The latter class may be considered as entirely eliminated by their mutual destruction when vast numbers are concerned, and the whole series of collected cases is so treated as

* Essai de Phys. Sociale, i. 57. Citing Hofacker and Sadler in corroboration.

to afford a single result. The same process also will in great measure destroy the effect of variable causes, if their variation be periodical in its law, and the observations be made indifferently in all the phases of their period. It is the peculiar property, however, of causes of this latter description, through whatever train of circumstances their action is propagated, ultimately to emerge to view in manifestations equally periodical with the causes themselves. In cases of dynamical action this peculiarity is susceptible of demonstration, and has been so demonstrated under the name of the "principle of forced vibrations;"* and experience abundantly proves its general applicability to every case of indirect action, whether physical or moral. To those, therefore, who assiduously watch the development of phenomena, and register effects as they arise with sufficient exactness, such causes will be detected, and their periods at the same time disclosed by the periodical fluctuations they occasion; or they may be searched for, if suspected to exist overlaid by accidental errors, by dividing the series of observed results into groups, differing in *phase* (*i. e.*, dividing the extent of the period suspected into several equal portions, and grouping the results observed in each together). The influence of the periodical cause suspected will then become apparent in the form of differences in the mean results of the several groups. Of this

* Encyclop. Metropol. Article Sound, § 323. *et seq.*

process every part of science teems with examples. In astronomy we owe to it the grand discoveries of aberration of light, the nutation of the earth's axis, the separation of the effects of the sun and moon on the tides, and an infinity of others; in meteorology, that of the diurnal and annual fluctuations of the barometer; in magnetism, the daily and annual changes in the direction and intensity of the magnetic forces; and in statistics, the annual oscillations observable in all the great elements of population, which the researches of M. Quetelet have placed in a distinct light.

But among accumulated masses of results, without any attempt at subdivision into *periodic* groups, the influence of periodical causes may start into evidence on a general inspection of the differences from a mean result, after a totally different manner. We have seen that these differences present *inter se* a definite and perfectly cognizable law of arrangement, so long as their causes are purely casual. Any deviation *from this law* among the differences of the observed values from the mean, then, becomes at once an indication of a determining tendency, and will very often, by the character of the deviation, lead to a well-grounded surmise of the nature of its cause. For instance, if a sudden falling off in the number of observed differences, beyond certain limits either way from the mean, *accompanied with some degree of improbable accumulation at or about those limits*, should be noticed, it may be taken as a certain indication of a periodical disturbing influence, having those limits for the maximum and minimum of its effect.

Again, if at any particular point in the scale of results arranged in order of magnitude we should notice a sudden and marked irregularity confined to a small extent, we may be sure that it arises from the action of some single, powerful, and exceptional influence. Thus, from the undue accumulation of conscript measurements below the standard height of 5 feet 2 inches, accompanied with a deficiency to the extent of 2275 cases in the two inches just above that standard, M. Quetelet is led to conclude that an influence foreign to the subject—in fact, a fraudulent practice, favouring the escape of the shorter men, has prevailed to that extent in the formation of the official returns he has employed as the basis of his calculations. (P. 98. *Transl.*)

Astronomy affords us a very remarkable example of this nature, which we adduce, by reason of a singular misconception of the true incidence of the argument from probability which has prevailed in a quarter where we should least have expected to meet it. The scattering of the stars over the heavens, does it offer any indication of law? In particular, in the apparent proximity of the stars called "double," do we recognize the influence of any *tendency to proximity*, pointing to a cause exceptional to the abstract law of probability resulting from equality of chances *as respects the area occupied by each star?* To place this question in a clear light, let us suppose that, neglecting stars below the seventh magnitude, we have measured the distance of each from its nearest neighbour, and calculated the squares of the sines of half these distances, which therefore stand to each other in the relative proportion of the areas occupied exclusively by each star. Suppose we fix upon a circular space of 4″ in radius as the unit of superficial area, and that we arrange all the results so obtained in groups, progressively increasing from 0 by the constant difference of one such unit. Now the fact, to which M. Struve originally called attention *, and on which we believe all astronomers are agreed, is, that the first of these groups *is out of all proportion richer than any of the others;* and that the numbers degrade in the groups adjacent with excessive rapidity; so that, for example, calculating on the numbers given by Struve †, we find the first group to contain 182 cases; the next three 68, or on an average 22 each; the next twelve 70, or 6 each on an average; and the next forty-eight only 94 in all, averaging 2 to each; while a general average ‡ would assign

* Catalogus Novus Stellarum duplicium, &c. Dorpati, 1827.

† *Ibid.*, p. xxxii., Introduction. Each of M. Struve's classes is doubled, since each constituent of a double star counts as a separate case.

‡ Taking 12,400 as the number of stars of the magnitudes and within the region of the heavens contemplated, viz. from the North Pole to 15° south declination, which number, for the reason in the foregoing note, has to be doubled.

only one star to 540,000 such units of area. The case, then, is parallel to that of a target of vast size, marked out into 6700 millions of equi-distant rings, riddled with shot marks in the bull's eye, and with a tolerable sprinkling in the first fifty or sixty rings, beyond which the whole area offers nothing for remark indicative of any particular local tendency, though *dotted all over with marks*, in the sparing manner above described. Any one who should view such a target, bearing in mind what is said above, must feel convinced that a totally different system of aiming had been followed in planting the interior and exterior balls.

Such we conceive to be the nature of the argument for a physical connexion between the individuals of a double star prior to the direct observation of their orbital motion round each other. To us it appears conclusive; and if objected to on the ground that every attempt to assign a numerical value to the antecedent probability of any given arrangement or grouping of fortuitously scattered bodies must be doubtful *, we reply, that if this be admitted as argument, there remains no possibility of applying the theory of probabilities to any registered facts whatever. We set out with a certain hypothesis as to the chances: granting which, we calculate the probability, not of one certain definite arrangement, which is of no importance whatever, but of certain *ratios* being

* London, Ed. and Dub. Philosoph. Magazine, &c. Aug., 1849.

found to subsist between the cases in certain predicaments, on an average of great numbers. Interrogating nature, we find these ratios contradicted by appeal to her facts; and we pronounce accordingly on the hypothesis. It may, perhaps, be urged that the scattering of the stars is *un fait accompli*, and that their actual distribution being just as possible as any other, can have no *à priori* improbability. In reply to this, we point to our target, and ask whether the same reasoning do not apply equally to that case? When we reason on the result of a trial which, in the nature of things, cannot be repeated, we must agree to place ourselves, in idea, at an epoch antecedent to it. On the inspection of a given state of numbers, we are called on to hold up our hands on the affirmative or negative side of the question, Bias or No bias? In this case who can hesitate?

Accidentally variable causes overlay altogether the evidence of regular action, so that the elimination of their influence is in all cases synonymous with the extension of knowledge. It is not, however, to this or to any other calculus that we can look for special rules of conduct in this part of inductive inquiry beyond the simple precept of collecting facts in great numbers, and employing mean results in lieu and to the exclusion of single observations wherever numerical magnitude is concerned. This precept is, however, of infinite use in all cases where we test the efficacy of a presumed cause by the numerical correspondence between its known energy and the amount of the observed effect. All Nature is full of such cases. That selected by M. Quetelet as an example is one of much agricultural and botanical interest, viz., the inquiry into those peculiarities of season on which its character as a forward or a backward one depends. The rudest observation suggests the prevalent *temperature* of the season as the element on which the difference in question mainly turns, though it may justly be inquired whether other meteorological elements, especially moisture, may not come in for their share in producing it; and should these prove to be but little influential, according to what laws, as regards the *distribution* of temperature over the period of vegetable activity, the arrival of a plant at any phase of its annual life is accelerated or retarded. This inquiry is not new. Reaumur, and after him Boussingault and the Abbé Cotte, taking the simplest possible view of the subject, maintained that the arrival of a plant at a definite stage of its growth is solely dependent on the *total amount* of temperature to which it has been subjected from the first movement of the sap in spring *, without regard to its distribution over the intervening time, or the extent of its variations. Such a law is unlikely in itself, and the experience of every one would lead him to doubt its universal applicability. It has, however, been adopted by M. Gasparin in a work

* Cotte assumed arbitrarily the 1st of April.

("Cours d'Agriculture") which has commanded considerable attention, an account of which, and of the arguments which may be adduced to show the inadequacy of this hypothesis, will be found in a paper by the Earl of Lovelace in the "Journal of the Royal Agricultural Society," vol. ix. part 2. M. Quetelet, who has independently arrived at a similar conclusion, proposes to substitute for the *total temperature* (estimated by the *sum of the daily mean temperatures*) the *sum of the squares of* such daily means, reckoned from the freezing point; assigning as a reason, that "the force exercised by the temperature is of the same nature as actual force. It is by the sum of the squares of the degrees, not by the simple sum of the degrees, that we must appretiate its action." Such an analogy is not calculated to produce much conviction; but there is good reason to presume that vegetation *is* accelerated in a higher ratio than that of the simple temperature, from the consideration, not only of the continual increase of dilatability by equal increments of heat which aqueous liquids undergo, but also from their much greater fluidity at high than at low temperatures, the one cause rendering circulation more free, the other producing a more rapid dilatation of the cellular tissue by the direct action of warmth. Pending the discovery of the true law of connexion between the phenomena (which cannot be that of the squares, if only for the simple reason that it would give equal efficacy to temperatures *below* and *above* the freezing point), M. Quetelet's as a provisional one has this advantage, that it affords scope for the influence of differences in the *distribution* of temperature, which that of Reaumur does not, and gives a better account of the rapid burst of vegetation which a few genial days produce in spring.

M. Quetelet has selected for observation the epoch of flowering, as more definitely observable than any other phase of vegetation; and as there are few things more agreeable to a country resident than watching and noting the commencement of flowering in the early spring flowers which adorn our gardens, fields, and hedgerows, this branch of botanical inquiry promises to become quite as popular as it is interesting in itself.

We can only afford room for his result as regards the common lilac. That beautiful ornament of our walks and shrubberies blossoms so soon as the sum of the squares of the *mean daily temperatures* (as indicated by the centigrade thermometer) amounts to 4264°, so that the mean time of its flowering at any given station may be at once determined from the meteorological records of its climate. At Brussels this mean date is the 27th or 28th of April. In other localities it occurs earlier or later by about three or four days for every degree of latitude south or north of Brussels, and about five or even six days later for every hundred yards of elevation above the level of that city, which is itself sixty-five yards above the sea:—

"To each plant" (thus he states his general conclusion) "is attached a constant, the square of a certain number of degrees of warmth necessary for the occurrence of 'inflorescence.' Whether a plant is found in such and such a latitude, at such and such a height, in the open air or in a greenhouse, it is the temperature" (so measured) "that must be considered. Thus are explained all the anomalies that present themselves in this kind of research. Geographical causes have no influence but by the variations they cause in temperature."—(*Transl.* p. 172.)

Among those branches of knowledge which are most effectually advanced by the consideration of mean or average results concluded from great masses of registered facts, to the exclusion of individual instances, statistics hold beyond all question the most important rank as regards the social well-being of man. To this subject M. Quetelet devotes the fourth and last division of his work; not, indeed, to the delivery of statistical tables or results, nor to the actual discussion of any particular class of documents, but to the points which it so much imports to have generally well understood of the methods and principles which ought to prevail in the collection and subsequent employment of such documents.

Whether statistics be an art or a science (a question to which he devotes a preliminary letter) or a scientific art, we concern ourselves little. Define it as we may, it is the basis of social and political dynamics, and affords the only secure ground on which the truth or falsehood of the theories and hypotheses of that complicated science can be brought to the test. It is not unadvisedly that we use the term Dynamics as applied to the mechanism and movements of the social body; nor is it by any loose metaphor or strained analogy that much of the language of mechanical philosophy finds a parallel meaning in the discussion of such subjects. Both involve the consideration of momentary changes proportional to acting powers,—of corresponding momentary displacements of the incidence of power,—of impulse given and propagated onward,—of resistance overcome,—and of mutual reaction. Both involve the consideration of time as an essential element or independent variable; not simply delaying the final attainment of a state of equilibrium and repose,—the final adjustment of interests and relations,—but, from instant to instant, pending the process of mutual accommodation, altering those relations, and, in effect, rendering any such final state unattainable. One great source of error and mistake in political economy consists in persisting to regard its problems as statical rather than dynamical in their character; confounding the propagation of impulse with a step towards equilibrium,—a state unattainable where the interests of masses of mankind are concerned. So long, indeed, as society is little developed, its movements fettered, its commercial activity sluggish, and all things go on leisurely, the distinction is one of small importance; a state of *acquiescence*, nearly approaching to that of equilibrium and final adjustment, being taken up from instant to instant, and following at a little distance, yet *pari passu*, the slow changes of the acting causes. It is otherwise under the increased facilities, excessive mobility, and excited energy which prevail under the high temperature and pressure of modern civilization. Friction (which has an equally real existence in both mechanisms) is diminished, the intensity of the active powers increased, the scale on which movements are carried on enlarged,—a state of things which finds its expression in the "over-speculation," "gluts," "panics," "reactions," *et hoc genus omne* of modern commerce and social change. The same must be the case whenever efficient causes, of whatever nature, act through a train of varying circumstances, and result in effects of which it can only be securely asserted that their momentary and infinitesimal changes stand under given circumstances in given relations. It may be true, for example, that capital tends to a common level of profit in the choice among its possible employments; but endless fallacies would be involved in any reasoning which should proceed on the assumption that it finds that level. Demand may tend to increase supply by stimulating exertion, but a supply proportionate to the demand, and steadily following its variations, is what no sound political economist will ever expect to see. The Rule of Three has ceased to be the sheet anchor of the political arithmetician, nor is a problem resolved by making arbitrary and purely gratuitous assumptions to facilitate its reduction under the domain of that time-honoured canon.

Number, weight, and measure are the foundations of all exact science; neither can any branch of human knowledge be held advanced beyond its infancy which does not, in some way or other, frame its theories or correct its practice by reference to these elements. What astronomical records or meteorological registers are to a rational explanation of the movements of the planets or of the atmosphere, statistical returns are to social and political philosophy. They assign, at determinate intervals, the numerical values of the variables which form the subject-matter of its reasonings, or at least of such "functions" of them as are accessible to direct observation; which it is the business of sound theory so to analyse or to combine as to educe from them those deeper-seated elements which enter into the expression of general laws. We are far enough at present from the actual attainment of any such knowledge, but there are several encouraging circumstances which forbid us to despair of attaining it.

The first of these is the exceeding regularity which is found to prevail in the annual march of statistical returns and the constancy of the ratios they indicate where great masses of population are concerned, where leading features of human nature are the obviously influential elements on which the observed results depend, and where temporary or periodical causes of disturbance (evidently such) do not visibly interfere. As instances might be cited the relative proportion in the births of the sexes already spoken of; the ratio of illegitimate to legitimate births in the same country and the same section of the population; nay, even the number of the still-born (with a distinct percentage for town and country), which M. Quetelet has ascertained to be so uniform in Belgium that, on a total number of nearly 6000 annual cases, the yearly deviation from the mean falls short of 140; the ratio of marriages to the whole population, of second marriages to the whole number of annual marriages, and, still more minutely, of widowers with widows, widows with bachelors, and widowers with spinsters; the relative ages of parties intermarrying; and innumerable other particulars; all which, free as air in individual cases, seem to be regulated with a precision, where masses are concerned, clearly proving the existence of relations among the acting causes so determinate, that there is evidently nothing but the intricacy of their mode of action to prevent their being subjected to exact calculation, and tested by appeal to fact. *Taken in the mass*, and in reference both to the physical and moral laws of his existence, the boasted freedom of man disappears; and hardly an action of his life can be named which usages, conventions, and the stern necessities of his being, do not appear to enjoin on him as inevitable, rather than to leave to the free determination of his choice.

Another encouraging feature in the aspect of

statistical documents, which shows them, when properly collected, to be trustworthy for the purposes to which we desire to apply them, and holds out a rational hope of their available application,—is their evident *sensitiveness* to the influence of real and unmistakable causes, which we are sure, *à priori*, ought to influence them. Thus we see the uniform march in the number of annual marriages, corresponding to an increasing population, visibly accelerated in years of prosperity and abundance, and visibly retarded in those of scarcity and public distress. Thus, too, we see in Bavaria laws restraining marriage result in an increased number of illegitimate births.* Wherever monthly returns, of whatever kind, are compared, the influence of season is marked by a more or less conspicuous annual maximum and minimum. Instances of this, of the most striking character, are adduced by our author in his "Essai de Physique Sociale." In these and similar cases, where we clearly perceive the existence of definite tendencies, or of a generally modifying cause pervading the whole field of their action, it is satisfactory and reassuring to find the result in correspondence with our views. For it must never be forgotten that tendencies only, not causes, emerge as the

* The vast multitude of illegitimate births in France would seem to be traceable in great measure to the difficulties thrown in the way of marriage by requiring the expressed consent of a great number of relatives of both parties to its celebration.

first product of statistical inquiry,—and this consideration, moreover, ought to make us extremely reserved in applying to any of the crude results of such inquiries the axioms or the language of direct unimpeded causation. The proportionality of cause to effect, for instance, is a principle rather emphatically repudiated in the history of the correspondence of increase of imposts with increase of revenue, and of profits as compared with prices.

"Population," says M. Quetelet, "is the statistical element, *par excellence:* it necessarily rules all others, since it relates, above all, to the people and the appretiation of their welfare and their wants. It would be vain to attempt to form statistics of value without taking as a basis the results of a census executed with all the care and precision which so delicate an operation requires. The other data have no real value, except in so far as they relate to the number of the population. A census carefully made sums, in a measure, the most important problems which can be proposed to a statist. The classification according to age allows of the establishment of tables of population, of forming correct ideas on mortality, on the forces at the disposal of the state in case of necessity, and of fixing the ratio between the useful fraction which contributes to the general well-being, and the fraction which yet requires assistance and support to become in its turn useful. The classification by professions, indicates the means by which the population provides for its subsistence and tends to augment its prosperity. . . . Those by civil condition, by origin, by education, furnish the administration with no less precious information to assure internal good order, and to facilitate the execution of the laws."—(*Transl.* p. 183.)

A well-organised system of civil registration ("*état civil*,") is therefore one of the first wants of an enlightened people. No man in such a people is above or beneath the obligation of authenticating his existence, his claims on the protection of his country, and his fulfilment of the duties of a citizen,—or of contributing his individual quota of information, in what personally concerns himself or his family, in reply to any system of queries which the Government in its wisdom may see fit to institute respecting them. Such information may be regarded as a poll-tax, which, in this form, a Government is fairly entitled to make, and which indeed is at once the justest and least onerous of taxes; or rather, it may be looked on as a mode of self-representation, by which each individual takes a part in directing the views of the legislature in objects of universal concern. Nothing, therefore, can be more unreasonable than to exclaim against it, or to endeavour to thwart the views of Government in establishing such a system,—nor anything more just than to guarantee its fidelity by penalties imposed on false returns or wilful omissions.

The analysis of the population returns of a great nation, or rather the drawing from that analysis, duly executed according to rational classifications, just and philosophical conclusions, is a task calling for the exercise of much acuteness and discrimination in appretiating the influence which the relative proportions between the classes, as to age, condition, calling, must necessarily have on national character and habits, and in weighing—with reference to future prospects—the probable influence on that character and those habits which is involved in even a very moderate observed change from time to time, in those proportions.

"The numerical tables of a population, when made with care and with all the development which science requires . . . form, in the annals of a people, the most eloquent page that a statesman can read, if he understand them well. In fact it only belongs to the practical observer completely to understand the language of figures, and not to go beyond what they can teach him. Censuses, well made, and which succeed one another on a uniform plan and at intervals sufficiently near, should present most precise notions of the physical and moral condition of a people,—of the degree of its power,—of its prosperity,—and of the tendencies which may compromise its future: they would teach much better than voluminous inquiries, which are often fettered by prejudices and private interests, what we ought to think of the retrograde state or the immoderate development of certain branches of industry."

Among the first results of such an analysis, are those general ones which our Continental neighbours technically understand by the "movement" of the population—its increase, that is to say, by the excess of births over deaths and emigrations, and the internal change in the proportions of those living at different ages corresponding to changes, if any, in the law of mortality as indicated by the ages of death. On this point M. Quetelet, in an earlier part of this work, has the following pertinent remark. : —

"The movement of a stationary population is often compared with that of a population increasing by an excess of births over deaths. However, this is a comparison of heterogeneous elements: all other things being equal, the latter population should have a greater mortality; for there are more children in it."

So far as this remark goes it is just, but it does not include the whole case, or exhibit fully the influence of the consideration in question. To judge of the extent of this influence it is only necessary to consider that, in a given population now existing, the individuals living at any assigned age are not the survivors of that age among a number equal to that born in the current year, but among a number born antecedently, when the population was less than at present, in a proportion easily calculated, the age being given, and the annual rate of increase known. Thus, supposing the population of a country to double in fifty years, a man fifty years old is the survivor of only half the number of cotemporary births, and of one hundred of only one-fourth those which would appear, on a comparison of the number actually born in a given year with those actually living at the age specified, in that year. Not only, therefore, are there more children in comparison with adults in an advancing population, but at the same time fewer old men. Now the ratios of the helpless, the active, and the meditative elements of a population to the entire mass and to each other, —of giddy youth and adult enterprise to mature experience, timid caution, and declining powers, must necessarily give rise to corresponding features of national character. A disproportion in this respect, influencing all the great lines of development of national activity and impressing the whole career of a people, cannot but make itself felt in every feature of their existence. It is only necessary to contrast the energy displayed by a nation whose population doubles in twenty-five years, as in the United States, with the sobriety of movement, not to say torpor, of another, where, as in Holland, it is nearly stationary, to perceive the connexion in question to be that of effect with cause.

"An exposition of the political condition belongs essentially to the statistics of a country. We do not, however, know how to express it in figures. The same may be said of information relative to the moral and intellectual condition. The simple recital of what has passed in a locality at a particular time sometimes better teaches the moral condition of a people than all the numerical tables possible."

Statistics, however, deals essentially with numbers. It may be difficult, or impossible to express numerically the degree of political freedom, the extent to which the institutions of a country fulfil the ends of their establishment and maintenance,— or the degree in which its fiscal regulations press upon its inhabitants,—yet these are nevertheless results capable of being estimated, and which it is of the last importance to estimate; and the estimation must ultimately rely, to a considerable extent, on the numerical exhibition of particulars. Thus, to say nothing of the statistics of elections in which numbers are easily and precisely attainable, or of those of crime, accurate returns may and ought to be obtained and published of a great variety of particulars relative to the administration of justice in our *civil* courts, by which our judgment as to their well or ill working may be influenced. As examples, we may specify the statistics of juries, common and special,— those of legal decisions in civil cases, more especially as regards the cases of new trials moved for and obtained, and their grounds; — of decisions appealed from to higher tribunals, and of the proportion of cases in which such new trials or such appeals have affirmed or reversed the former decision,— points of great interest as concerns the confidence with which the decision of a civil court may be relied on by its suitors, but of which, if any official returns exist in this country, we have been unable, after some considerable amount of inquiry, to procure them. This is to be regretted, because the application of the theory of probabilities to judicial decisions with this very view (that of determining, from the amount of self-contradiction existing among them, their value as tests of truth) has been expanded by Laplace and Poisson into a very elaborate theory, which the latter especially has applied to the statistical returns of the French tribunals, civil as well as criminal. It may be worth while here to mention the conclusions deduced by the last-mentioned geometer from the consideration of 17,157 cases adjudicated on in French courts of civil appeal, during the years 1831, 1832, 1833. Of these the judgment of the inferior tribunal was confirmed in 11,747 cases, or in 685 cases out of 1000,—a percentage certainly not calculated to inspire a high degree of *primâ facie* confidence in the efficacy of a resort to a court of justice for the redress of a civil injury, though it must be admitted that appeals would chiefly take place in cases where the original decision was obviously contrary to common sense at least, if not to law. Setting out with these data, and taking into consideration the peculiar circumstances of the French institutions, in which three judges are required to pronounce, by a majority of voices, a "*jugement de première instance*," and seven in a court of appeal, M. Poisson concludes the probability that a confirmatory decision will be a just one, to be 0·948, or about 19 to 1, in its favour, and 0·641, or about 16 to 9, that a reversal of the former decision will be so. With respect to the probability that a second appeal will confirm the decision of a previous one, be that in favour or not of the original decision, he assigns it at 0·7466, or about 3 to 1 in favour of its doing so.*

* Recherches sur la Probabilité des Jugemens. Paris, 1837

Taxation, too, is an element of political condition easily enough represented in figures, but in which it is hardly possible to get two persons to agree in their interpretation. M. Quetelet sums up his few and cursory remarks on this subject with a dictum which, notwithstanding Mr. Norman, most Englishmen must feel to be intended for their peculiar consolation. "It has been justly remarked," he says, "that those are the most civilized countries who [which] pay proportionally the most to the government."

The chief difficulty to be encountered in aiming at correct results in the collection of agricultural, industrial, and commercial statistics is, that it —

"Requires the intervention of persons who are almost always interested, or think they have an interest, in disguising the truth. When the government collects them, it is generally opposed by the manufacturer, who supposes it done with fiscal views. The desire to obtain freedom for his industry, and to obtain what are called protecting laws . . . . almost always tends to exaggeration in one direction or another. Governments also publish documents on importations and exportations. These tables, which are useful to consult, nevertheless often contain very vague returns: they are generally confined either to the fixing of prices from faulty valuations or of quantities without considering either price or quality. In the official valuations, moreover, we only know a part of the truth: it is especially here that information not susceptible of reduction to numbers becomes necessary, in order to determine the probable quantity which escapes the legally stated values."

Owing to these causes of jealousy and partial presentation, many important statistical elements, relating to matters of pecuniary concern, can hardly be collected by official intervention. It is here that a Statistical Society may render most valuable service by setting on foot systematically, yet amicably and unobtrusively, local and private inquiries, with the guarantee of personal veracity for their answers, and the purely scientific and truth-loving spirit of such a body of enlightened inquirers for their fair presentment.

"The statistics of the moral and intellectual condition of a people," he goes on to observe, "presents still greater difficulties; for the appreciation can only be founded on facts much more contestable than those given by industry and commerce. When we say that a province produces so many quarters of corn or so many gallons of oil, we know that the figures may be more or less in error; but we understand the nature of the unit. It is not the same when we say that a province produces annually so much crime. . . . Infinite precaution and sagacity are necessary to read with success the statistics of tribunals, for the documents they contain are very complex in their nature, and almost always incomplete." . . . .

"What a mass of errors have we not accumulated in treating of pauperism! To probe this leprosy of society we have had recourse to lists of the poor, and very often without inquiring if these lists were complete and comparable in different countries or even within the limits of the same country. Real poverty is nearly always very different from the poverty officially returned. . . . *In Belgium a man will enter his name on the list of paupers to escape serving in the civic guard or to obtain other advantages, without receiving a farthing of public benevolence*" [! !].

With such difficulties in the way of exhibiting fairly, and interpreting truly, statistical facts, arises a necessity for laying down precautionary rules for the guidance of those to whom is confided the important task of their collection and registry—for checking their correctness when collected—and for their legitimate employment in aid of legislative or administrative purposes. On each of these heads M. Quetelet gives us a letter—short, indeed, and somewhat desultory; but abounding in useful and sensible remarks. Each of them would, in fact, require a treatise for its complete illustration.

A fool can ask questions, but only a wise man pertinent ones; and it often takes a wiser man to ask than to answer. After recommending to the statist a due and ample course of preparatory study of the subject in hand, our author goes on to observe, on the collection of statistical information:—

"The principal considerations which should guide an administration as to the questions to be asked are the following:—

"1. Only ask such information as is absolutely necessary, and as you are sure to obtain.

"2. Avoid demands which may excite distrust, and wound local interests or personal susceptibility as well as those whose utility will not be sufficiently appreciated.

"3. Be precise and clear, in order that the inquiries may be everywhere understood in the same manner, and that the answers may be comparable. Adopt for this uniform schedules, which may be filled up uniformly.

"4. Collect the documents in such a way that verification may be possible.

. . . . . .

"Simplicity and clearness of demand, together with uniformity in the forms to be filled up, are essential conditions to obtain comparable results. Without them, no statistics are possible. When the question relates to ages, professions, or diseases, it is of the greatest importance to employ classifications perfectly identical, in order that the general information may be compared even to the slightest detail. The most perfect unity should reign throughout the whole. It is to establish a unity like this that in certain states, such as Belgium and Piedmont, central commissions have been formed to collect and arrange the different elements which should be included in the national statistics. The necessity of such institutions is particularly shown when we see in very enlightened countries the principal departments sometimes publish very different numbers to express the same things, or make classifications which render comparison impossible." — (*Transl.* pp. 196, 197.)

Not to secure facility for the verification of the documents we collect is to miss one of the principal aims of the science. Statistics are only of value according to their exactness, without which they can serve but to establish error. Every statistical document requires a twofold examination — a moral and a material one, the former being, in all cases, by far the most important, as it involves the inquiry into the influence under which it has been collected — a point on which the whole colouring of the document essentially depends: —

"During the war of independence, the United States carefully misrepresented the true number of their population: they exaggerated considerably the numbers of inhabitants in maritime cities, in order to put the enemy on the wrong scent. Assuredly no good appreciation of the American population could be founded on the documents of this period."—(*Transl.* p. 202.)

Every statistical document ought to carry on the face of it, the exceptions, exemptions, and limitations, under which its entries are made. In respect of the use which may be made of it, negligence in this respect may amount in effect, if not in culpability, to a falsification.

"Thus, by means of official numbers, M. Sarauw pretended to prove that in the island of St. Croix, in the Danish Antilles, the mortality of the black slaves was less than that of white men even in Europe; and this assertion might appear so much the more imposing, as M. Sarauw resided in the island in question."

This result (which was arrived at in good faith) rested solely on the omission of negro children dying before attaining their first year from the register of births, such children being exempt from poll-tax, and therefore their omission being deemed of no importance.

The material examination of statistical documents rests chiefly on the internal evidence they may offer of self-consistency. It is singularly aided by diagrams. A simple line, properly laid down from a consecutive series of numbers, by what is called graphical projection, enables us to appretiate at a glance the continuity and regular progression of their succession; and, what is of still more importance, to apprehend correspondences between two series so projected, which often afford immediate conviction of a relation between them, such as the most subtle mind would find it difficult to perceive without such aid. They give to the study of phenomena the same advantage which algebra has introduced into calculation—they generalize and allow of abstraction; and they enable us at once to detect and often to rectify errors which, if undetected, would affect mean results, and throw everything into confusion. We are glad to find M. Quetelet strong in his advocacy of this mode of dealing with a series of observations which the generality of French *savans* affect, very unwisely, to despise as inconsistent with their notions of mathematical rigour.

There is nothing more indicative of a man's fitness or unfitness for the duties of a legislator and a statesman than his manner of dealing with statistical documents. When appealed to, as they too commonly are, for the purpose of establishing extreme positions, or of lending support to party views, or to particular interests, we are continually reminded of the doctrine of one long accustomed to listen to such arguments. "Nothing can be more fallacious than theories—except facts!" Those who use them in this manner will be found invariably to sin against truth and common sense in one or other of the following ways, viz.:—

"1. By 'having preconceived ideas of the final result.'

"2. By 'neglecting the numbers which contradict the result they wish to obtain.'

"3. By 'incompletely enumerating causes, and only

attributing to one cause what belongs to a concourse of many.'

"4. By 'comparing elements which are not comparable.'"

To which we may add a 5th, the most common of all and the most inexcusable, viz.: singling out the extreme partial results which tell on the side to be defended, and ignoring all the rest.

With such eclecticism we may find in statistics the means of defending almost every position. In politics, especially, they

"Become a formidable arsenal, from which the belligerent parties may alike take their arms. . . . Some figures, thrown with assurance into an argument, have sometimes served as a rampart against the most solid reasoning; but when closely examined, their weakness and nullity have been discovered. Those who allow themselves to be frightened by such phantoms, instead of looking to themselves, prefer rather to accuse the science than to confess their blind credulity, or their inability to combat the perfidious arms opposed to them.

"We see persons profoundly convinced of a truth, seek to establish it directly by the authority of figures, and give, as they think, a mathematical demonstration. However, by means of the statistical documents which they unskilfully employ, they most frequently produce an opposite effect to that which they desired. Thus we cannot reasonably doubt that enlightenment contributes to man's happiness, by illuminating his intellect and fortifying his morals. In the attempt to demonstrate this what has been done? It has been thought necessary to establish that the number of crimes is inversely as the number of children sent to school — as if the number of crimes, even were it known, had as its only cause the greater or less development of the intellect; and as if the development of intellect were measured by the number of children sent to school. What has been the result of this? It has been found, after well examining statistical documents, that the number of crimes is more generally in a *direct* proportion to the number of children sent to school, than in the *inverse* proportion. The conclusion is exactly the opposite of what was at first desired — a new error, which some have, with the same levity admitted."—(*Transl.* p. 214.)

The necessary incompleteness of all statistical documents is sometimes urged as a general argument against trusting implicitly to conclusions drawn from them. The argument is valid, in so far as we have reason to believe that the unenumerated cases differ systematically, *i. e.*, in some essential point of classification, from the enumerated; so as to render the proportions in which the several classes are represented in the returns different from what they would be were the enumeration complete. But granting their incompleteness — and granting even that the incompleteness is such as to affect injuriously the proportionate numbers in classified results — this does not preclude the drawing of many sound and valuable conclusions from such documents, if only we are assured that in comparing similar ones for several successive years, or under circumstances otherwise different, the same causes of incompleteness prevailed and continued to affect the several classes in an invariable ratio.

This position M. Quetelet illustrates by a reference to the Criminal Statistics of Belgium. — Prior to 1830 the official returns gave only the number of crimes *known* and *prosecuted*, but for the seven years from 1833 to 1839 they included also the number of crimes known, but which were not prosecuted because the authors were unknown. Now it was found that this latter number proceeded from year to year with even more regularity than that of crimes prosecuted. No doubt, therefore, the number of crimes altogether unknown to justice, could it have been made a matter of registry, would have presented a similar constancy. Of known crimes against person, two thirds were regularly prosecuted, and one third escaped, the authors being undiscovered. In the case of crimes against property the proportions were reversed, and were nearly those of one fourth and three fourths; the graver crimes being those most sure of detection. On the whole it would appear from these records that out of 1154 crimes annually known to justice in Belgium, only 416, or little more than one third, formed subjects of prosecution. Assuming, then, that the number of unknown crimes is equal to that of known (this would hardly be admissible for crimes against person), the amount of prosecuted crimes in Belgium would not exceed one sixth of those actually committed.

"I am absolutely ignorant and shall never know whether the crimes on which the tribunals have to pass judgment form the sixth or seventh or any other part you will of the total number of crimes. What is important for me to know is that this ratio does not vary from year to year. On this hypothesis I can judge *relatively* whether one year has produced more or less crimes than another."

Admitting that this ratio remains invariable from year to year, and that justice pursues criminals with the same activity, two countries or two provinces of the same country might be compared in respect of morality. But as the latter condition almost certainly does not hold good under different administrations, it becomes impossible, from the official returns of prosecutions, fairly to institute such a comparison between nations. Even should the same legislation, the same repression, and the same activity to bring criminals to justice, subsist, if the result be made to depend on a comparison of the number of *condemnations*, instead of those of *prosecutions*, a difference in the mode of trial would alone suffice to destroy the comparability of the cases.

"We know, in fact, that the establishment of the jury in Belgium has doubled the number of acquittals." — *Transl.* p. 227.)

On the subject of Medical Statistics, M. Quetelet has a brief, digressive, and somewhat pungent letter, and presents what must be confessed to be rather a deplorable picture of the actual state of this branch of the general subject.

"All reasonable men," he says, "will, I think, agree on this point, that we must inform ourselves by observation, collect well recorded facts, render them rigorously comparable before seeking to discuss them with a view of declaring their relations, and methodically proceeding to the appreciation of causes. Instead of this what do we see? Observations incomplete, incomparable, suspected, heaped up pell-mell, presented without discernment, or arranged so as to lead to the belief of the fact which it is wished to establish; and nearly always it is neglected to inquire whether the number of observations is sufficient to inspire confidence."

This is, no doubt, the impression which the perusal of the generality of medical books and dissertations leaves on the mind. The fact is, that in a science like medicine the statistical method of inquiry is not the most natural and obvious. Under circumstances of excessive complication in any line of research, and more especially in one in which success leads so directly to celebrity and fortune, men usually look for what Bacon terms "instantiæ luciferæ," those "luminous instances" where the result of a single experiment, the striking issue of a novel process, makes its way at once to the inductive instinct without being subjected to the scrutiny of reason. The comparison of multitude with multitude, the destruction of errors by mutual collision, and the slow emergence of truth from the conflict by its outstanding vitality, belong to a maturer age of science than that in which medicine had its origin or attained its present importance. Yet there have not been wanting in its walks men of philosophical views, who have both seen themselves and recommended to others this course of procedure.*

* The following striking passage occurs in Dr. Holland's "Medical Notes and Reflections:"—"A very especial advantage has been the application of numerical methods and averages to the history of disease; thereby giving it the same progress and certainty which belong to statistical inquiry on

Neither is the deficiency so absolute as M. Quetelet's expressions would lead us to suppose. So far at least as the statistics of *disease* are concerned, some material progress may be reported. Medical science, imperfect as it is, has at least succeeded to a certain extent in classifying diseases under more or less general heads, and identifying them with sufficient distinctness to attribute to each something like its due share in contributing to the total annual mortality. This is a great step. It enables us at once to compare the prevalence of particular disorders (in that degree of intensity at least which leads to a fatal termination) with that of other statistical elements or with meteorological registers, and so to work our way by sure though perhaps slow degrees from the detection of tendencies in some certain atmospheric conditions, food, habits, &c., to their production, up to a knowledge of their proximate

other subjects. Averages may in some sort be termed the mathematics of medical science. The principle is one singularly effectual in obviating the difficulties of evidence already noticed; and the success with which it has been employed of late by many eminent observers affords assurance of the results that may hereafter be expected from this source. Through medical statistics lies the most secure path into the philosophy of medicine. The inquiries which so greatly distinguish M. Louis as a pathologist may be noted as eminent examples of this method, which is now pursued with great success by many physicians in our country."—*On Medical Evidence*, vol. i. p. 5.

The dissertations of the late Sir Gilbert Blane abound with statistical statements well collected and ably reasoned on, to the attainment of most important results.

or remote causes, and thus to devise measures of an administrative kind, not indeed for their cure in particular cases, but for their general mitigation and possible final extinction (as in the case of the sea scurvy); and doubtless much greater progress might be made in this direction, would medical practitioners agree (or were it made incumbent on them as a condition of their *status*) to forward classified returns of the cases under their treatment to some common sanitary centre, — the form of classification and nature of the entries to be prescribed on uniform and well-considered principles, and the results authoritatively published at stated intervals. Publicity indeed is the *sine quâ non* of statistical science, and the grand condition of its useful application, not merely by reason of the openings thereby afforded for the detection of error and the exposure of unfairness of registry, but, what is of infinitely more consequence, letting in the broad good sense of the thinking part of mankind on the subjects themselves abstractedly presented to them — than which nothing so effectually tends to clear away *professional* prejudices and errors, and to bring professions themselves (*as every profession ought to be brought, for its own sake as well as the public's) under the watchful inspection of its laity.*

The statistics of cure are necessarily more imperfect than those of disease. Excessive difficulties must lie in the way of tabulating the medical treatment of cases upon anything like uniform and intelligible principles of classification and registry, owing to the multitude of particulars to be embraced, the difficulty of recognizing diseases in their earlier stages, the necessity of continually swerving from a uniform preconceived system of treatment in accommodation to age, sex, habits of life, and constitutional peculiarities—the absurd system of administering *mixtures* of *mixtures* of medicaments, so as to render it next to impossible to say what quantities of the *prima medicamenta* have been really swallowed, and all the thousand and one causes which conspire to render medical practice tentative and uncertain, and the statements of its degree of success untrustworthy. Supposing these difficulties overcome, if not in all, yet in selected classes of disease, supposing every essential particular intelligibly registered, and the result candidly stated, it has still to be borne in mind that such registers must necessarily exclude all cases in which nature has been left to her own unaided resources, and nearly all in which the natural remedies of rest, regulated diet, ventilation, cleanliness, &c., may have been alone resorted to. It would require a physician of no common forbearance to abstain in fifty out of each hundred cases from the use of *all active medicines*, and of no common candour and defiance of professional censure to declare that he had done so, and to put on record the *failures* of this line of treatment.

"To judge," says M. Quetelet, "of the advantages which

therapeutics may present, we must commence by inquiring what would become of a man afflicted with such a malady if abandoned to the force of nature only. Perhaps we might be led to conclude that in doubtful and difficult cases it is better to give up the patient to the efforts of nature than to the remedies of art, confining ourselves to the use of a careful diet. Different kinds of treatment have less influence on mortality than is generally supposed. A respected and learned man, Dr. Hawkins, thus expresses himself: — 'A friend took private notes on the comparative mortality under three doctors in a hospital. The one was *eclectic*, the second pursued the *expectant* system, and the third the *tonic* regimen. The mortality was the same; but the duration of indisposition, the character of the convalescence, and the chances of relapse very different.' Thus the mortality was the same. We might draw the same conclusions from the documents collected in the principal hospitals of Europe: the mortality varies between very narrow limits, and depends more "on the general maintenance and supervision of the hospitals," [de la *tenue* des hôpitaux,— most incorrectly translated on the *principals* of the hospitals"] "than on the therapeutic means employed. . . . Did I not fear being taxed with exaggeration, I should say that a good administration saves more patients in hospitals than the science of the most skilful doctors."—(*Transl.* p. 235.)

We have just had occasion to notice a serious mistranslation, throwing upon an individual the responsibility of the general success or failure of an establishment, contrary to the plain meaning of the passage in the original French, and we wish it had been possible for us to conclude this article without further remark on the manner in which the translator of the work before us has executed his task. It is full of such misrenderings, which betray a palpable ignorance of the language of the original, issuing in expressions which are neither French, English, nor sense. Thus we have "*revoquer en doute*" (to call in question) continually rendered by "*to revoke in doubt*" (p. 2. &c.); "*exceptionnel*" (p. 18.) is rendered by "*exceptionable;*" "*temps affreux,*" shocking weather (p. 23.), by "*frightful times;*" "*modeste,*" moderate (p. 28.), by "*modest;*" "*parties,*" (p. 34.), games, by "*parts;*" "*lunettes*" (telescopes), by "*lunettes;*" "*hasardés*" (precarious), by "*hazarded;*" "*siècles*" (ages), by "*centuries,*" — the definite for the indefinite sense,—giving an almost puerile air to the passage in which it occurs: "Our planet is but a very secondary body, a grain of dust lost in immensity, and yet *centuries* have been required to bring it to the state in which we now see it." (P. 133.) In p. 147. we have the idiomatic phrase, "On aurait lieu de plaindre un pays" (a country would be to be pitied), perverted into, "He would have to complain of a country." Again (p. 228.) we have "all *indistinctly*" (*indistinctement*, indiscriminately) "collect statistics, but some confide their results to their memories, others to paper; some even collect them unwittingly, *like* M. Jourdain *does* prose," ("*comme* M. Jourdain *faisait* de la prose,")—as Molière's M. Jourdain (with whom we should have thought every one at all conversant with the language must be familiar) *used to make* prose.

"Who can affirm," says the translator, "that this principle" (the law of gravity) "is not a particular case of a much more general law, or that the results deduced from it are not values *sufficiently* approximative, *since* the neglected quantities *are not* appretiable in the present state of science." M. Quetelet's expression is, "ne sont pas des valeurs *suffisamment* approximatives *pour que* les quantités négligées ne soient pas apprétiables," &c. (are not mere approximations, *sufficiently such*, however, that the quantities neglected shall be inappretiable, &c.). Obvious errors, and misprints too, in the original, are transferred uncorrected into the translation. Thus, in p. 81., we have the important and mischievous misprint ·01 instead of 0·1 twice repeated. M. Quetelet, with the usual laxity of a foreigner, is privileged to misspell our English names, but it does not become an English writer on Probabilities to acquiesce in the transformation of the honoured name of Stirling into Stierling. We must add, too, that the manner in which the French metrical system used in the original is converted into British equivalents in the translation is such as to interfere materially with a clear understanding of the purport. Thus, in the table of the limiting heights of giants, tall and short men, and dwarfs, in p. 103., the limits are given in the original to millimetres, while in the translation they are stated only to the nearest inch, and that in one instance erroneously. We notice these blemishes, not in the spirit of cavil, but in order that they may be removed in a subsequent edition.

The letters on the use of statistics to the administration and on the ulterior prospects of this branch of science, though they can hardly be said to contain anything very new or striking, yet come opportunely at a period like the present, when vast changes, both legislative and economical, are in progress, and when opportunities are lapsing of seizing *in transitu* results which will one day be most valuable for future comparison. Steam, railroads, and free-trade principles are making such inroads into all that used to be considered fixed or slowly alterable, that it will be of the utmost interest to have secured points of departure in the new career which opens on society.

"Statists should be eager to register, from this time forward, all the facts which may assist in the study of this vast transformation in the social body, which is in process of accomplishment.

"A government in modifying its laws, especially its financial laws, should collect with care documents necessary to prove, at a future state, whether the results obtained have answered their expectation. *Laws are made and repealed with such precipitation that it is most frequently impossible to study their influence.*"

These words deserve to be written in letters of gold. They point to an evil whose tendency is to degrade social policy from the list of sciences of observation and experiment to the rank of an empirical art. *Avant nous le Chaos! Après nous le Déluge!* should be the motto of that statecraft which, under a momentary sense of pressure from those whom even the uneasiness of change makes restless and impatient, urges on the social movement faster than a sound philosophy can count the revolutions of its mechanism or register the work accomplished; or of that which, by the simultaneous alteration of every condition, makes the separate estimation of any single effect hopelessly impracticable.

NOTE on p. 400.—Sir Joshua Reynolds, in his celebrated Lectures to the Royal Academy has laid it down as the fundamental principle of the pictorial art, that beauty of form and feature consists in their close approximation to the mean or average conformation of the human model. Were this the case, ugliness ought to be extremely rare, and the highest degrees of beauty those of the most ordinary occurrence, a conclusion contrary to all experience. (H. 1857.)

AN ADDRESS

DELIVERED BY J. F. W. HERSCHEL, ESQ., PRESIDENT OF THE ASTRONOMICAL SOCIETY OF LONDON, ON THE OCCASION OF THE DISTRIBUTION OF THE HONORARY MEDALS OF THAT SOCIETY, ON APRIL 11. 1827, TO FRANCIS BAILY, ESQ., LIEUT. W. S. STRATFORD, R.N., AND COL. MARK BEAUFOY.

GENTLEMEN,

THE ordinary business of the evening being now terminated, it remains to fulfil the object for which we are especially convened this night, which is one of no less interest than the distribution of the Honorary Medals awarded by your Council, in pursuance of the principle of encouraging works of great labour, high practical utility, and steady perseverance in astronomical observation, and in redemption of the pledges held out in the Address circulated at the origin of this Society, explanatory of its objects.

On former similar occasions when we have been called on to witness the execution of this important duty, it has frequently been our good fortune to acknowledge and applaud the claims of foreign merit, and to prove by our awards, that no mean jealousies, or narrow and mistaken views of national honour, are capable of blinding our judgment or biassing our decision; but that he who, whatever be the spot of earth he inhabits, most promotes the cause of Astronomical science, is most our brother and our countryman. Yet, I am sure it will be gratifying to you to know that, on this occasion, ample scope has been found for selection in the merits of our own compatriots, and in the home list of our members. It is not that great and important Astronomical works have not emanated from our continental neighbours: on the contrary, the spirit of research and discovery appears to have prevailed with extraordinary activity; and the last year has even witnessed the addition to our system of another of those singular bodies, the discovery of which has conferred so much lustre on the names of Halley and Encke. No less than three independent claimants to the almost simultaneous disclosure of this interesting fact may be enumerated; and this circumstance, while it marks the spirit of the age more forcibly perhaps than any trait which could be produced, must obviously render it impossible for this Society to interfere or decide on the priority and rank of the competitors. But though unmarked by any tangible memorial of our approbation, the names of Biela, Clausen, and Gambart will not the less be cherished among us, and enrolled by posterity in the choicest and most permanent annals of Astronomical celebrity.

It is however for labours of a very different kind that our medals are this day to be conferred:

labours, if less brilliant, yet more vital; if less associated with lofty speculations on the nature of the universe, yet more intimately linked with the practical uses of this world. The first award of your Council is that of a gold and silver medal respectively to your late excellent President Mr. Baily, and your indefatigable Secretary Mr. Stratford, for their joint labours in the construction of the Catalogue of 2881 principal fixed stars, which forms the Appendix to the second volume of the Memoirs of this Society.

A catalogue of stars may be considered in two very distinct lights, either as a mere list of objects placed on record, to fix on them the attention of astronomers, and to afford them matter for observation, or as a collection of well-determined zero points, offering ready means of comparing their observations with those of others, and of detecting and allowing for instrumental errors. In this light only I shall now consider it as chiefly of importance to the practical astronomer. It is for his uses that an amount of pains, labour, and expense, both national and individual, has been bestowed on the perfection of such catalogues, which on a superficial view must appear in the last degree lavish, but which yet has been no more than the necessity of the case demands. If we ask to what end magnificent establishments are maintained by states and sovereigns, furnished with master-pieces of art, and placed under the direction of men of first-rate talent, and high-minded enthusiasm sought out for those qualities among the foremost in the ranks of science: — if we demand *cui bono?* for what good a Bradley has toiled, or a Maskelyne or a Piazzi worn out his venerable age in watching? the answer is, — not to settle mere speculative points in the doctrine of the universe; not to cater for the pride of man, by refined inquiries into the remoter mysteries of nature,— to trace the path of our system through infinite space, or its history through past and future eternities. These indeed are noble ends, and which I am far from any thought of depreciating; the mind swells in their contemplation, and attains in their pursuit, an expansion and a hardihood which fit it for the boldest enterprize.—But the direct practical utility of such labours is fully worthy of their speculative grandeur. The stars are the land-marks of the universe; and amidst the endless and complicated fluctuations of our system, seem placed by its Creator as guides and records, not merely to elevate our minds by the contemplation of what is vast, but to teach us to direct our actions by reference to what is immutable in his works. It is indeed hardly possible to over-appretiate their value in this point of view. Every well-determined star, from the moment its place is registered, becomes to the astronomer, the geographer, the navigator, the surveyor,— a point of departure which can never deceive or fail him,—the same for ever and in all places, of a delicacy so extreme as to be a test for every instrument yet invented by man, yet equally adapted for the most ordinary purposes; as available for regulating a town clock, as for conducting a navy to the Indies; as effective for mapping down the intricacies of a petty barony, as for adjusting the boundaries of transatlantic empires. When once its place has been thoroughly ascertained and carefully recorded, the brazen circle with which that useful work was done may moulder, the marble pillar totter on its base, and the astronomer himself survive only in the gratitude of posterity; but the record remains, and transfuses all its own exactness into every determination which takes it for a ground-work, giving to inferior instruments, nay even to temporary contrivances, and to the observations of a few weeks or days, all the precision attained originally at the cost of so much time, labour, and expense.

To avail ourselves of these records, however, we must first have the means of disentangling the observed places of the stars at any moment, from the regularly progressive effect of precession, and from a variety of minuter periodical inequalities arising from the nutation of the earth's axis, and from the aberration of light, of which the genius of theoretical, no less than the industry of practical, astronomers has at length succeeded in developing the laws and fixing the amount, so as to leave little probability of any material change being induced by future researches.

The calculations, however, required for this purpose, if instituted for each particular star at the time it is wanted, are so numerous and troublesome as to become a very serious evil; the effects of which have been severely felt in Astronomy in the discouragement it has offered to the reduction of observations, owing to which the labour of many an industrious observer's life has been in great measure thrown away. Indeed, a lamentable picture might be drawn of the waste of valuable labour traceable to this cause. The want of tables therefore to facilitate the reduction of particular stars was early felt. I shall not, however, enter into any historical detail of the attempts hitherto made from time to time to supply this desideratum. A well drawn up and concise account of them is given in Mr. Baily's Preface to the Catalogue, which renders superfluous all I could say on the subject. Indeed, useful as they have been, and considerable as has been the pains bestowed on them, they are all so far surpassed by this work of Mr. Baily, that it ought rather to be considered as belonging to a new class, than to be compared in any way with preceding ones, which must eventually all be superseded by it.*

It is time now to speak more particularly of the Catalogue itself. Its whole plan and arrangement, the selection of the stars, the preparation and revision of the formulæ, the choice of the co-effici-

* From this sentence, however, I ought to except special tables for the daily reduction of a certain number of select stars, whose use is no way superseded by the general Catalogue, being destined for continual, as the latter is only for occasional, reference.

cients, and the discussion of the terms to be retained or rejected, we owe to Mr. Baily, who has stated every particular relating to it in a most elaborate preface, which may indeed be regarded as a compendium of all that is known on the subject of the corrections, and is remarkable at once for its precision and perspicuity. A great portion of the computation has been gratuitously performed by Mr. Stratford, checked by a computer engaged for that purpose. From this very severe labour, however, he was unfortunately compelled to desist, I regret to say, by ill-health, and his place supplied by a professional computer: but the hardly less laborious task of comparing and checking the computations of his assistants, and, what is as important in all such cases as accuracy of computation, the careful superintendance of the press, and repeated revision of the whole work, has entirely devolved on him; and never, I must say, was task performed with more diligence and exactness.

The selection of the stars has been made from the Catalogues of Flamsteed, Bradley, Lacaille, Mayer, Piazzi and Zach, so as to include all stars down to the 5th magnitude, wheresoever situated in the heavens,—all of the 6th within 30° of the equator, and all the stars to the 7th magnitude inclusive, within 10° of the ecliptic. Almost all of them, however, are to be found in the Catalogues of Bradley or Piazzi, from which they have been reduced to 1830 (the epoch adopted), by formulæ given by Bessel. Their number is so considerable, that in whatever part of the heavens we may be observing, one or more are sure to be within a moderate distance; so that no one provided with this Catalogue can possibly be at a loss for a zero-point to check his observations, and ascertain the state of adjustment of his instrument. To its convenience and utility in this respect, I can speak from individual experience. It is indeed become my sheet anchor, and has infused into a series of observations wholly dependent on such aid, a degree of exactness which, without it, I should hardly have expected to attain.

The formulæ employed for calculating the corrections are almost entirely those of Bessel, who has laboured with such diligence and perseverance on this department of Astronomy, as to make the subject almost his own. In adopting them, however, Mr. Baily has taken nothing for granted, even from such high authority. He has gone over the whole subject anew; and the slight inaccuracies which he has detected and corrected in several of the results of this profound geometer, although almost insensible in a numerical point of view, are valuable, as proving at once the general accuracy of his investigations and the minuteness of the scrutiny they have undergone.

The most delicate part of the whole operation, however, was the choice of the several co-efficients, which, if erroneously assumed, would render the whole subsequent work of no value. In making this assumption, Mr. Baily has exercised a degree of judgment which I feel convinced will unite the suffrages of astronomers. Taking a comprehensive view of the results afforded by all former investigations, he has uniformly adhered to the principle, to steer clear of extreme quantities, and to adopt only such as not only rest on the greatest number of the best observations, but agree in their values nearly with the average of all. Thus, in the case of the aberration, the value adopted is the mean of the almost miraculously coincident results of Brinkley and Struve, and agrees within two hundredths of a second with that of the extreme values assigned by Bradley and Bessel, so that this datum may be regarded as one of the best established in Astronomy. In the same cautious manner has Mr. Baily proceeded with the other co-efficients. That of precession he has taken entirely from Bessel's elaborate investigations compared with those of Laplace, in which the only remaining source of uncertainty, is that arising from our ignorance of the mass of Venus; the influence of which cannot possibly produce an error, however, of a tenth of a second in the precession. The nutation he has taken as it results from Dr. Brinkley's observations, which (like his aberration) justify this partiality by holding almost exactly an average value among all the different results of Bradley, Mayer, Maskelyne, Laplace, and Lindenau, and can hardly be considered as more than a tenth of a second in error.

This judicious choice will secure the present tables from a possibility of ever sharing the fate of preceding labours of this sort. They can never be superseded by others of greater accuracy, nor fall into disuse or grow obsolete till the apparent places of the stars shall have become so much altered by the effect of precession as to render the computations inexact, for which a very long series of years will be required.

But the distinguishing characteristic of this work, is the adoption throughout of Professor Bessel's capital improvement in the system of applying the corrections, by arranging the formulæ in such a manner that all that is peculiar to each star, and permanent in magnitude, shall stand distinctly separated from all that is ephemeral, or varying from day to day; and *that* in such a manner that a short ephemeral table, capable of being compressed into a single page, shall serve, not only for these stars, but for every star in the heavens. The convenience of this method, the brevity it introduces into the computations, the distinctness it gives to all the process of reduction, requiring neither thought nor memory on the computer's part, give it an incalculable advantage over every other. To reduce any observation, no other book need be opened. The work occupies four lines, and is done in half that number of minutes. If we compare this with the tedious and puzzling operation required by former processes, we shall fully agree with Mr. Baily that "those only who are versed in such calculations can appretiate the labour, the risk of error, and the loss of time incurred in their several operations;" all which are saved by the present arrangement.

These considerations will amply justify the award of your council in your eyes and those of the world. They will justify a great deal more. At no time was the necessity of pressing on the attention of astronomers the utility, I may say the duty, of uniformity in their systems of reduction more urgent than at present*, when hardly a nation in Europe is unprovided with a good observatory, and when rival astronomers in all quarters of the globe are contending for the palm of accuracy and diligence. So long as they persist in continuing to reduce their observations by different systems, their merits can never be fairly compared. Each may boast the perfection of his instruments, and vaunt himself in the security of his pre-eminence. Each may promulgate his standard Catalogue, which will be adhered to in his own nation, and rejected by all others; thus dividing astronomers into sects and parties,—a state of things which ought surely not to continue. The only remedy is to agree to speak one language, to adopt one system. It matters little, in the present advanced state of science, whether that system be still open to infinitesimal corrections. Let astronomers only consent to use

* This applies with equal or greater force to the correction for refraction; a common table for which ought to be agreed on and adhered to by all.

it as, like all human works, confessedly imperfect, and in process of time to be corrected: but not at the caprice of each individual who may think one co-efficient a tenth of a second too small, or another as much too great; but after full consideration, when the necessity and amount of correction shall have become certainly known and generally agreed on.

Meanwhile, a fair opportunity is offered to rival astronomers throughout the world, to try their strength, in an arena of ample extent, and where every part of the honourable contest will be brought distinctly into sight. In giving this Catalogue to the world, we invite their examination to its errors (for such it must contain), and call on them to lend their aid to its perfection, by determining, with all the exactness their resources afford, the mean places of the stars it comprises. For this, its arrangement affords every facility, and those who observe have no excuse for neglecting to reduce. Let us hope then, that instead of lavishing their strength in fruitless attempts to give superhuman precision to fifty or a hundred select objects, the formation of a standard Catalogue of nearly 3000 will be deemed of sufficient importance to fix the attention of astronomers; and that not only those to whom the direction of great national observatories is confided, but even private individuals, if such there be, who feel themselves in possession of the means required, may take a share in this glorious, but at the same time arduous undertaking.

(The President then, delivering the Gold Medal to Mr. Baily, addressed him as follows:—)

Mr. Baily,

Accept this Medal, which the Astronomical Society bestows on you, by an award which every astronomer in Europe will confirm. The work you have accomplished will identify you with the future progress of that Science, into almost every department of which it is calculated to infuse new life; since every practical astronomer has in it to thank you for an accession of power. It is needless for me to accompany this testimony of the sense the Society entertains of your distinguished merits, with the expression of a hope that your exertions in the cause of Astronomy will continue. You could not struggle against nature so far as to desist from pursuits, which, demanding of ordinary men a total devotion of their time, and concentration of their whole intellectual powers, have been to you a relaxation from the most active business. Possessing thus within yourself a source of pure and exalted enjoyment, enhanced by the consciousness of public utility, and a certainty of the approval and admiration of those whom you esteem, we can only add our wishes that length of years, and continuance of health, may render your distinguished talents, and rare zeal for the promotion of your favourite science, as useful to Astronomy as it is honourable to yourself.

(*The President next presented the Silver Medal to Mr. Stratford, addressing him at the same time in these words:* —)

MR. STRATFORD,

The Medal which, in the name of the Astronomical Society, I now deliver to you, though "less fine in carat" will, I trust, be to you "more precious" than gold, as proving how highly we estimate your devoted and persevering attention to the work you have so happily brought to a conclusion. Those only who have actually entered into the details of a work of this nature can possibly understand the overwhelming and soul-sickening labour of such a task; but the pile of volumes now lying on the table, a great portion of which you have yourself penned, and the whole of which you must, in the course of your undertaking, have repeatedly read over, figure by figure, will serve to give some idea of it. In executing this arduous duty, you have had no other inducement than your zeal for the progress of science, and that devotion to the interest of this Society which is so conspicuous in every part of your conduct, and which would not suffer you to tolerate the idea of any incorrectness, anything unworthy the importance of the subject emanating from it. The habits of correctness in numerical computation, and systematic fidelity of detail indispensable for such a work, you possess, though in perfection, yet in common with many: but the enthusiasm in the cause of abstract science, which could carry you successfully through the task thus voluntarily imposed on yourself, you share with few. You have, however, the satisfaction of knowing that so much labour has not been bestowed in vain; for, if there be anything on which we can calculate with certainty, it is that the work you have been mainly instrumental in completing must exercise a powerful influence on the future destinies of Astronomy.

(*The President then resumed his Address to the Members in general, as follows:—*)

GENTLEMEN,

We have still another, and a very interesting part of the business of this meeting to perform, in the delivery, to Colonel Beaufoy, of a Medal for his valuable series of observations of eclipses of Jupiter's satellites, communicated to this Society, and in part already printed in the first part of the second volume of our Memoirs; in part recently read at a late meeting; and completed up to the present time, by the paper you have heard read to-night.

The subject of the eclipses of Jupiter's satellites is one of singular interest in the history of Astronomy. The discovery of these bodies was one of the first brilliant results of the invention of the telescope; one of the first great facts which opened the eyes of mankind to the system of the universe, which taught them the comparative insignificance of their own planet, and the superior vastness and nicer mechanism of those other bodies, which had before been distinguished from the stars only by their motion, and wherein none but the boldest thinkers had ventured to suspect a community of nature with our own globe. This discovery gave the holding turn to the opinions of mankind respecting the Copernican system: the analogy presented by these little bodies (little however only in comparison with the great central body about which they revolve) performing their beautiful revolutions in perfect harmony and order about it, being too strong to be resisted. As if to confirm this analogy beyond dispute, Kepler lived just long enough to witness the discovery, and to demonstrate* the extension of the same general law to their periods which he had found to obtain among those of the primary planets about the sun. The conclusion was irresistible; and the full establishment of the Copernican system must date from the discovery of the satellites of Jupiter.

This elegant system was watched with all the curiosity and interest the subject naturally inspired; and the eclipses of the satellites speedily attracted attention, and the more when it was discerned, as it immediately was, by Galileo himself, that they afforded a ready method of determining the difference of longitudes of distant places on the

* According to Delambre this extension of Kepler's law is due to Vendelinus.

earth's surface by observations of the instants of their disappearances and re-appearances simultaneously made. Thus, the first astronomical solution of the great problem of the longitude, — the first mighty step which pointed out a connexion between speculative Astronomy and practical utility, and which, replacing the fast dissipating dreams of astrology by nobler visions, showed how the stars might really and without fiction be called arbiters of the destinies of empires, we owe to the satellites of Jupiter; to those atoms, imperceptible to the naked eye, and floating like motes in the beam of their primary — itself an atom to our sight, — noticed only by the careless vulgar as a large star, and by the philosophers of former ages as something moving among the stars, — they knew not what, nor why; perhaps only to perplex the wise with fruitless conjectures, and harass the weak with fears as idle as *their* theories.

No wonder now that the eclipses of the satellites were watched with anxious earnest interest; they were soon to afford matter for yet greater wonder and deeper contemplation. Roemer's discovery of the velocity of light from the retardation of their eclipses, about the end of the 17th century, was the next in order, and the sublimest truth they were destined to be the means of unfolding; a truth so amazing, so overwhelming to human faculties, that (not to mention the feebler names of Cassini, Maraldi, and Fontenelle,) even the comprehensive genius of a Hooke quailed before it, and refused to admit the existence of a motion so little short of infinite in a finite system like our own. The discovery of the aberration of light by Bradley, however, more than forty years afterwards, confirmed it in its full extent; and no truth in the circle of physical science is either more astonishing or better established than this.

We are not yet come to the end of the long catalogue of useful and admirable results afforded to science and to mankind by the discovery of these bodies. We have hitherto regarded only obvious results; broad and evident conclusions from apparent facts. Let us now trace them in the quiet succession of their convolutions, in the unfolding of their periodical inequalities, in the slowly accumulating amount of their mutual action, in the influence of the oblate figure of their primary on their orbits; in short, through all the mazy intricacies of their perturbations. The lessons they have thus whispered to the intellect of man, over the midnight lamp, have not been less instructive, less fraught with wonder and utility, than those which they have blazoned to his senses. It is to that powerful and gifted genius, * now so recently gathered in an illustrious grave; on whose ashes the tears of mourning science are yet warm, — to him, whose revered name so freshly sanctified by death, I am unwilling to pronounce, that we owe the complete development of their theory. His penetrating mind saw all the advantages likely to

* Laplace.

accrue to the general theory of the planetary perturbations from the study of this miniattre system, where years are represented by days, and ages by years, and where inequalities, which in the planetary theory have a character approaching to secular, can be traced in their increase and on their wane. Aided, therefore by his powerful analysis, he succeeded in applying the law of gravitation to the minute investigation of all their inequalities; and the result has been not merely another triumph of the Newtonian theory in the complete explanation of all their complicated irregularities, but the formation of tables even more perfect than observation itself*: and in addition, a mass of most valuable and instructive information on the general nature of planetary perturbations amply repaying all the labour of the inquiry, and adding fresh lustre to the already imperishable glory of his name.

This slight sketch of the history of the satellites of Jupiter may serve to show how intimate is the connexion of distant parts of science with each other, and that in it we are to regard nothing as trivial and nothing as great in itself, but in respect of the instruction we may draw from it;—to show, in fine, how deep are the foundations and how

* Than any single observation.—*Delambre.*

† We owe yet another important piece of information to these satellites. The comet of Lexell passed on the 23rd of August, 1779, among them, without in the smallest degree disturbing their motions, thus proving the minuteness of its own mass. (H. 1857.)

wide spread the ramifications of that tree of knowledge which, in the poet's words,

> . . . . "quantum caput ardua ad astra
> Attollit—tantum radice in Tartara tendit."

which draws its increments from small beginnings and matters of speculative curiosity, and ends in becoming the ornament, the shelter, and the support of society.

It is by observations of the eclipses of the satellites alone that their theory can be compared with nature, their apparent distances from the planet being too small and its change too slow to admit of micrometrical measurements precise enough for the purpose, though perhaps the modern improvements both in the telescope and micrometer may authorize a hope that this may not long be an insuperable difficulty. Accordingly, from the time of Roemer downwards, a series of eminent astronomers have occupied themselves with observations of these phænomena, and it is on no less than two thousand of such observations that Delambre, improving on the tables of Wargentin by the aid of the profound theory just alluded to, succeeded in calculating the first series of tables laying claim to precision.

The longitude is so much better ascertained now by lunar distances and occultations, that these observations are less resorted to than heretofore for that purpose. Nevertheless they are occasionally used, especially those of the first and second, whose eclipses not only happen much more frequently, but are much more definite, than those of the exterior ones. Indeed, the observations of the latter have been declared by high authority, utterly useless. It is not always good, however, to trust to authority; and Mr. South, by a comparison of his own with Colonel Beaufoy's observations, has arrived at a very different conclusion, at least for the cases when both the beginning and end of the eclipse can be seen. Still, however, it is highly desirable that they should continue to be assiduously observed, not merely to furnish corresponding observations, but to afford the means of further perfecting the tables, so as ultimately to enable us to dispense with corresponding observations altogether.

Colonel Beaufoy has for many years past been a most careful and assiduous observer of these eclipses, and indeed of all occasional phænomena; such as occultations, eclipses both solar and lunar, and of late of that very useful and important class, the transits of moon-culminating stars, of which one of his recent communications contains an extensive and highly interesting series. His observations of the immersions and emersions of the satellites communicated to this Society, amount to no less than 180, all (with the exception of two or three of the earlier ones) being made in the interval from 1818 to 1826 inclusive;—a fine series, indeed a surprising one, when the comparative rarity of the phænomena is considered, not more than about forty visible at Greenwich occurring annually on an average, and when the great drawback on observations of this sort from unfavourable weather in this anti-astronomical climate is taken into the account. What chiefly adds to their value as a series, however, is the circumstance of their being all made by one observer, and with one telescope, — a fine five-feet achromatic of Dolland, and with the same magnifying power 86. In no class of astronomical observations is uniformity in this respect of such importance, since the variations in the times of appearance and disappearance, when observed at the same spot, simultaneously, by different observers with different telescopes, is found to amount not merely to few seconds but to whole minutes.

It must be a matter of deep regret to us all, both for his own sake and for that of astronomy, that so valuable and interesting a series of observations should sustain, what I trust however will prove only a temporary interruption from the severe illness of Colonel Beaufoy, which prevents him from receiving in person the mark of our approbation adjudged him by your Council. At his request, therefore, I will hand it to our worthy secretary.

(Here the President delivered the medal to Mr. Stratford, as proxy for Colonel Beaufoy, at the same time thus addressing him:) —

MR. STRATFORD.

When you shall transmit this medal to Colonel Beaufoy, accompany it with the assurance of our warmest approbation of the useful and excellent example he has set, in thus steadily prosecuting from year to year, a train of observations so important in itself and requiring so much patient and persevering attention: an example we trust to see emulated by others, since it shows how much, how very much, may be done with moderate instrumental means, by regular, systematic, and well directed observation. He has succeeded in rendering his name conspicuous among astronomers, and his observatory a standard point of reference,— one of those zero points on earth which, like the standard stars in the heavens, will serve for the determination of innumerable others. Already we are furnished with a conspicuous instance of its use in this respect, in the determination of the longitude of Madras by Mr. Goldingham, which has this night been read to the Society, in which that important element is derived from a very moderate number of corresponding observations made at the two stations, with considerable presumption of exactness. Nor can we suppose that this will prove a solitary instance. Assure Colonel Beaufoy how much we consider science as practically benefited by his labours:—assure him too of lively grief and sympathy for his present sufferings, and our earnest wishes and prayers that he may be speedily restored to the full enjoyment of health, to his friends, and to his favourite astronomy.

AN ADDRESS

DELIVERED BY J. F. W. HERSCHEL, ESQ., PRESIDENT OF THE ASTRONOMICAL SOCIETY OF LONDON, ON THE OCCASION OF THE DELIVERY OF THE HONORARY MEDALS OF THAT SOCIETY, ON FEB. 8. 1828, TO LIEUTENANT-GENRAL SIR T. MACDOUGAL BRISBANE, K.C.B., AND JAMES DUNLOP, ESQ.

GENTLEMEN.

IN pursuance of the award of your Council, which you have just heard, I have now to call your attention to the subject of the honorary marks of this Society's approbation, which it is part of our business at this meeting to bestow. The selection of objects on which such distinction may most deservingly and most usefully be conferred, has been, in this instance, of much interest and some difficulty,—not from a paucity of claims, but from their variety and magnitude. On all sides, both abroad and at home, the spirit of Astronomical research and discovery has been diligently alive. The great work which has been commenced on the Continent, for the determination of the places of all the stars of our hemisphere in zones, has been continued with a patient ardour to which no words can do justice.—The heavens have been ransacked for double stars; and the results of the search, developing a most rich and unlooked-for harvest of striking discoveries, being the first fruits of the great telescope of Fraunhofer, have been consigned to immortality, in a work which does honour to its age and nation, and which has already been brilliantly rewarded in another quarter. The ingenuity of one of our own countrymen has placed new, simple, and powerful means in the hands of observers for verifying the stability of their instruments, and determining their fluctuations. And in every quarter, to go no further in this detail, an activity worthy of the high ends and dignity of our science, has been remarkably displayed. Among so many important labours, however, some of which are yet awaiting their final completion, or receiving the last touches of their authors, the attention of your Council has been fixed, by the imposing mass of valuable observations which has emanated during a series of years, from the Observatory at Paramatta, established by the late Governor of the Colony of New South Wales, Sir Thomas Macdougal Brisbane, one of our Vice-Presidents, long distinguished among us by his ardent love of Astronomy, and an intimate familiarity both with its theory and practice.

Nothing can be more interesting in the eyes of an European astronomer, especially to those whose field of research, like our own, is limited by a considerable northern latitude,— than the southern hemisphere, where a new heaven, as well as a new earth, is offered to his speculations; and where the distance, the novelty, and the grandeur of the scenes thus laid open to human inquiry, lend a character almost romantic to their pursuit.

A celestial surface equal to a fourth part of the whole area of the heavens, which is here for ever concealed from our sight, or whose extreme borders at least, if visible, are only feebly seen through the smoky vapours of our horizon,—affords to our antipodes the splendid prospect of constellations different from ours, and excelling them in brilliancy and richness. The vivid beauty of the Southern Cross has been sung by poets, and celebrated by the pen of the most accomplished of civilized travellers; and the shadowy lustre of the Magellanic clouds has supplied imagery for the dim and doubtful mythology of the most barbarous nations upon earth. But it is the task of the Astronomer to open up these treasures of the southern sky, and display to mankind their secret and intimate relations. Apart, however, from speculative considerations, a perfect knowledge of the astronomy of the southern hemisphere is becoming daily an object of greater practical interest, now that civilization and intercourse are rapidly spreading through those distant regions,—that our own colonies are rising into importance,—and that the vast countries of South America are gradually assuming a station in the list of nations, corresponding with their extent and natural advantages. It is no longer possible to remain content with the limited and inaccurate knowledge we have hitherto possessed of southern stars, now that we have a new geography to create, and latitudes and longitudes without end, to determine by their aid. The advantages, too, to be obtained, even for the perfect and refined astronomy of the north, by placing nearly a diameter of the globe between the stations of observatories, and taking up the objects common to both hemispheres, in a point of view, and under circumstances so every way opposite to those which exist here, have been strongly pointed out by a venerable and illustrious member of this Society, in an elaborate paper published in its Memoirs, and would alone suffice to justify a high degree of interest, as due to every well conducted series of observations from that quarter. The observations of Halley at St. Helena had made known the places of a moderate number of the brighter southern stars; but the only catalogue of any extent and accuracy, which existed previous to the establishment of the observatories of the Cape and Paramatta, was that of Lacaille, who spent three years at the Cape of Good Hope, and the Isles of France and Bourbon; and, though with very inadequate instrumental means, yet, by dint of the most indefatigable industry, succeeded in observing and registering upwards of 10,000 stars. But by far the greater part of these observations have never been reduced; a selection only from them of 1942 of the principal ones, not amounting to a fifth of their whole number, having been formed into a catalogue, and published by this meritorious astronomer. It must be admitted, however, that the degree of accuracy stated by Lacaille himself to have been probably attained by him, is hardly such as to make us now very deeply regret their want of reduction, especially as the observations themselves are printed with every requisite for that purpose, when required. Still, however, from his method of observing, which was with a fixed telescope and rhomboidal network, his observations have what may be termed a dormant value, as they most probably give correct differences for each night's work; and when a catalogue of standard southern stars shall be completed, Lacaille's observations will become available, by regarding these as zero points, and referring all the rest to them.

Such was nearly, with little improvement, the state of the astronomy of the southern hemisphere, when Sir Thomas Brisbane was appointed Governor of the Colony of New South Wales. The intention of our Government to found an observatory on the largest scale, at the Cape of Good Hope, was, indeed, already fixed , and the observer, a member of this Society, supplied with instruments sufficient for the purpose of constructing a preliminary catalogue, occupied himself with the necessary observations, while awaiting the arrival of those ultimately destined to adorn that establishment, and the building of his observatory. The approximate catalogue so constructed and reduced, containing all the southern stars observed by Lacaille, down to the fifth magnitude, is already printed by the Royal Society in their Transactions.

Sir Thomas Brisbane's attachment to Astronomy has ever been a prevailing principle of his mind, and one which even amidst the distractions of a military life of no ordinary degree of activity and adventure, he found means to indulge; and which never deserted him, however the calls of his country might demand his services in a different and more splendid career.

His appointment to the important office of Governor of New South Wales, however, put it in his power to execute to their fullest extent and under the most favourable circumstances, plans of astronomical investigation, which to a private individual would be utterly impracticable. The opportunity was embraced with eagerness. The best instruments,—consisting of an excellent transit of 5½ feet focal length, by Troughton; a mural circle of two feet in diameter, the workmanship also of Troughton, and said to have been the model on which that of Greenwich was constructed, and which had long been in his possession; and a fine 16-inch repeating circle of Reichenbach,—were destined for this service: and two gentlemen engaged as assistants at considerable salaries; the one a foreigner of high estimation as a mathematician and calculator, the other Mr. Dunlop, of whom I shall presently have occasion to say much more. It ought to be mentioned, that this noble equipage was furnished entirely from Sir Thomas's private fortune, and maintained wholly at his own expense. Immediately on his arrival in the colony in 1821, and so soon as an observatory could be erected, and the instruments established, the work of observation commenced, and continued with little interruption under the immediate superintendence and direction of Sir Thomas Brisbane himself, who, though the pressing and important duties of his high office would of necessity seldom admit of his devoting any material proportion of his time to actual observation, yet frequently took a personal share in the labours of the observatory, as a relaxation from higher duties, and in particular, a great portion of the transits were observed by himself.

The first fruits of this enterprise were the observations of the December solstice of 1821, which were published in the Astronomical Notices of Schumacher; in which work also appear those of both the solstices of 1822, and a number of detached and occasional observations, which reached Europe at different times by a variety of channels, and found their way into that valuable collection. The solstices of 1823 were communicated by Sir Thomas Brisbane to this Society, in a letter to our late worthy president, together with a considerable extensive series of observations of principal stars, chiefly those visible in both hemispheres, and which have undergone a careful reduction and close scrutiny in the hands of Dr. Brinkley, the details of which, as well as the original observations, are printed in the first part of the second volume of the Transactions of this Society, and which justify, in the eyes of that experienced observer, as they must in those of every practical astronomer, a decided opinion of the great care and skill with which they have been made.

A great number of occasional observations,—such as eclipses, occultations, and observations of the planets Venus and Uranus, near their conjunctions and oppositions, and of comets from the same source,—are also printed in the same volume. One of the most remarkable single results we owe to the establishment of Sir Thomas Brisbane's observatory consists in the re-discovery of the comet of Encke in its predicted place, on the 2nd June, 1822. The history of this extraordinary body is well known to all who hear me; and as its re-discovery at Paramatta by Mr. Rümker, has already been, on a former occasion, distinctly noticed and rewarded by this Society, there is no occasion that I should here enlarge on it; and yet I cannot help pausing a moment to figure the delight its celebrated discoverer must have experienced to find the calculations, on whose exactness he had pledged himself, thus verified beyond the gaze of European eyes; and this strange visitant gliding, as if anxious to elude pursuit, into its primitive obscurity, thus arrested on the very eve of its escape, and held up to mankind,—a trophy at once of the certainty of our theories and the progress of our civilization.

Observations of the length of the pendulum were not neglected by Sir Thomas Brisbane; and the determination of this important element at Paramatta, forms the subject of a highly interesting and valuable communication made by him to the Royal Society, and printed by them in their Transactions for 1823, and discussed by Captain Kater with his usual care and exactness.

The remainder, and indeed the great mass of the observations made with the mural circle and the transit instrument, have at different periods been communicated to the Royal Society, and are for the present deposited in its archives. Forming our judgment only upon those of which an account has been publicly read at the meetings of that illustrious body, but which are understood to constitute only a comparatively small part of the whole, — they form one of the most interesting and important series which has ever been made, and must ever be regarded as marking a decided era in the history of Southern Astronomy.

It is for this long catalogue of observations — whether scattered through the journals of Europe, printed in our own Transactions, or deposited as a precious charge in the care of a body so capable of estimating their merits; but still more for the noble and disinterested example set by him in the establishment of an observatory on such a scale, in so distant a station, and which would have equally merited the present notice, had every observation perished in its conveyance home,—that your Council have thought Sir Thomas Macdougal Brisbane deserving the distinction of a medal of this Society, which, as he is unable personally to attend this meeting, I will now deliver to his proxy, Mr. South.

Mr. South,

We request you to transmit to Sir Thomas Brisbane this medal, accompanied with the strongest expressions of our admiration of the patriotic and princely support he has given to Astronomy, in regions so remote. It will be a source of honest pride to him while he lives, to reflect that the first brilliant trait of Australian history marks the era of his government, and that his name will be identified with the future glories of that colony in ages yet to come, as the founder of her science. It is a distinction truly worthy of a British governor. The colonial acquisitions of other countries have been but too frequently wrested from unoffending inhabitants, and the first pages of their history blackened by ferocious conquests and tyrannical violence. The treasures of gold and silver they have yielded—the fruits of rapine—have proved the bane of those who gathered them; and in return, ignorance and bigotry have been the boons bestowed on them by their parent nations. Here, however, is a brighter prospect. Our first triumphs in those fair climes have been the peaceful ones of science; and the treasures they have transmitted to us, are imperishable records of useful knowledge, speedily to be returned with interest, to the improvement of their condition and their elevation in the scale of nations.

(The President then resumed his address to the Members, as follows:—)

I have now to call your attention, Gentlemen, to the award of another medal, to Mr. Dunlop, who accompanied Sir Thomas Brisbane in capacity of his assistant, and who, since the middle of the year 1823, when his companion Mr. Rümker left the observatory, remained in the sole charge of the instruments; and up to the period of the departure of his principal from the colony, continued an uninterrupted series of observations with a care and diligence seldom equalled, and never surpassed. In such cases it is not only the head which plans, but the hand which faithfully and promptly executes, that claims our applause. The most liberal provision of instrumental means would have been comparatively unavailing, had the spirit of him who supplied them, been seconded by any ordinary zeal on the part of his assistants. The records of this Society already alluded to, bear sufficient testimony to the merits of Mr. Rümker, and to our sense of them. In Mr. Dunlop were combined qualities rendering him, of all others, the very individual fitted for the duties imposed on him—zealous, active, ready—but above all (and the combination is not an ordinary one), industrious and methodical. In the vast mass of observations made and registered by him, all is equable and smooth, as if the observations had all been made at a sitting:

> . . . . . "Servatur ad imum
> Qualis ad incepto processerit"—

no lacunæ—no long intervals of inactivity—nothing hurried or sketchy; but the same painstaking, laborious filling-in, pervading the whole,—marking that the observer's whole heart and soul were in his work, and that each individual observation possessed its own peculiar, though momentary, interest. Nor is this wonderful. The heavens visible to Europeans, have been so thoroughly examined, and their contents so carefully registered, that there is not the slightest rational probability of anything new or uncommon offering itself to instruments of moderate power in the ordinary course of observation. Here, however, all was new;—for the optical power of Lacaille's telescope was far too feeble to afford much insight into the physical constitution of the objects determined with it: and thus all the excitement of discovery was maintained during every step in the progress of the work.

But to be susceptible of this excitement, so maintained, the observer must be animated by the true spirit of the Astronomer; and few have possessed this spirit in a greater degree than Mr. Dunlop. In a scientific point of view, therefore, he must be regarded as the associate, rather than the assistant of his employer; and their difference of situation becomes merged in their unity of sentiment and object.— These considerations alone would have rendered it impossible to your Council to disunite in any expression or mark of their approbation, individuals who have thus, each in his sphere, gone hand in hand together, towards the perfection of Southern Astronomy, even had the labours of Mr. Dunlop been confined to the ordinary business of an observatory, or to observations with fixed instruments. But this is very far from having been the case. The nebulous, as well as the sidereal heavens, have occupied his attention; and in the prosecution of this most delicate and difficult branch of Astronomy, he has availed himself entirely of his own resources, in the most literal sense,—the instrument which he used being not simply his *own*, but the work of his own hands; and the observations being performed by him after the departure of Sir Thomas Brisbane from the colony, at a personal sacrifice of his private interests, and in the face of difficulties which would have deterred any one not animated with a real and disinterested love of science from their prosecution. The results of these observations have been the description and determination of the places of upwards of 600 nebulæ and clusters of stars. And when it is recollected that Lacaille was able to observe not more than about forty or fifty of these curious objects, we may form some idea of the extent of this labour. In addition to these interesting results, Mr. Dunlop has amassed a copious and valuable collection of Southern double stars, which he is at present occupied in reducing and arranging; and a variety of interesting and curious particulars relative to the magnitudes, colours, and other peculiarities of all the more conspicuous single ones.

Shut out as we are by our geographical situation from the actual contemplation of these wonders, the astronomers of Europe may view, with something approaching to envy, the lot of these their more fortunate brethren. The feeling, if an unworthy, is, however, but a passing one, and merges in that of admiration of their zeal and enterprize, and of gratitude for the information they have afforded us. In testimony of that admiration and that gratitude, on the part of this Society, towards Mr. Dunlop, I beg of you, Mr. South, to transmit to *him* also, this our medal, and to accompany it with the assurance that wheresoever his future fortunes may lead him — whether in the land which has already witnessed his meritorious labours — to complete and extend them, or in his native country, which is both able and willing to appretiate his value, to put the finishing stroke to the noble fabric he has been mainly instrumental in raising, by taking a leading part in the less exciting, but not less useful or indispensable work, of reducing the observations already made:—in either case he will be attended by our best wishes for his prosperity and happiness, and our confidence that science will continue to benefit by his exertions.

AN ADDRESS

DELIVERED AT THE ANNIVERSARY MEETING OF THE ASTRONOMICAL SOCIETY OF LONDON, FEB. 13. 1829, ON PRESENTING THE HONORARY MEDALS TO THE REV. W. PEARSON, PROFESSOR BESSEL, AND PROFESSOR SCHUMACHER, BY J. F. W. HERSCHEL, ESQ., V.P.R.S. F.R.S.E. M.R.I.A. AND PRESIDENT OF THE SOCIETY.

GENTLEMEN,

I HAVE now to claim your attention to the subject of our Annual Prizes. In the distribution of such honours there is frequently a difficulty of choice, from the number and strength of competing claims; and the past year has produced matter of much interest to astronomers. The comet of Encke has revisited our sphere,—not, like a forgotten thing, to take us unawares, but as an old, familiar friend, giving due notice of his arrival, and strict to his appointment. Of all the wonders that astronomy has disclosed to us, there is none more astonishing than to see this dim, misty, all-but-incorporeal thing, whose parts can have no more cohesion than the floating particles of the lightest fog, borne along by its inertia, and commanded by its gravity, like the denser planets with which it must henceforth be associated. We look to this comet as the revealer of many secrets, such as whether there exist a ponderable, or at least material ether inthe planetary spaces, or any vestige of unabsorbed, nebulous matter in our system, susceptible of being caught up by it, and thus diminishing its speed and retarding its progress.

The list of papers which has been read to you, and which has been stated in the Report of your Council, will prove the zeal and diligence of our British astronomers and our associates; and the same spirit which has now for many years been prevalent on the Continent still animates it, and continues to produce the happiest results. At home we have further to congratulate the astronomical public on the attention that has been drawn to the improvement of the achromatic telescope, whether by the use of different media from the usual ones, or by a different combination of those commonly employed; and on all hands we find no reason to regard astronomical discovery as having relaxed in its progress.

The first medal which I shall this day have the honour to present, according to the award of your Council, is a gold medal to Dr. Pearson, for his book, entitled "An Introduction to Practical Astronomy;" one of the most important and extensive works on that subject which has ever issued from the press. The treatises existing on theoretical astronomy are numerous and extensive, as befits the dignity and abstruseness of the subject; but works on the practical department, explaining the best methods of observing, and the precautions necessary, condensing and bringing together the Tables most compendious or more generally adopted, and describing in detail the various instruments employed,—the manner of using and adjusting them,—and the species of observations for which they are adapted;—such works were altogether wanting, or existed at best in a very insufficient state. We have here, however, the desideratum supplied. Dr. Pearson, in his first volume, has brought together all the Tables which the practical astronomer is ever likely to want. Nor is this a mere compilation. Nearly half the volume consists of new Tables, and the rest are for the most part either extended or improved in their arrangement. The most considerable and striking part of Dr. Pearson's work, however, is the second volume, now just published. This is destined to a description of instruments, and is accompanied with a series of plates, so complete as almost to render description superfluous. It is no small advantage in such a case, to know that the instruments described have been for the most part in the actual possession of their describer, and used by him in actual observation. Such has been the case in this instance.

(*The President then, addressing Dr. Pearson, continued thus:*—)

DR. PEARSON.

You have applied your splendid collection of instruments to a most excellent use. You have not frittered your time away in commencing and breaking off a series of observations with one and with another, at the impulse of the moment; but, by studying their peculiarities, noting down their adjustments, and delineating their forms, you have given a most useful aid to those of less experience than yourself, and have added a value to every instrument you have described, by your labours. In dedicating this second volume of your book to our admirable Troughton, you have raised him a monument which must be peculiarly gratifying to his feelings. In your pages he may see how large a place he occupies in his art, and in the science on which it depends; and we all rejoice that he has lived to receive this applause at your hands. For yourself, sir, be assured, this Society, who know the sacrifices you have made, and the liberality with which you have set at nought all calculations of profit and loss in producing a work of such vast expense and such small probable return, congratulate you heartily on thus bringing your labours to a close, and mark the epoch by presenting you with this medal.

(*The President then resumed his address to the Members, as follows:—*)

GENTLEMEN,

The next medal which has been awarded by your Council is a gold medal to Professor Bessel, for his observations of the stars in *zones*, made by him at the Royal Observatory of Königsberg;—a vast undertaking, and one which would alone suffice to confer immortal honour on a name which has already so many other independent claims to astronomical distinction. The attention of astronoiners, in fixed national observatories, up to a late period, was almost exclusively confined to observations of the sun, moon, and planets, and a moderate number of the principal fixed stars. The smaller stars, the minor host of heaven, were systematically neglected, and the conspicuous ones only deemed worthy of being observed in any other than a desultory way. Their utility for the purposes of nautical astronomy might of course be expected to draw upon the most remarkable ones a proportionate attention; but astronomers, like the vulgar, had been too much influenced by appearances and by glitter, and had fallen into habitual neglect of the rest, or contented themselves with rough approximations to their places, sufficient to mark them down in maps, or include them in lists and approximate catalogues; but inadequate to the determination of any delicate question as to their proper motions, parallax, &c. To this, however, one splendid exception must be made in the Catalogue of Piazzi. This record of the places of more than 7000 stars of all magnitudes, determined with an excellent instrument, with all the care of a diligent and cautious observer, and from several observations of each, is one of the finest monuments of astronomical research. Nor ought the labours of Lalande to be forgotten. His examination, indeed, was extended to an enormous list, to no fewer than 50,000, and was conducted, like Professor Bessel's, in zones. It has been rendered available, also, to astronomers, by tables of reduction, of the simplest possible kind, published by Professor Schumacher, and is, indeed, a most useful and valuable collection. It labours, however, under the disadvantage of a great inferiority in an instrumental point of view, and therefore can be nowise regarded as superseding or anticipating the more refined inquiries of Professor Bessel.

It would be quite superfluous to speak here of the general merits of Professor Bessel as an astronomer, or of the excellence of the observations regularly made in the observatory under his direction. We know and appretiate them : but they are not to be made the subject of our remarks or our praise on this occasion. The observations for which your medal is awarded to him were commenced in 1821, and have been continued with little intermission ever since, at the Royal Observatory at Königsberg, with the meridian circle of Reichenbach, having a magnifying power of 106 applied to a most excellent telescope. This instrument being confined to a zone of about two degrees in breadth, is made to oscillate or sweep up and down continually, while the heavens pass in review before the observer by their diurnal motion, and all stars, down to the ninth magnitude, which pass the field, are taken at once in right ascension and declination, and read off by the clock and limb of the circle. This mode of observing presents two capital advantages, — viz., multitude of objects, and facility of reduction. Of the former we may judge by the fact, that in some of the zones we find between three and four hundred objects observed at a sitting: with respect to the latter, a little table, of the simplest use and most compendious form, is attached to each zone, and by its aid the readings of the clock and limb are at once reduced (by a calculation comprised in three lines) to the mean right ascensions and declinations of the objects at a fixed epoch, freed from instrumental error, and ready for the catalogue. Those only, who know by experience the labour of reducing observations not made on this system, can imagine the saving of toil and drudgery thus arising. Nay, more — it renders the observation book itself available as a catalogue; for, by the system of indexing the zones, any point in the heavens may at once be referred to, and every object there observed at once reduced, without need to turn over the book, to enter into any inquiry, or in any way refer beyond the page before us and the table of reductions in the beginning of each volume. This is the perfection of astronomical book-keeping.

This course of observations was commenced, as I have already said, in 1821; and you may judge of the industry and perseverance with which it has been prosecuted, by the fact, that, by the end of 1824, the whole equatorial belt, of thirty degrees in breadth, of the heavens, had been swept, and between 30 and 40,000 stars observed. But this did not satisfy the zeal, or exhaust the patience of M. Bessel. He has since continued the work northward with unabated ardour, and is extending his zones to the forty-fifth degree of northern declination: thus embracing, in the whole, sixty degrees of the finest part of the heavens.

A great many double stars, some of them very delicate ones, have been detected in these sweeps; they are included in M. Struve's splendid catalogue of these objects. Nor is it at all improbable that many new planets may have been seen, and, on a repetition of the observations, will be found missing. In a word, we have, in this collection, one of those great masses of scientific capital laid up as a permanent and accumulating fund, the interest of which will go on increasing with the progress of years. It is a harvest sown, and already springing, but of which the ripened produce is destined for after generations. Yet the crop, if a remote, is a sure one. It will neither be uprooted by political convulsions, nor stinted by neglect, nor spoiled by premature gathering in. The language of such a record is like that of prophecy. It is written, but we cannot yet read it. It is full of truth, but not for us. It contains the statement of a vast system, but future generations must develope it. Could it be permitted us to look forward and draw aside the veil which a few centuries interpose between us and its interpretation, we might expect to see all the great questions which agitate astronomers set at rest, and new ones, more refined and grounded on their solution arising. Some minute and telescopic atom will perhaps have become the stepping-stone between our system and the starry firmament—its parallax will mark it for our neighbour — and either its fixity will demonstrate the equilibrium of our immediate sidereal system, or its proper motion reveal to us the nature and extent of the forces which pervade it. The orbits of those remarkable stars which are ascertained to be really *erratic*, or which have a proper motion too large to be overlooked, such as 61 *Cygni* and μ *Cassiopeiæ*, will become known. They will be seen to deviate in their paths from great circles of the heavens—their convexity or concavity will mark the directions, and their changes of velocity the intensities, of the forces which urge them. Already, since the date of the first catalogue of fixed stars, the former of these wonderful objects has moved over no less than four degrees of the heavens. Had it been accurately observed but once in a century, what might we not have known! Let this consideration stimulate astronomers to follow up the splendid example Professor Bessel is setting, and complete and pursue the gigantic task he has carried on so far, but which is beyond the power of any one man to go through, much less to repeat. How much is escaping us! How unworthy is it of those who call themselves philosophers to let these great phenomena of nature — these slow, but majestic, manifestations of the power and glory of God — glide by unnoticed, and drop out of history, beyond the power of recovery, because we will not take the pains to note them in their unobtruding and furtive passage; because we see them in their every-day dress, and mark no sudden change; and conclude that all is dead, because we will not look for the signs of life; and that all is uninteresting, because we are not impressed and dazzled.

We must not, however, be hasty in our reproaches. There is a general sense afloat among the continental astronomers, of the necessity of laying a foundation for future sidereal astronomy, as deep and as wide as the visible constituents of the universe itself. Nothing less than ALL will be enough — *quicquid nitet notandum.* To say, indeed, that every individual star in the milky way, to the amount of eight or ten millions, is to have its place determined and its motion watched, would be extravagant; but at least let samples be taken — at least let monographs of parts be made, with powerful telescopes and refined instruments, that we may know what is going on in that abyss of stars, where, at present, imagination wanders without a guide. Let us at least scrutinize the interior of sidereal clusters. Who knows what motions may subsist, what activity may be found to prevail, in those mysterious swarms? Or if we find them to be composed of individuals at rest among themselves — if we are to regard them as quiescent societies of separate and independent suns, bound by no forcible tie like that of gravity, but linked by some more delicate and yet more incomprehensible cause of union and common interest—the wonder is all the greater. We walk among miracles, and the soul yearns with an intense desire to penetrate some portion of those secrets, whose full knowledge, after all, we must refer to a higher state of existence, and an eternity of sublime contemplation.

Another gold medal, gentlemen, has been awarded to M. Schumacher for the important services rendered by him to both practical and theoretical astronomy by the publication of his various Astronomical Tables, and of his work entitled "Astronomische Nachrichten."

Astronomy is a science peculiarly in unison with the German national character. The persevering industry which forms so striking a feature in it, is the quality, of all others, requisite for an astronomer—that diligence which never wearies, and which, working slowly, and destroying nothing that is done *, goes on adding grain by grain to the mass of results, and accumulating them with a kind of avarice to swell the heap;—that painstaking scrutiny which penetrates through all details, and will not be satisfied till perfection is attained. And, on the other hand, an enthusiasm seemingly

* Beschäftigung die nie ermattet, die langsam wirkt doch nie zerstört, &c.—*Schiller*.

incompatible with this plodding turn, yet often co-existing with it in the same mind; a love of systems for their own sake; a spirit of speculation, sometimes bordering on wildness; and an ardent inherent love of the vast and wonderful. Among minds of this turn it is no wonder that astronomy should flourish—with enough of sublimity and mystery to exhaust the wildest imagination, and enough of laborious detail to keep in employment the most patient industry. Accordingly, Germany has always been fruitful in astronomers, and (regarding as Germans all who are bound in the common family tie of language and manners) German astronomy has at present reached a pitch of eminence, which only national pride prevents our acknowledging to be unexampled in the history of the science—whether we consider the researches of their theorists, the activity of their computers, or the number and importance of their national observatories: or those of Russia, several of which are manned (so to speak) with directors and assistants who have been educated in the German school, and transplanted from German observatories, and from the personal tuition of their most illustrious men, who have worked with them as their friends and pupils, rather than as mere assistants, and who look up to them with the veneration of the scholar to his master.

Among all these, and among those numerous and talented individuals throughout the continent, and in England, who are attached to astronomy professionally, or from love of the science, the "Astronomische Nachrichten" of Professor Schumacher establish a point of concourse—a complete bond of union: we have there a theatre of discussion of whatever is most new and refined in the theory and practice of astronomy—the utmost delicacies of computation and scrupulous investigations of instrumental errors are given by those most competent to supply and to judge of them. To its pages observations of every kind find their way, especially those which depend for their utility on corresponding observations, or which lose their interest and importance by long suppression. Not a comet appears but *there* we find its elements handed in from all quarters with emulous rapidity—occultations—moon culminating observations—computations of longitudes and latitudes—disquisitions on practical points—descriptions, advertisements, and prices of instruments—in a word, everything which can awaken and keep alive attention to the science—everything that can facilitate the contact of mind with mind. Every one who has attended to the progress of knowledge in recent times, must feel all the importance of such an engine. But it cannot be kept in action without a strong presiding power. In any inferior hand it would languish, and soon fall into disrepute and inaction. Professor Schumacher is, of all men, that one whom the voice of Europe would have fixed on for the conduct of such a work: an excellent astronomer himself, and presiding over an observatory in which every thing is delicate and exquisite, he possesses that practical and theoretical knowledge which commands respect, and gives his acceptance or rejection of contributions a weight from which there is no appeal. He has, moreover, the eminent but merited good fortune to possess the full and effective support of a Government deeply impressed with the importance of astronomical science. With this powerful aid, which would have been accorded to no other, he has been enabled to establish sure and regular communications with every part of the civilized world —and to face an expenditure which, under similar circumstances, no private individual would have ventured to undertake. He has thrown his whole weight into the scale of advancing science; and the effect has been, the establishment of a great European astronomical republic, with a common feeling, and a sense of common interests.

But the services rendered by M. Schumacher to astronomy are not limited to this publication. A numerous and useful collection of Tables has been edited by him, under the title of "Hülfstafeln," or Assistant Tables, and others. One of these volumes is devoted to facilitate the reduction of the observations of Lalande in the "Histoire Céleste," on the same plan with those used for the reduction of Bessel's zones. This truly useful work rescues from oblivion the labours of Lalande, and renders his observations available to science. M. Schumacher, liberally assisted, in a pecuniary point of view by the Royal Danish Hydrographic Office, has also followed up the example set by the Coimbra Ephemeris, of the publication of lunar distances from the planets,—thus rendering available a new branch of nautical astronomy, and hastening the period when observations of the planets at sea would have naturally been called for.

In the computation of the Assistant Tables, M. Schumacher has had most active assistance from several accomplished Danes; of whom I may mention Hansen, Clausen, Ursin, Nissen, Nehus, Zahrtmann, and Petersen. In honouring the principal we honour the accessaries; and we trust that the tribute of this passing notice will not be displeasing to them and their coadjutors.

Captain Smyth,

As you are kind enough to act as proxy for Professors Bessel and Schumacher, receive for them these their respective medals; and, in transmitting them, take care to convey to them the expression of our gratitude and admiration for the services they have rendered to our science, and our wishes that their brilliant and useful career may be prolonged yet many years, with increase of glory, and with health and prosperity to enjoy it.

AN ADDRESS

TO THE ROYAL ASTRONOMICAL SOCIETY ON THE OCCASION OF PRESENTING THEIR GOLD MEDAL TO M. PLANA, DIRECTOR OF THE ROYAL OBSERVATORY OF TURIN, FEB. 1840.

GENTLEMEN.

THE award of our medal for this year to Signor Plana is an act, as it may at first sight appear, of somewhat tardy justice. Those great works on the lunar theory (for which that award is made), and on the perturbations of the planets, especially of Jupiter and Saturn, have now been so long before the public, that it may almost appear as if, in the dearth of matter of sufficient interest of later date, your Council had been ransacking the annals of modern astronomy to find something on which they might rely in a kind of inglorious safety for a justification of their award.

This would be a very erroneous view indeed to take of this subject. So far from experiencing a lack of matter to choose from,—so far from a deficiency of interest in the subjects which have shared the consideration of the Council in coming to the conclusion they have done—there have been, in fact, on probably no occasion, such powerful countervailing claims—and so far from seeking, in this award, a merely safe and justifiable course, it has required no common share of boldness and decision in four judges to put aside those claims, in favour of M. Plana's,—of that boldness I mean which is based on justice and a longsighted view of public utility.

Before I proceed, therefore, to state the reasons which have weighed with the Council to take the step they have done, it will be right for me to mention, at least in general terms, two of the subjects which have chiefly divided their attention on the occasion; and this I am fortunately enabled to do, infinitely better than I could pretend to do it on my own knowledge and reading, by the aid of most excellent reports on those subjects laid before the Council by Professor Airy and Mr. Main, the one on the subject of Professor Hansen's general researches on physical astronomy, the other on Professor Bessel's and Mr. Henderson's observations on the parallax of those remarkable double stars, 61 Cygni and α Centauri—observations which it would appear, beyond question, have brought us to the very threshold of that longsought portal which is to open to us a measurable pathway into regions where the wings of fancy have hitherto been overborne by the weight, or baffled by the vagueness of the illimitable and the infinite.

Mr. Hansen's researches on the lunar and planetary theories are every way most remarkable, and seem likely to lead to results of the utmost generality and importance. He has attacked the great problem of three bodies (extended, in the conception and application of his methods, to the mutual perturbations of *four*) by a method entirely novel in its idea, although based on and starting from Lagrange's idea of the variation of the elements. Of this method, it would not be easy, in words unaided by symbolic expression, to give any distinct account, but its principle may be stated in general terms, as assuming not the elliptic *elements*, but the elliptic *time*, to be subject to perturbation, or, in other words, as considering the perturbed co-ordinates each to arise from the combination of invariable *elements* with a varied or perturbed *time*, the amount of whose variation shall exactly account for all that the variation of the elements accounts for in Lagrange's method. The mere mention of this refined and abstruse mode of conceiving the problem must suffice to show, that, to carry it into effect, must require at every step a contention of mind, a degree of intellectual effort, far surpassing what is required for the mere management of algebraic symbols and developments, however intricate.

Whatever be the skill and dexterity, however, exhibited by the author of this truly original conception, and whatever promise it must be considered as holding out for the future advancement of our knowledge in this intricate research, it can hardly yet be regarded as having attained that extent of development which it will require to supersede in the construction of Tables, and the actual calculation of the lunar and planetary perturbations, the methods already in use, which the researches of Clairaut, Laplace, Lagrange, Poisson, Damoiseau, and Plana, have wrought up to such a pitch of practical perfection. Hansen's theory appears to afford what, in the actual state of our knowledge, must be regarded as most precious—a new handle by which to seize this refractory problem—one of universal applicability and gigantic power and *purchase*, but of which the management is not yet fully reduced to practice, and of which even the author himself can scarcely yet be said to have acquired the entire mastery. In the theory of Jupiter and Saturn, indeed, the final numerical results are obtained, and Tables calculated; but in the lunar theory, which (in the words of Mr. Airy) "must be considered as the ground of his chief analytical triumph, there exist at present only what may be termed the *foundations* for such a theory." "No man living" (I continue to use the words of the eminent geometer last mentioned), "no man living, probably, except Mr. Hansen himself, could work it into a complete lunar theory; and the exhibition of numerical results is here, therefore, still distant." Let us hope that he will not long leave them so.

On the other subject to which I alluded—the parallax of the fixed stars—it would be doing an injustice to the valuable report of Mr. Main, which, as a beautiful specimen of astronomical history, I hope to see ere long adorning our Transactions, if I were to avail myself more largely on this occasion than is absolutely necessary. It has long been understood by astronomers, that the research of parallax ought not to be confined to the largest stars, but that, in order to determine our choice of stars for this research, other *prima facie* grounds for suspecting a proximity to our system ought to be taken into consideration; such as great proper motion, or, in the case of a double star, great apparent dimensions of the orbits described about each other. In the case of the double star 61 Cygni, both these indications combine to point it out as deserving inquiry. In that of α Centauri they also conspire: for it is well known that this fine double star has a considerable proper motion: and my own observations prove, that the mutual orbit described by its individuals about each other, is of unusually large angular dimension. The great brilliancy of the star also, and its situation in a region of the heavens in which the stars, generally speaking, seem to be less remote than in others, all favour the expectation of a measurable parallax being detected in it: and such Mr. Henderson, from his own observations, assigns to it. I am not about to criticise this result; on the contrary, I am disposed to attribute much weight to his conclusion, but it is only on a very long series of observations, on *absolute places*, affected as they are by instrumental error and uncertainty of refraction, that any conclusion of this kind can rest with security.

Bessel has attacked the question in a different way, by measuring at all times of the year the angular distance of the stars composing the double star 61 Cygni from two small stars visible in the same field of view, and within limits adapted to secure micrometrical measurement. The method is unexceptionable, the measurements conducted with consummate skill, and their reduction executed with all possible regard to everything likely to influence the result. And this result is, to assign a minute, it is true, but perfectly unequivocal amount of parallax, in a way so striking as hardly to allow a doubt of its reality. Such is the impression on merely reading the numerical statement: but, put in the light in which Mr. Main has placed it, by the graphical projection of the measures, the conclusion seems quite irresistible:

> "Segnius irritant animos demissa per aurem,
> Quam quæ sunt oculis subjecta fidelibus, et quæ
> Ipse sibi tradit spectator."

It may now be reasonably asked, if all this be so, why have your Council hesitated to mark this grand discovery with that distinct stamp of their conviction and applause, which the award of their annual medal would confer? A problem of this difficulty and importance solved, so long the cynosure of every astronomer's wishes, the ultimate test of every observer's accuracy — the great landmark and *ne plus ultra* of our progress, thus at once rooted up and cast aside, as it were, by a *tour de force*, ought surely to have commanded all suffrages. It is understood, however, that we have not yet all M. Bessel's observations before us. There is a second series, equally unequivocal (as we are given to understand) in the tenour, and leading to almost exactly the same numerical value of the parallax, and not yet communicated to the public. Under these circumstances, it became the duty of your Council to suspend their decision. But should the evidence finally placed before them at a future opportunity justify their coming to such a conclusion, it must not be doubted that they will seize with gladness the occasion to crown, with such laurels as they have it in their power to extend, the greatest triumph of modern practical astronomy.

M. Plana is well known to the astronomical world as the director of the Observatory at Turin, from which have emanated some valuable series of observations. In conjunction with M. Carlini, he also carried on that extensive and important triangulation of the Savoy Alps, which have made his name celebrated as a geodesist. His works, too, on many other subjects, both astronomical and purely analytic, are of great importance, particularly his investigations on the subject of refraction prefixed to the Turin observations, from 1822 to 1825, published in 1828; those on the motion of a pendulum in a resisting medium, &c. But it is of his researches on the lunar theory for which our medal has been actually awarded, and of these it behoves me now to speak; and I cannot do so in more clear, concise, and discriminating terms, than those used by Mr. Airy, in his report already alluded to: —

"The method pursued by Plana in his 'Théorie de la Lune,' is slightly, but not importantly, different (I mean in the fundamental equations) from those of his predecessors, Clairaut, Laplace, and Damoiseau. He first starts with the method of variation of elements, and pursues it to such an extent as to ascertain generally the form of the expressions connecting the longitude, the latitude, and the time. He then reverts to Clairaut's equations; and as these equations require for the successive substitutions an approximate expression for time in terms of longitude, he adopts a peculiar form (suggested by the variation of elements) for the principal part of it, and attaches to that principal part a subordinate part marked with the prefix δ. The same thing is done for the latitude. The process then is tolerably direct, and is almost similar to that of antecedent writers. In the fundamental algebra, therefore, there is no very great originality in the plan; but the mode followed in the detail of the work is beyond all praise. In the whole of the analytical combinations of this immense work, every part arising from the combination of any one term (however small) with any other term, is given separately in such a form as to leave no difficulty in the detection of error to any careful examiner. The terms of peculiar difficulty (as, for instance, that depending on twice the distance between the node and the perigee) are made the subject of special discussion; and, in some instances, the origin of discordance between the author's results and those of Laplace, is investigated with the same clearness which prevails through the other operations.

"In one respect, the plan of investigation differs much from those of his predecessors, as well as from Hansen's. The investigation is wholly symbolical: no numerical value is introduced, and no consideration of relation of values entertained, till the final substitutions are made. As an example of theory, there can be no doubt of the beauty of this process. As a subject for practical accuracy, it may be not so certain whether it is advisable. The convergence of the series is sometimes extremely slow. As far as I can observe, the accuracy of this method is exactly and properly that of successive substitution; but in various parts of the lunar theory (in all places where the terms rise two orders by integration) the method of successive substitution is not sufficient: in fact, it is necessary to assume a term in order to find its correct value. Adopting this method, however, the author has pushed it as far as, probably, it will ever be carried. The whole is worked to the fifth order, and some parts to the seventh order.

"Finally, the author has determined from observations the principal constants which require to be substituted in the symbolical expressions, and has substituted them, and has thus produced a set of numerical expressions, which may immediately be used for the formation of lunar Tables.

"In terminating the remarks on the works of these two authors, Plana and Hansen, I must again express my very great admiration for both. But their merits are of very different kinds. The theory of Hansen is undoubtedly of the higher order, but it can hardly yet be said to be practical (at least in the lunar theory): many years will yet elapse before it will influence the lunar Tables. The theory of Plana is very good, and probably adequate in all respects: it is eminently practical in form: it has already influenced the investigations of other writers, and will probably soon influence the Tables."

There is but one thing more to add to this clear and powerful summary, and I will supply it by a quotation from the work itself:

"Je n'ai pu me faire aider par personne; j'ai dû traverser seul cette longue chaîne des calculs, et il n'est pas étonnant si par inadvertence j'ai omis quelques termes qu'il fallait introduire pour me conformer à la rigueur de mes propres principes."

When we look at the work itself, there seems something almost awful in this announcement.

A very important memoir of M. Plana, on the theory of the planetary perturbations has adorned the Transactions of this Society. The points of which it treats are miscellaneous, and some of them, perhaps, not of the highest importance, except in one point of view, and that, perhaps, the most important of all. Everyone who is at all conversant with these researches, must be impressed with the enormous interval which separates — I will not say the mere differential equations of the planetary motions — but their integrals, after much and intricate development, from the final numerical results on which their Tables are to be constructed; that is to say, the computed values of the co-efficients of terms having the same argument, when assembled from all the points whence they arise in the algebraic processes, and amalgamated together. M. Plana appears to have proposed to himself the gigantic task of revising and correcting not only those algebraic developments, but the actual numerical calculations of the whole "Mécanique Céleste;"

and this paper contains many examples sufficiently proving the necessity of such revision, and leading the way to those further and more elevated researches on the theory of Jupiter and Saturn, to which the latter part of this memoir must be considered as having given occasion; and which are further developed in several other memoirs published in various academical and other collections.

Neither the time nor the nature of this occasion would allow of my entering into any history of the controversy to which the revision thus set on foot, and the discordant results arrived at in this memoir, gave occasion. Suffice it to say, that errors — venial, no doubt, and such as it would be miraculous did they not exist, were discovered on all sides, and the absolute necessity established not merely of a thorough revision of every part of these immense computations, but of printing and publishing the steps in that regular and methodical form, which alone can put it in the power of subsequent calculators to lay their finger on the precise point where error shall have crept in, and to resume the calculations from that point without sacrificing the whole of what precedes.

It is this methodical clearness, this letting in of the light on every dark corner of every intricate combination and heart-breaking numerical calculation, which may be regarded as marking from this time a new era almost in the planetary theory itself. In the "Mécanique Céleste," we admire the elegance displayed in the alternate interlinking and development of the formulæ, and exult in the power of the analytical methods used; but when we come to the statement of numerical results, we quail before the vast task of filling in those distant steps, and while cloud rolls on after cloud in majesty and darkness, we feel our dependance on the conclusions attained rather to partake of superstitious trust, or of amicable confidence, than of clear and demonstrative conviction. Let me not be misunderstood, as, by these expressions casting any reflection on the conduct of that immortal work. The surest proof of its title to such immortality which can be given, is that microscopic examination subsequently lavished on every point embraced in its immense outline. It is no disparagement to the agriculturist whose energies have extirpated the wilderness, and established in its place cultivation and wealth, that a period shall arrive, when his furrows shall, in their turn, be replaced by the garden, and his system of culture by a measured and calculated succession. Neither would I be understood to lay the whole stress of our applause of M. Plana's researches on the luminousness of their statement. His analysis is always graceful, his combinations well considered, and his conceptions of the ultimate results to be expected from them perfectly just, and justified by the results when obtained.

It cannot but be agreeable to this meeting to know that our award is duly appreciated by M.

Plana himself, and regarded by him in the light which it is ever most desirable it should be, as a stimulus to fresh researches, and further exertions of his powerful talents in the same line where they have already reaped so rich a harvest. No sooner had the Council decided on their award, than, as in private regard no less than in public duty bound, I communicated to him the result; and his reply, which breathes the warmest spirit of attachment to the Astronomical Society, and of undiminished zeal in his own peculiar line of research, is now before me. In the absence of any personal friend to receive it for him, I shall now, therefore, present our medal to Mr. Rothman (in the absence of our Foreign Secretary, Captain Smyth), in his name, and request him to forward it to him, with our best wishes for his health and happiness.

ADDRESS

OF THE PRESIDENT, SIR J. F. W. HERSCHEL, BART., ON THE PRESENTATION OF THE GOLD MEDAL OF THE ROYAL ASTRONOMICAL SOCIETY, TO PROFESSOR BESSEL, AT THE ANNIVERSARY MEETING, FEBRUARY 12. 1841, FOR HIS OBSERVATIONS AND RESEARCHES ON THE PARALLAX OF 61 CYGNI.

GENTLEMEN.

THE Report of the Council has placed before you so ample a view of the state of the Society, of its labours during the last year, of the accessions to its members, and of the many and severe losses it has had to deplore, that little is left for me to add, except my congratulations on its continued and increasing prosperity. It would be inexpressibly gratifying to me if I could persuade myself that my own exertions in its chair had contributed, even in a small degree, to that prosperity; but, alas! I have felt only too sensibly how very feebly and inefficiently, especially during the last year, owing to a variety of causes, but chiefly to residence at a distance from London, I have been able to fill that most honourable office.

The immediate object of my now addressing you, gentlemen, is to declare the award by your Council of the gold medal of this Society to our eminent associate, Mr. Bessel, for his researches on the annual parallax of that remarkable double star 61 *Cygni*,—researches which it is the opinion of your Council have gone so far to establish the existence and to measure the quantity of a periodical fluctuation, annual in its period and identical in its law with parallax, as to leave no reasonable ground for doubt as to the reality of such fluctuation, as something different from mere instrumental or observational error: an inequality, in short, which, if it be *not* parallax, is so inseparably mixed up with that effect as to leave us without any criterion by which to distinguish them. Now in such a case, parallax stands to us in the nature of a *vera causa*, and the rules of philosophizing will not justify us in referring the observed effect to an unknown and, so far as we can see, an inconceivable cause, when this is at hand, ready to account for the whole effect.

I say, in the nature of a *vera causa*, since each particular star must of necessity have some parallax. Every *real existing material* body, must enjoy that indefeasible attribute of body, viz., *definite place*. Now place is defined by *direction* and *distance* from a fixed point. Every body, therefore, which does exist, exists at a certain definite distance from us and at no other, either more or less. *The distance of every individual body in the universe from us is, therefore, necessarily admitted to be finite.*

But though the distance of each particular star be not in strictness infinite, it is yet a real and immense accession to our knowledge to have *measured* it in any one case. To accomplish this has been the object of every astronomer's highest aspirations ever since sidereal astronomy acquired any degree of precision. But hitherto it has been an object which, like the fleeting fires that dazzle and mislead the benighted wanderer, has seemed to suffer the semblance of an approach only to elude his seizure when apparently just within his grasp, continually hovering just beyond the limits of his distinct apprehension, and so leading him on in hopeless, endless, and exhausting pursuit.

The pursuit, however, though eager and laborious, has been far from unproductive even in those stages where its immediate object has been baffled.

The fact of a periodical fluctuation of *some kind* in the apparent places of the stars was recognized by Flamsteed, and erroneously attributed to parallax. The nearer examination of this great phenomenon with far more delicate instruments, infinitely greater refinement of method, and clearer views of the geometrical relations of the subject, rewarded Bradley with his grand discoveries of aberration and nutation, and enabled him to restrict the amount of possible parallax of the stars observed by him within extremely narrow limits.

Bradley failed to detect any appretiable parallax, though he considered 1″ as an amount which would not have escaped his notice. And since his time this quantity has been assumed as a kind of conventional limit, which it might be expected to attain but hardly to surpass. But this was rather because, in the best observations from Bradley's time forward, 1″ has been a tolerated error; a quantity for which observation and mechanism, joined to atmospheric fluctuations and uncertainties of reduction, could not be held rigidly accountable even in mean results; than from any reason in the nature of the case, or any distinct perception of its reality. If parallax were to be detected at all by observations of the absolute places of the stars, it could only emerge as a "residual phenomenon," after clearing away all the effects of the uranographical corrections as well as of refraction, when it would remain mixed up with whatever uncertainties might remain as to the co-efficients of the former, with the casual irregularities of the latter, and with all the forms of instrumental and observational error. Now these have hitherto proved sufficient, even in the observation of zenith stars, quite to overlay and conceal that minute quantity of which astronomers were in search.

It is not my intention, gentlemen, to enter minutely into the history of the attempts of various astronomers on this problem, whether by the discussion of observations of one star, or by the combination of those of pairs of stars opposite in right ascension; nor with the occasional gleams of apparent success which, however, have always proved illusory, which have attended these attempts. For such a history, and, indeed, for a complete and admirably drawn up monograph of the whole subject, I must refer to a paper lately read to this Society by Mr. Main, and which is now in process of publication in the forthcoming volume of our "Memoirs." In whatever reference I may have to make to the history of the subject, I must take this opportunity to acknowledge my obligations to the author of this paper, as well as for his exceedingly luminous exposition of the results of those more successful attempts on the problem by Henderson, Struve, and Bessel, which I shall now proceed more especially to consider.

It would be wrong, however, not to notice that the first indication of some degree of impression beginning to be made on the problem seems to be found in Struve's discussion of the differences of right ascension of circumpolar stars in 1819, 20, and 21. The only *positive* result, indeed, of these observations is, that in the case of twenty-seven stars examined, none has a parallax amounting to half a second. But *below* this, there certainly do seem to be indications in the nature of a real parallax, which might at least suffice to raise the sinking hopes of astronomers, and excite them to further efforts.

But the time arrived when the problem was to be attacked from a quarter offering far greater advantages, and exposed to few or none of those unmanageable sources of irregular error to which the determinations of absolute places are liable. I mean by the measurement of the distances of such double stars as consist of individuals so different in magnitude as to authorize a belief of their being placed at very different distances from the eye; or, as Struve expresses it, *optically* and not *physically* double. This, in fact, was the original notion which led to the micrometrical measurements of double stars; but not only was anything like a fair trial of the method precluded by the imperfections of all the micrometers in use until recently, but the interesting phenomena of another kind, which began to unfold themselves in the progress of those measurements, led attention off altogether from this their original application, which thus lay dormant and neglected, until the capital modern improvements, both in the optical and mechanical parts of refracting telescopes, and the great precision which it was found practicable, by their aid, to attain in these delicate measurements, revived the idea of giving this method, what it never before had, a fair trial. The principle on which the determination of parallax by means of micrometrical observations of a double star turns, is extremely simple. If we conceive two stars very nearly in a line with the eye, but of which one is vastly more remote than the other, each, by the effect of parallax, will appear to describe annually a small ellipse about the mean place as its centre. These two ellipses, however, though similar in form, will differ in dimension; that described by the more remote star being comparatively much smaller: consequently, the apparent places being similarly situated in each, their apparent distance on the line joining these apparent places will both oscillate in angular position and fluctuate in length, thus giving rise to an annual relative alternate movement between the individuals both in position and distance, which is greater the greater the difference of the parallaxes.

Thus it is not the absolute parallax of either, but the difference of their parallaxes, which is effectively measured by this method; *i. e.*, by repeating the measurements of their mutual distance at all times of the year. But, on the other hand, aberration, nutation, precession, and refraction, act equally on both stars, or so very nearly so as to leave only an exceedingly small fraction of these corrections bearing on the results. And when the stars are very unequal in magnitude, there is a presumption that the difference of their parallaxes is very nearly equal to the whole parallax of the nearer one.

The selection of a star for observation involves many considerations. In that pitched on by M. Bessel (61 *Cygni*), the *large star* so designated, is in fact a fine double star: nay, one that has been ascertained to be physically double. It is in every respect a highly remarkable star. The mutual distance of its individuals is great, being about 16¼″. Now this being necessarily less than the axis of their mutual orbit, affords in itself a presumption that the star is a *near one*. And this presumption is increased by the unusually great proper motion of this binary system, which amounts to nearly 5″ per annum, and which has been made by Sir James South the subject of particular inquiry, and found to be *not* participated in by several small surrounding stars, *which, therefore,* are not physically connected with it. Moreover, the angular rotation of the two, one about the other, has been well ascertained.

Now, it fortunately happens, that of these small surrounding stars there are two very advantageously situated for micrometrical comparison with either of the individuals of the binary star, or with the middle point between them. The one of these (*a*), at a distance of 7′ 42″, is situated nearly at right angles to the direction of the double star; the other (*b*) at a distance of 11′ 46″, nearly *in* that direction. Considering (*a*) and (*b*) as fixed points then, and measuring at any instant of time their distances from (*c*), the middle point of the double star, the situation of (*c*) relative to (*a*) and (*b*) is ascertained; and if this be done at every instant, the relative *locus* of (*c*), or the curve described by it on the plane of the heaven with respect to the fixed base-line *a b*, will become known.

Now, on the hypothesis of parallax, that locus ought to be an ellipse of one certain calculable eccentricity and no other. And its major and minor axes ought to hold with respect to the points, *a b*, certain calculable positions and no others. Hence it follows that the distances *a c* and *b c* will each of them be subject to annual increase and diminution; and *that,* 1st, in a given and calculable ratio the one to the other; and, 2ndly, so that the maxima and minima of the one distance ($a\,c$) shall be nearly contemporaneous with the *mean* values of the other distance $b\,c$, and *vice versâ*.

Thus we have, in the first place, several particulars independent of mere numerical magnitudes; and, in the second place, several distinct relations *à priori* determined, to which those numerical values must conform, if it be true that any observed fluctuations in these distances ($a\,b$) ($a\,c$) be really parallactic. So that if they be found in such conformity, and the above-mentioned maxima and minima do observe that interchangeable law above stated; and if, moreover, all due care be proved to have been taken to eliminate every instrumental source of annual fluctuation; there becomes accumulated a body of probability in favour of the resulting parallax, which cannot but impress every reasonable mind with a strong degree of belief and conviction.

Now, all these circumstances have been found by M. Bessel, in his discussion of the measures taken by him (which have been very carefully and rigorously examined by Mr. Main in the paper alluded to, as have also M. Bessel's formulæ and calculations, for in such matters nothing must remain unverified), to prevail in a very signal and satisfactory manner. Not one case of discordance, in so many independent particulars, have been found to subsist; and this, of itself, is high ground of probability. But we may go much farther. Mr. Main has projected graphically the deviations of the distances ($a\,c$) and ($b\,c$) from their mean quantities (after clearing them of the effects of proper motion and of the minute differences of aberration, &c.). Taking the time for an abscissa, and laying down the deviations in the distances so cleared as ordinates, two curves are obtained, the one for the star (a) the other for the star (b). Each of these curves ought alternately to lie for half a year above, and for half a year below, its axis.—*It does so.* Each of them ought to intersect its axis at those dates when the maximum and minimum of the other above and below the axis occurs. With only a slight degree of hesitation at one crossing—*it does so.* The points of intersection with the axis ought to occur at dates in like manner calculated *à priori;* and so they do within very negligible limits of error. And, lastly, the general forms, magnitudes, and flexures of the curves ought to be identical with those of curves similarly projected, by calculation on an assumed resulting parallactic co-efficient. This is the final and severe test: Mr. Main has applied it, and the results have been placed before you:—*oculis subjecta fidelibus.* If all this does not carry conviction along with it, it seems difficult to say what ought to do so.

The only thing that can possibly be cavilled at is the shortness of the period embraced by the observations: viz., from August 1837 to the end of

March 1840. But this interval admits of five intersections of each curve with its axis; of two maxima and two minima in its excursions on either side; and of ample room for trying its agreement in general form with the true parallactic curves. Under such circumstances, it is quite out of the question to declare the whole phenomenon an accident or an illusion. *Something* has assuredly been discovered, and if that something be not parallax, we are altogether at fault, and know not what other cause to ascribe it to.

The instrument with which Bessel made these most remarkable observations is a heliometer of large dimensions, and with an exquisite object-glass by Fraunhofer. I well remember to have seen this object-glass at Munich before it was cut, and to have been not a little amazed at the boldness of the maker who would devote a glass, which at that time would have been considered in England almost invaluable, to so hazardous an operation. Little did I then imagine the noble purpose it was destined to accomplish. By the nature and construction of this instrument, especially when driven by clock-work, almost every conceivable error which can affect a micrometrical measure is destroyed, when properly used; and the precautions taken by M. Bessel in its use have been such as might be expected from his consummate skill. The only possible apparent opening for an annually fluctuating error seems to be in the correction for temperature of its scale. But this correction has been ascertained by M. Bessel by direct observation, in hot and cold seasons, and applied. Nor could this cause destroy the evidence arising from the simultaneous observation of the two companion stars, since a wrong correction for temperature would affect both their distances proportionally, leaving the apparent parallactic movement still unaccounted for.

The resulting parallax is an extremely minute quantity, only thirty-one hundredths of a second; which would place the star in question at a distance from us of nearly 670,000 times that of the sun! * Such is the universe in which we exist, and which we have at length found the means to subject to measurement, at least in one of its members, probably nearer to us than the rest.

It becomes necessary for me now to refer to two series of researches on this important subject, which have been held by your Council to merit very high and honourable mention; though neither of them, separately, for reasons which I shall state, would have been considered as carrying that weight of probability in favour of its conclusions, which would justify any immediate decision of the nature which they have come to in the case of M. Bessel's. I allude to M. Struve's inquiries, by the method of micrometric measures, into the parallax of α *Lyræ*; and to Mr. Henderson's, by that of meridian observations on the parallax of α *Centauri*.

* The orbit described by the two stars of 61 *Cygni* about each other will, therefore, be about 50 times the diameter of the earth's about the sun, or $2\frac{1}{2}$ times that of *Uranus*.

α *Lyræ* is accompanied by a very minute star, at the distance of about 43″. That this star is unconnected with α by any physical relation, is clear from the fact ascertained by Sir James South and myself, that it does not participate in the proper motion of the large star. The mutual angular distance of these stars has been made by M. Struve the subject of a very extensive series of micrometric measures with the celebrated Dorpat achromatic, bearing this object steadily in view, and working it out to a conclusion of the very same kind, and, though materially inferior in the degree and nature of its evidence to that of Bessel, yet certainly entitled to high consideration. M. Struve's observations on this star, and for this purpose, extend from Nov. 1835 to Aug. 1838, and are distributed over sixty nights, averaging twenty per annum; and from their combination according to the principle of probabilities he concludes a parallax of 0″·261. Mr. Main has subjected these observations to an analysis and graphical projection, precisely similar in principle to those I have explained in the case of 61 *Cygni*. The curves so projected have been subjected to your inspection, and that inspection certainly does leave a very strong impression of a real and tolerably well-ascertained parallax *having* been detected in this star. But at the same time an impression no less decided, owing to irregularities in the march of the curve, when compared with the true parallactic curve, is created,— that the errors of observation are far from being eliminated,—that, on the contrary, they bear such a proportion to the parallax itself as to leave room for some degree of hesitation, and to justify an appeal to a longer series of observations, and to concurrent evidence from other quarters, before declaring any positive opinion. The evidence of this kind, in short, is not equal to that afforded by the similar projection of Bessel's observations of *either* of his two comparison stars. And to this it must be added, that only one star of comparison existing in the case of *α Lyræ*, the possible effect of temperature and *annual* instrumental variation is not eliminated from the result in the way in which it is from the measures of 61 *Cygni;* while all that great mutual support which the observations of parallaxes of the two comparison stars afford each other in the latter case, is altogether wanting in the former. These considerations, without any under-estimation of the great importance and value of M. Struve's researches yet formed essential drawbacks on the immediate admission of his results.

In a word, I conceive the question of discovery as between these illustrious, but most generous and amicable rivals, may be thus fairly stated. M. Struve's meridian observations in 1819–1821 seem to have made the first impression on the general problem, but too slight to authorize more than a hope that it would yield at no distant day. His micrometric measures of *α Lyræ* commenced more than a year earlier, and have extended altogether over a longer period than M. Bessel's of 61 *Cygni.* From their commencement they afford indications of parallax, and these indications accumulating with time have amounted to a high degree of probability, and rendered the supposition of parallax more admissible than that of instrumental or casual errors producing the same influence on the measures. On the other hand, M. Bessel's measures commencing a year later, and continued on the whole through somewhat less time, have exhibited a compact and consistent body of evidence drawn from two distinct systems of measures mutually supporting each other, and so steadily bearing on their object as to leave no more reasonable doubt of its truth than in the case of many things which we look upon as, humanly speaking, certain. And this conviction once obtained, reacts on our belief in the other results, and induces us to receive and admit it on the evidence adduced for it; which, without such conviction so obtained, we might hesitate to do until after longer corroboration of the same kind.

The other series of observations to which I must now call your attention are those of Mr. Henderson, made at the Cape of Good Hope, on the great star α *Centauri*, the third star in brightness which the heavens offer to our view. It is a magnificent double-star consisting of two individuals, the one of a high and somewhat brownish orange, the other of a fine yellow colour, and each of which I consider fairly entitled to be classed in the first magnitude.*

* I have seen *both* their images projected on a screen of three

Their distance is at present about 15″ asunder, but it is rapidly diminishing, and in no great lapse of time they will probably occult one another, their angular motion being comparatively small. Their apparent distance was formerly much greater: how much we cannot say for want of observations, but probably the major axis of their mutual orbit is little short of a minute of space. They, therefore, afford strong indications of being very near our system. Add to which their proper motion is very considerable, and participated in by both, which proves their connexion as a binary system; and an additional presumption in favour of their proximity may be drawn from their situation in what, from general aspect, I gather to be the nearest region of the milky way, among an immensity of large stars.

Mr. Henderson observed these stars with great care both in right ascension and declination with the very fine transit, and (in spite of certain grievous defects in the axis) the otherwise really good and finely divided mural circle of the Royal Observatory in that colony. Since his return to England, he has reduced these observations with a view to parallax, and the result is the apparent existence of that element to what, after what has been said, we must now call the great and conspicuous amount of a full second. Mr. Main, to whom I am so largely indebted for allowing me to draw freely on his labours, has also discussed these results, and

thicknesses of stout paper, the eye being on the opposite side of the screen from that on which the images were depicted.

comes to the conclusion that (as might, perhaps, be expected) the right-ascension observations afford a trace, but an equivocal one, of parallax, but that in declination (I use his words) "The law of parallax is followed remarkably well. There is scarcely an exception to the proper change of sign, according to the change of sign of the co-efficients of parallax. This is quite as much as can reasonably be expected in a series of individual results obtained from any meridional instrument for observing zenith distances. We cannot expect to find the periodical function regularly exhibited by the differences. On the whole, therefore, we should say that, in addition to the claims of α *Centauri* on our attention with relation to its parallax, arising from its forming a binary system, its great proper motion, and its brightness,—it derives now much additional importance, in this point of view, from the investigation of Mr. Henderson. This we are at least entitled to assume until some distinct reason, independent of parallax, shall have been assigned for the changes in the declinations. Such I do not consider impossible, having before my eyes the results which Dr. Brinkley derived, in the cases of certain stars, from the Dublin circle. For the present it must be considered that the star well deserves a rigorous examination by all the methods which the author himself has so well pointed out; and that, in the event of a parallax at all comparable with that assigned by Mr. Henderson being found, he will deserve the merit of its first discovery, and the warmest thanks of astronomers, as an extender of the knowledge which we possess of our connexion with the sidereal system."

With this view of Mr. Henderson's labours I fully agree, and await with highly excited interest the result of Mr. Maclear's larger and complete series of observations on this star both with the old circle and with that more perfect one with which the munificence of government has recently supplied the observatory. Should a different eye and a different circle continue to give the same result, we must, of course, acquiesce in the conclusion; and the distinct and entire merit of the *first* discovery of the parallax of *a* fixed star will rest indisputably with Mr. Henderson. At present, however, we should not be justified in so far anticipating a decision which time alone can stamp with the seal of absolute authenticity.

Gentlemen of the Astronomical Society, I congratulate you and myself that we have lived to see the great and hitherto impassable barrier to our excursions into the sidereal universe; that barrier against which we have chafed so long and so vainly —(*æstuantes angusto limite mundi*)—almost simultaneously overleaped at three different points. It is the greatest and most glorious triumph which practical astronomy has ever witnessed. Perhaps I ought not to speak so strongly—perhaps I should hold some reserve in favour of the bare possibility that it may be all an illusion—and that further researches, as they have repeatedly before, so may now fail to substantiate this noble result. But I confess myself unequal to such prudence under such excitement. Let us rather accept the joyful omens of the time, and trust that, as the barrier has begun to yield, it will speedily be effectually prostrated. Such results are among the fairest flowers of civilization. They justify the vast expenditure of time and talent which have led up to them; they justify the language which men of science hold, or ought to hold, when they appeal to the governments of their respective countries for the liberal devotion of the national means in furtherance of the great objects they propose to accomplish. They enable them not only to hold out but to redeem their promises, when they profess themselves productive labourers in a higher and richer field than that of mere material and physical advantages. It is then when they become (if I may venture on such a figure without irreverence) the messengers from heaven to earth of such stupendous announcements as must strike every one who hears them with almost awful admiration, that they may claim to be listened to when they repeat in every variety of urgent instance that these are not the last of such announcements which they shall have to communicate,—that there are yet behind, to search out and to declare, not only secrets of nature which shall increase the wealth or power of man, but TRUTHS which shall ennoble the age and the country in which they are divulged, and by dilating the intellect, react on the moral character of mankind.

Such truths are things quite as worthy of struggles and sacrificcs as many of the objects for which nations contend, and exhaust their physical and moral energies and resources. They are gems of real and durable glory in the diadems of princes, and conquests which, while they leave no tears behind them, continue for ever unalienable.

It must be needless for me to express a hope that these researches will be followed up. Already we have to congratulate astronomy on the resolution taken by one of our great academic institutions to furnish its observatory with an heliometer of the same description as Bessel's; nor can we fear but that the research will speedily be extended to other stars, offering varieties of magnitude and other indications to draw attention to them.

On the whole, then, the award of our medal, which the Council have agreed on, seems to me, under the circumstances, fully justified. I will now request the foreign secretary to convey it to our distinguished associate; and in so doing I will add our hope that, in the painful and distressing visitation with which it has pleased Providence recently to try him, he may find occasion to withdraw his mind awhile from that melancholy contemplation to receive with satisfaction such a tribute to this his last and perhaps his greatest achievement, accompanied as it is by the truest regard for his private worth and the most respectful sympathy for his present distress.

MEMOIR OF FRANCIS BAILY, ESQ.,

D.C.L. OXFORD AND DUBLIN.

[From the Monthly Notices of the Royal Astronomical Society, Vol. VI. November, 1844.]

In the performance of the melancholy duty imposed on me by the wishes of the Council, that I should endeavour, on this occasion to place before the assembled members of this society, a sketch of the scientific life and character of our late lamented President, I have been careful both to examine my own competency to the task, and to consider well the proper limits within which to confine myself in its execution. In the first of these respects, indeed, though tolerably familiar with some of the leading subjects which I shall have to touch upon, there are others on which I have seriously felt the want of a longer interval for preparation. On these, of course, I shall take care to express myself with becoming diffidence; and in so vast a field of laborious inquiry and of minute, yet important research as I shall have to range over, it may easily be supposed I have more than once found occasion to wish that the duty had fallen into abler hands. A duty, however, it is, and a very sacred one, which we owe to departed merit, to society, and to ourselves, to fix as speedily as possible, while its impress is yet fresh and vivid upon us, its features in our minds with all attainable distinctness and precision, and to store them up beyond the reach of change and the treachery of passing years.

As respects the limits within which I feel it necessary to confine myself on this occasion, it is to astronomers to whom I have to speak of an astronomer — to members of a large and, in the simplicity of truth I may add, a highly efficient public body — of an officer to whom, more than to any other individual, living or dead, it owes the respect of Europe. To make what I have to say complete as a biography, however interesting to us all, however desirable in itself, is very far, either from my intention or my power. Nor is the time fitting for the attempt. The event is too recent, the particulars which can be collected at the present moment too scanty, the grief of surviving relations too fresh, to admit of that sort of close and pertinacious inquiry into facts, anecdotes, documents, and evidence, which personal biography requires to be satisfactory. In this respect, therefore, a mere sketch is all that I can pretend to give.

Francis Baily was born on the 28th of April, 1774, at Newbury, in the county of Berks. His father was Mr. Richard Baily, a native of Thatcham, in the same county, who became established as a banker at Newbury. He married Miss Sarah Head, by whom he had five sons and two daughters. Francis, who was the third son, received his education at the school of the Rev. Mr. Best, of Newbury, an establishment of considerable local reputation, where, although probably little of an abstract or mathematical nature was imparted, the chief elements of a liberal and classical education were undoubtedly communicated. From his early youth he manifested a propensity to physical inquiry, being fond of chemical, and especially of electrical experiments, — a propensity sufficiently marked (in conjunction with his generally studious habits) to procure for him, among his young contemporaries, the half-jesting, half-serious *sobriquet* of "The Philosopher of Newbury."

It does not appear that he received any further instruction beyond the usual routine of an establishment of the kind above mentioned, so that in respect of the sciences, and especially of that in which he attained such eminent distinction, he must be regarded as self-educated. This taste for and knowledge of electricity and chemistry, were probably acquired from Dr. Priestley, with whom, at the age of seventeen, he became intimately acquainted, and of whom he always continued a warm admirer. But that his acquaintance with the subject was considerable, and his attachment to it permanent, may be concluded from the fact that Mr. Welsh, the organist of the parish church of Newbury, who had a very pretty electrical apparatus, and at whose house I remember myself to have first witnessed an electrical experiment is stated to have imbibed his taste for that science, and to have acquired its principles, from his example and instructions at a somewhat subsequent period.

He quitted Mr. Best's school at fourteen years of age, and, having chosen a mercantile life, which accorded with the views of his parents, he was sent to London, and placed in a house of business in the city, where he remained till his twenty-second year; when, having duly served his time, and either not feeling an inclination to the particular line of business in which he had commenced his life, or being desirous of the general enlargement of mind which travel gives, or from mere youthful love of adventure and enterprise, he embarked for America on the 21st October, 1795, which, however, he was not destined to reach without twice incurring the most imminent danger from shipwreck, both on our own coast, under most awful circumstances, on the Goodwin Sands, and off New York, which he was prevented from reaching, being driven to sea in a gale, and after endeavouring in vain to reach Bermuda, was driven into Antigua, whence he subsequently embarked for Norfolk, in Virginia.

In America he remained one or two years, travelling over the whole of the United States, and through much of the Western country, in which travel he experienced, at various times, much hardship and privation, having, as I remember to have heard him state in conversation (and which must have referred to this period of his life), passed eleven months without the shelter of a civilized roof. During his residence in America, he was not unmindful of his intellectual and social improvement, having not only read much and observed much, as a copious journal* which he transmitted home proves, but formed the acquaintance of some eminent persons, among whom may be mentioned Mr. Ellicot, the Surveyor-General of the United States, from whom he obtained some curious information bearing on the periodical displays of meteors on the 12th November, of which that gentleman observed a superb instance in 1799, and from whom it is not impossible he may have acquired a taste for observations of a more distinctly astronomical and geographical nature.

Whatever may have been the more direct object of this journey, if indeed it had any other than to gratify a youthful inclination for travel and adventure, it does not appear to have exercised any material influence on his after life, since, on his return to England, in place of immediately entering into business, he continued to reside for some time with his parents at Newbury, which, however, at length he quitted for London, to engage in business as a stockbroker, being taken into partnership by Mr. Whitmore, of the Stock Exchange. The exact date of this partnership I have not been

* This interesting record has been edited by Mr. De Morgan, under the title "Journal of a Tour in unsettled parts of North America in 1796 and 1797, by the late Francis Baily, Esq. F.R.S., President of the Royal Astronomical Society: with a memoir of the author. London, Baily Brothers, Royal Exchange Buildings, 1856."

able to learn. I believe it to have been 1801 *; but that it must have been prior to 1802 may be concluded from the subject of his first publication, which appeared in that year, viz., "Tables for the Purchasing and Renewing of Leases for Terms of Years certain, and for Lives, with Rules for determining the Value of the Reversions of Estates after any such Leases." This work (as well as the next) is preceded by a highly practical and useful Introduction, and followed by an Appendix, which shows that, at the age of twenty-eight, he had become well versed in the works of the English mathematicians, and had also consulted those of foreign ones. It speedily attained a standard reputation on account of its intrinsic utility, and went through several editions. His next work, a pamphlet in defence of the rights of the Stock-Brokers against the attacks of the City of London, printed in 1806, at all events shows him at that time to have become identified in his feelings and interests with that body of which he lived to be an eminent and successful member. A similar conclusion may be drawn from his next publication, which appeared in 1808, "The Doctrine of Interest and Annuities Analytically Investigated and Explained," a work than which no one more complete had been previously published, and which is still regarded as the most extensive and standard work on compound interest. It was speedily followed by

* It was about the end of 1799.

other works on the same subject, viz., in 1810, by "The Doctrine of Life Annuities and Insurances Analytically Investigated and Explained;" to which, in 1813, he added an Appendix. This is a work in many ways remarkable, and its peculiarities are of a highly characteristic nature; method, symmetry, and lucid order being brought in aid of practical utility in a subject which had never before been so treated, and old routine being boldly questioned and confronted with enlarged experience. A friend of great mathematical attainments and extensive practical acquaintance with subjects of this nature, thus characterizes it: — "It is not easy to say too much of the value of this work in promoting sound practical knowledge of the subject. It was the first work in which the whole of the subject was systematically algebraized; the first in which modern symmetry of notation was introduced; and the first modern work, since Price and Morgan, in which the 'Northampton Tables' were not exclusively employed, and in which the longer duration of human life was contended for; and the first in which some attempt was made to represent by symbols the various cases of annuities and assurances, afterwards more systematically done by Mr. Milne." In the Appendix to this work, a method originally proposed by Mr. Barrett of forming the Tables, by which cases of temporary and deferred annuities, formerly requiring tedious calculations, become as easy as the others, and which, in the improved form subsequently given to it by Mr. Griffith Davies, has come into very general use in this country was, by the penetration of Mr. Baily, given to the public, but for which it would probably have been altogether lost. It may serve to give some idea of the estimation in which this work was held, that when out of print its copies used to sell for four or five times their original price. A chapter of this work is devoted to the practical working of the several life assurance companies in London, containing some free remarks on several points of their practice. Mr. Babbage has subsequently followed in the same line (as he has also advocated extending the estimation of the duration of life to still more advanced ages). However unpleasing it may be to public bodies, especially commercial ones, to see practices of whose injustice they may perhaps have been unaware, convicted of it, and made matter of public animadversion, there can be no doubt that criticisms of this kind, when really well grounded and expressed with temperance and moderation, are both salutary to the parties concerned, and merit, in a high degree, the gratitude of the public. A higher praise is due to the candour and boldness of openly entering the lists on such occasions, and despising the anonymous shield of which so many avail themselves.

But while devoting his attention thus assiduously to matters of direct commercial interest, he could yet find time for other objects of a more general nature. Astronomical pursuits had already begun to assume in his eyes that attraction which was destined ultimately to draw him aside entirely from business, and to constitute at once the main occupation and the chief delight of his life. As every thing to which he turned his thoughts presented itself to them, if I may use the expression, in the form of a palpable reality, a thing to be turned and examined on all sides — to be reduced to number, weight, and measure — to be contemplated with steadiness and distinctness, till everything shadowy and uncertain had disappeared from it, and it had moulded itself, under his scrutiny, into entire self-consistency, the practical branches of astronomical calculation early became, in his hands, instruments of the readiest and most familiar application, as the touchstones of the truth of its theories and the means of giving to them that substantial reality which his mind seemed to crave as a condition for their distinct conception by it. His first astronomical paper, on the celebrated solar eclipse, said to have been predicted by Thales, which was written in November, 1810, and read before the Royal Society on the 14th March, 1811, affords a remarkable instance of this. That eclipse had long been a disputed point among chronologists. It was easy to perceive, and accordingly all had perceived, that an eclipse of the sun, so nearly central as to produce great darkness, being a rare phenomenon in any part of the globe, and excessively so in any precisely fixed locality, must afford a perfectly certain means of determining the date of a coincident event, if only the geographical locality be well ascertained, and some moderate limits of time within which the event must have happened be assigned, and provided the means were afforded of calculating back the moon's place for any remote epoch. In this case, both the locality and the probable historical limits were sufficiently precise; and the account of Herodotus, which agrees only with the character of a total and not of an annular eclipse (as Mr. Baily was the first to remark) still further limits the problem. But the tables of the moon employed by all prior computists were inadequate * to carry back her place with the requisite exactness, nor was it till the publication of Burg's "Lunar Tables" that the means of doing so were in the hands of astronomers. The course of Mr. Baily's reading at this period (being then, no doubt, employed in collecting the materials for the Chronological Tables in his "Epitome of Universal History," which appeared not long after,) brought him necessarily into contact with the subject. He perceived at once both the uncertainty of all former calculations of this eclipse, and the possibility of attacking it with a fresh prospect of success. None, however, but a consummate astronomical calculator would have ventured on such an inquiry, which involved the computation of all the solar eclipses during a period of seventy years, six

* Recent improvements in the lunar tables have shown that this question must be re-opened. See a remarkable paper on the subject by Mr. Airy, *Phil. Tr.* 1853.

centuries before the Christian era. These calculations led him to assign, as the eclipse in question, that of September 30, B.C. 610, which was central and total, according to these tables, at the very point where all historical probability places the scene of action.

Most men would have regarded such a result, obtained by so much labour, with triumphant complacency: not so Mr. Baily. His habit of examining things on all sides, instead of permitting him to rest content with his conclusion, led him on to further inquiry, and induced him to calculate the phenomena of another total eclipse recorded in ancient history, that of Agathocles, which happened August 15, B.C. 310, an eclipse of which neither the date nor the locality admits of any considerable uncertainty, and which, therefore, appeared to him well fitted to test the accuracy of the tables themselves. Executing the calculation, he found indeed a total eclipse on the year and day in question, and passing near to the spot, *but not over it.* An irreconcileable gap of about 3°, or 180 geographical miles, remains between the most northerly limit of the total shadow, and the most southerly supposable place of Agathocles's fleet. Although this may justly be looked upon as a wonderful approximation between theory and historical fact (indicating, as it does, a correction of only 3′ in the moon's latitude, for an epoch anterior by more than 21 centuries to that of the tables), yet it did not escape Mr. Baily's notice, nor did his love of truth permit him to conceal the fact, that no presumed single correction of the tabular elements will precisely reconcile *both* eclipses with their strict historical statement. There seemed, however, no reason to doubt that the eclipse of 610 B.C. is, in fact, the true eclipse of Thales. It seems extraordinary that neither Professor Oltmanns, who investigated the eclipse of Thales about two years subsequently, and who came to the same conclusion, nor M. Saint Martin, who read an elaborate memoir on the same subject to the French Institute in 1821, should have made any mention of this very remarkable paper of Mr. Baily.

The "Epitome of Universal History," of which mention has already been made, was published in 1813, and intended to accompany an "Historical Chart" published the year before, an extension and improvement of Dr. Priestley's, in which the political alterations of territory are represented through the whole of history. It is an easy and useful work of reference, in which the number and accuracy of the dates, and the utility of the appended tables, are especially valuable. There can be little doubt that the object of this work was much less to produce a book than to systematise and concinnate the author's own knowledge. When such a task is undertaken by a mind at once vigorous in its grasp, and simple, practical, and natural in its points of view, it can hardly fail to result in a picture of the subject where all the parts are truly placed, and easily apprehended by the general reader. The chart with its explanation, forming a distinct work, was in considerable request, and went through three editions in five years.

About the 22nd of January, 1814, occurred the celebrated fraud of De Beranger, that being the assumed name of an impostor employed to bring important but false intelligence from the scene of war abroad, for the purpose of influencing the price of the British funds. The imposture was so adroitly managed that many bargains were made on the strength of this intelligence, and much confusion caused. In the detection and exposure of this fraud, Mr. Baily had a considerable share, and was appointed by the committee of the Stock Exchange to get up the evidence against the perpetrators,—a task which he is said to have performed in so masterly a manner, that no more complete and conclusive chain of evidence was ever produced in a court. The result of these inquiries and the steps taken in consequence, were made the subject of three Reports of the above-mentioned committee, drawn up by him, and printed in that and the subsequent year.

From this time, astronomy appears to have been continually engaging more and more of his attention. The subject of eclipses and occultations with their connected calculations, together with that of the improvement of the Nautical Almanac, which, whatever might be said on specific points, had certainly, at that time, begun to fall considerably behind the requisitions of astronomical, and even of nautical science, were those with which he may be said to have commenced his more active astronomical career. But I wish to call attention at present to two pamphlets which he published in 1818 and 1819 respectively, which will afford occasion for some remarks of moment. The first of these is a notice of the annular eclipse of September 20. 1820, whose path lay along the whole medial line of Europe from north to south. Two points in this tract merit our attention. In it he adopts a practice, which he subsequently on a great many occasions adhered to, of introducing in the way of prefatory statement a brief but very clear sketch of the history of the subject, and the observations of former astronomers. These little historical essays are for the most part extremely well drawn up and highly interesting, and show a perfect knowledge of the subjects treated of, drawn from very extensive reading. The next point, and one of more importance, is the studious consideration shown to observers possessed of slender instrumental means, in pointing out to them modes and forms of observation by which those means might be rendered available and useful. At no period of his life himself possessing any large and elaborate instrument or luxurious appliances, one of his constant aims was to render astronomical observation popular and attractive by showing that much of a highly useful character might be accomplished with even moderate instruments. There is no question more frequently asked by the young astronomer who has possessed himself of one or two tolerably good instruments which he desires to employ his time upon, than this, "How can I make myself useful?" nor any which can be more readily answered by a reference to the innumerable notices on almost every point of practical astronomy which Mr. Baily from this time forward for many years continued to scatter profusely to the public, and which have probably done more to create observers, and to cherish and foster a taste for practical astronomy among Englishmen, than any single cause which can be mentioned.

In 1819 he printed for private distribution a translation of Cagnoli's memoir on a "Method of deducing the Earth's Ellipticity from Observations of very Oblique Occultations," with an appendix recommendatory of the method, which is precisely such as requires for its perfect execution only a sufficient telescope, a moderately good clock, and an observer diligent in watching opportunities. This was, no doubt, Mr. Baily's chief reason for translating and distributing it, and for subsequently following it up by his chart and catalogue of the Pleiades, through which the moon had to pass at each lunation in 1822 and the following years, thereby affording admirable opportunities for applying the principle in question. I should not, however, have thought it necessary, in the midst of so many claims on our notice, to draw especial attention to this work, but for one passage in it deeply interesting to all of us. I mean that in which he alludes to the formation of an Astronomical Society as an event earnestly to be desired.

"It is much to be regretted," he observes, "that in this country there is no association of scientific persons formed for the encouragement and improvement of astronomy. In almost all the arts and sciences, institutions have been formed for the purpose of promoting and diffusing a general knowledge of those particular subjects. . . . . the beneficial effects of which are too evident to be insisted on in this place. But astronomy, the most interesting and sublime of the sciences . . . . cannot claim the fostering aid of any society . . . . The formation of an Astronomical Society would not only afford this advantage, but would in other respects be attended with the most beneficial consequences," &c. &c.

It is thus that coming events cast their shadows before them. But looking back from this point, as it were, to the then embryo state of our corporate existence, it would be ungrateful not to associate with the name of Francis Baily that of Dr. Pearson, as having at or about the same time made the same suggestion. It was happily and speedily responded to, and on Wednesday, the 12th of January, 1820, a preliminary meeting of the fourteen founders of our Institution took place, which resulted in its final establishment, and in which, during the first three years of its existence, Mr. Baily filled the office of secretary, in other words, undertook and executed the more laborious and essential duties. The establishment of this Society may, indeed, be considered as a chief and deciding epoch in his life, and to have furnished, though not the motive, yet, at least, the occasion, for the greater part of his subsequent astronomical labours. Looking to it, as every one must do, as a most powerful instrument for the advancement of the science itself, and the propagation of a knowledge of and a taste for it among his countrymen, he yet appeared to regard it as something more than simply as a means to an end. He made it an object of personal attachment and solicitude, which led him to watch over its infant progress with parental care, and to spare no exertion in its behalf. As years passed on, and as the Institution flourished (as every institution must do which is constituted on sound principles, whose members are loyal to those principles, and willing to work heartily in its cause), this sentiment, so far from diminishing, seemed to grow upon him till he regarded its welfare and interests as identical with his own. I shall reserve a more distinct statement of our obligations to him for a more advanced period of this notice; but in a narrative of his life it becomes impossible from this epoch to separate the Astronomical *Society* from astronomical *science*, in our estimate of his views and motives, or to avoid noticing the large and increasing devotion to its concerns of his time and thoughts. To the Transactions of the new Society he became, as might be expected, a frequent and copious contributor. In the interval between the first establishment of the Society and the year 1825 (the reason for this limit will presently be seen), he contributed five papers, viz.: "On the Meridian Adjustment of the Transit Instrument;" "On the Determination of Time by Altitudes near the Prime Vertical;" "On the Solar Eclipse of September 7. 1820;" "On the Mercurial Compensation Pendulum;" and "On the Determination of Longitudes by Moon-culminating Stars." The two first-mentioned of these turn on somewhat elementary points of astronomical observation, and contain tables, and suggest facilities, which he had found useful in his own practice. The eclipse was observed by him at Kentish Town, where, not being annular, he must have felt severely the sacrifice, imposed probably by the calls of business, of the opportunity of witnessing by a short continental trip, a phenomenon which had engaged so much of his thoughts. His paper on the "Mercurial Pendulum," though practical in its object, was of a much more elaborate kind than any thing which had previously emanated from him, with exception of his memoir on the eclipse of Thales. It contains a minute and excellent view of the whole subject of this most useful compensation; is prefaced (*more suo*) with a clear synoptic view of the then actual state of the subject, and goes into the whole subject of the expansion of the materials, the formulæ for determining with more precision than heretofore the proportional length of the mercurial column, and the mode of adjustment both for rate and compensation. This paper must certainly be regarded as a very valuable one, and an astronomer can hardly be said thoroughly to understand his clock who does not possess it. The object of the paper on moon-culminating stars is to recommend, facilitate, and render general, that most useful and widely available method of determining the longitude on land.

About this period, also, Mr. Baily began, and thenceforward continued, to be a frequent contributor to the "Philosophical Magazine," published by Messrs. Tilloch and Taylor, of articles interesting in a great variety of ways to the practical astronomer. These articles are so numerous, and so miscellaneous in their subject matter, that it would be vain to attempt any detailed account of them, within such limits as I must confine myself to. Nor, indeed, is it requisite to do so; as many of them, however useful at the time, have now ceased to present any especial interest, apart from their general object, which was that of diffusing among the British public a knowledge of the continental improvements in the art of observing, and the practice of astronomical calculation, and placing in the hands of our observers and computers a multitude of useful tables and methods, which, though sure to work their way ultimately into use, were undoubtedly accelerated in their introduction into English practice by coming so recommended.

More especial objects were those of recommending to general attention and use certain eminently practical methods, such as those of determining latitudes by the pole-star, longitudes by moon-culminations and occultations, copious lists of which were, on several occasions, either procured from abroad and reprinted here, or calculated by himself for the purpose.

The circulation of notices, also, of other remarkable expected phenomena, with a view to procuring them to be observed,— the description of newly invented foreign instruments, or of such as had been long known but little used in England,— the analysis of foreign astronomical publications, — every thing, in short, which could tend to excite curiosity, to cherish emulation, and to render the British astronomical mind more excursive and more awake than heretofore, found a place in these contributions; of which so constant and copious a fire was kept up, as may well excite our surprise at the industry which sustained, no less than our admiration of the zeal which prompted it.

A volume of astronomical tables and formulæ, printed in 1827 for private distribution (as was frequently his custom), and then largely circulated, but since published with corrections, is of the utmost convenience and value, and will be highly prized by every astronomer who may be fortunate enough to possess a copy, as a work of ready and continual reference for all the data and coefficients of our science. A series of zodiacal charts was also commenced by him, but I am not able to say if more than one plate was engraved.

One of the most practically important and useful objects, however, to which Mr. Baily's attention was about this period turned, was the facilitating, by tables properly contrived for the purpose, the reductions of apparent to mean places of the fixed stars. It seems almost astonishing that these computations, which lie at the root of all astronomy, and without which no result can be arrived at, and no practical observer can advance a single step, should have remained up to so late a period as the twentieth year of the nineteenth century, in the loose, irregular, and troublesome state which was actually the case, and *that* not from their theory being ill understood, but from their practice not having been systematized. Each of the uranographical corrections had to be separately computed by its own peculiar tables, and with coefficients on whose magnitude no two astronomers agreed. The latter evil, indeed, might be tolerated at a time when the tenth of a second of space was not considered of so much consequence as at present, but the calculations were formidable and onerous in the extreme to private astronomers, whatever they might be rendered in public establishments by habit and the use of auxiliary tables. So far as the fundamental stars were concerned, the subject had for some time attracted attention, and had begun to receive its proper remedy by the publication, by Professor Schumacher in Denmark, of their apparent places for every tenth day; and by the laudable exertions of Sir James South in our own country, who, for some years, prepared and circulated similar tables for every day, not without urgent representations of the necessity of taking it up as a public concern, which was at length done. But for stars out of this list, except about 500 somewhat facilitated by Zach, there was no provision of any kind, nor any auxiliary tables to have recourse to; so that sidereal astronomy, beyond the bounds of this favoured list, might be almost said to be interdicted to the private astronomer, owing to the excessive irksomeness of these calculations. This was precisely the sort of case for Mr. Baily to take pity on. He perceived a desert where, with a moderate expenditure of capital, a plentiful harvest might be made to grow, and forthwith proceeded to remedy the evil. Accordingly, with the aid of Mr. Gompertz, he investigated the subject generally, and succeeded in devising a method of arranging the terms of the corrections for aberration, solar and lunar precession, adapted to the purpose, and identical in principle with that adopted by M. Bessel, who, on his part, was at the same time, and, actuated by the same motives, engaged on the subject unknown to Mr. Baily. The latter had actually proceeded to the computation of his tables, when the labours of Bessel reached his knowledge, who had, moreover, included the precession under the same general mode of expression. Mr. Baily, with characteristic frankness and candour, immediately acknowledged this as an improvement in advance of his own idea, and at once adopted and recommended it for general use. He did more, he carried out the idea into a wide and most useful field; and in the Catalogue of the Astronomical Society he has put the astronomical world in possession of a power which may be said, without exaggeration, to have changed the face of sidereal astronomy, and must claim for him the gratitude of every observer. It detracts nothing from the merit of Mr. Baily, or from his claim to be considered the author of this precious work, that the numerical computations were chiefly executed by Mr. Stratford, and the expenses borne by the Astronomical Society. The conception was all his own, and the work prefaced, explained, and superintended, in every stage of its progress, by himself alone. The gold medal of this Society was awarded to him for this useful work.

On the 22nd of February, 1821, Mr. Baily was elected a Fellow of the Royal Society. He was also a member of the Linnean and Geological Societies, but I am unable to state the precise date of his election in either.

In 1825 he retired from the Stock Exchange, after a career in which his consummate habits of business, his uprightness, intelligence, and prudence, had established his fortune, and might, if continued, have led him on to any eminence of worldly wealth. But there was that in his disposition which the mere acquisition of wealth could not satisfy. All that he had before done for his favourite science seemed only preparatory to what he might do; and with the best years of his intellectual life before him, and with objects worthy of his efforts now opening to his view in that direction, he resolved henceforward to devote himself to their pursuit, though at the sacrifice of prospects whose attractions always prove irresistible to minds of a lower order. In thus calmly measuring the relative worth of intellectual and worldly pursuits, and stopping short in the full career of success, when arrived at a point which his undazzled judgment assured him to be the right one, he afforded an example of self-command as uncommon as it was noble. In the satisfaction which the decision afforded him, and the complete fulfilment of those aspirations which led him to form it, we have one proof (if proofs be wanting) how entirely a well-chosen and elevated scientific pursuit is capable of filling that void in the evening of life, which often proves so intolerably irksome to men who have retired early from business from mere love of ease or indolence. On no occasion did he ever appear to regret the sacrifice he had made, or even to regard it as a sacrifice.

No desire of listless ease or self-indulgence, however, could by possibility have mixed with Mr. Baily's motives in taking this step; for immediately on doing so he entered on a course of devoted and laborious exertion, which continued without interruption during the remainder of his life, and of which the history of science affords few examples. The mass of work which he got through, when looked at as such, is, in fact, appalling, and such that there seems difficulty in conceiving how it could be crowded into the time; the key to which is, however, to be found in his admirably conceived methodical arrangement of every piece of work which he undertook, and his invaluable habit of finishing one thing before he undertook another.

At this epoch, or very shortly subsequent to it, he purchased and took up his permanent residence in his house in Tavistock Place, excellently adapted in every respect both to his future comfort and convenience as a place of abode, and for those important and delicate researches of which it was destined to become the scene; standing, as it does, insulated in a considerable garden, well enclosed on all sides, and, from the nature of the neighbourhood, free from any material tremor from passing carriages. A small observatory was constructed in the upper part, for occasional use and determination of time, though he never engaged in any extensive series of observation. The building in which the earth was weighed and its bulk and figure calculated, the standard measure of the British nation perpetuated, and the pendulum experiments rescued from their chief source of inaccuracy, can never cease to be an object of interest to astronomers of future generations.

In endeavouring, according to the best of my ability, to give some account of the astronomical labours of Mr. Baily subsequent to this period, it will no longer be advisable to adhere, as I have hitherto done, to the chronological order in which they were undertaken and executed. It will rather be preferable (with exception of a few memoirs and publications of a miscellaneous nature) to consider them under distinct heads, according as they refer to one or other of the following subjects, viz.: —

1. The Remodelling of the "Nautical Almanac;"
2. The Determination of the Length of the Seconds-Pendulum;
3. The Fixation of the Standard of Length;
4. The Determination of the Density of the Earth;
5. The Revision of Catalogues of the Stars;
6. The Reduction of Lacaille's and Lalande's Catalogues; and,
7. The Formation of a new Standard Catalogue.

The Nautical Almanac. — The end of the 18th and the commencement of the 19th century are remarkable for the small amount of scientific movement going on in this country, especially in its more exact departments. It is not that individuals were not here and there busied in extending the bounds of science, even in these, but they met with little sympathy. Their excursions were limited by the general restriction of view which had begun to prevail, and by a sense of loneliness and desertion (if I may use such an expression) arising from that want of sympathy. Mathematics were at the last gasp, and astronomy nearly so; I mean in those members of its frame which depend upon precise measurement and systematic calculation. The chilling torpor of routine had begun to spread itself largely over all those branches of science which wanted the excitement of experimental research. I know that I have been blamed on a former occasion for expressing this opinion, but it is not the less true, though we may now happily congratulate ourselves that this inanimate period has been succeeded by one of unexampled activity. To break the dangerous repose of such a state, and to enforce that exertion which is necessary to healthy life, there is always need of some degree of friendly violence, which, if administered without rudeness, and in a kindly spirit, leads at length the revived patient to bless the disturbing hand, however the urgency of its application might for a moment irritate. It is in this light that we are to regard the earnest and somewhat warm remonstrances of Mr. Baily on the deficiencies which had long begun to be perceived and felt in the "Nautical Almanac," in its capacity of an astronomical ephemeris.

The subject once moved gave rise to a great deal of discussion from more than one quarter, which was from time to time renewed for some years; but as I have no intention to make this notice an occasion of dilating on any matter of a controversial nature, I shall merely add that, on the dissolution of the late Board of Longitude, followed almost immediately by the death of Dr. Young, on whom the charge of its superintendence rested (the new Berlin Ephemeris, by Encke, having also recently appeared, in which many of the principal improvements contended for were adopted), it seemed fitting to the Lords Commissioners of the Admiralty to place unreservedly before the Astronomical Society the subject of a complete revision and remodelling of that great national work— a high proof of confidence, which speaks volumes for the good sense, prudence, and activity which had continued to pervade its administration during the ten years which had now elapsed since its first institution.

It is hardly necessary to add that this important business received the most unremitting attention from Mr. Baily, as well as from every other member of the Committee, in all its stages. To him also was confided the task of drawing up the final report of the Committee appointed to carry out the wishes of the Admiralty, which will be found in the fourth volume of our "Memoirs," and which is a model of good sense, clearness, and lucid arrangement. The Report was immediately acted upon by Government, and the result was the present British "Nautical Almanac;" a work which, if it continue to be carried on, as I trust it ever will, on the principles which prevailed in its reconstruction, will remain a perpetual monument to the honour of every party concerned in it.

The Pendulum.—The seconds-pendulum having been constituted the legal source from which, in the event of the loss of the national standard of length, the yard might at any time be recovered, it may be easily imagined with what intensity of interest the announcement was received among all conversant with these fundamental determinations, that a very material correction had been entirely overlooked in the reduction of the experiments, on which the Act of 5 Geo. IV. c. 74. was founded. This correction is, in fact, no other than the correction due to *the resistance of the air*, and, placed in this light, it would seem somewhat wonderful that such an oversight could have been committed; but it had been customary to consider the effect of resistance on the time of vibration to be wholly confined to its influence in diminishing the arc, and this secondary effect being allowed for in the formulæ employed to compute what is called the correction for the arc of vibration, the primary or direct effect of resistance dropped altogether out of notice, or, rather (owing to an entire misconception of the nature of the mechanical process by which resistance is operated), had been supposed to be altogether inappretiable in its amount. The real effect of resistance, though under a somewhat confused statement as to its nature, had, however, been long before noticed, and its amount even ascertained with tolerable correctness, by the Chevalier Buat, in 1786; but his experiments and theory had so entirely fallen into oblivion as to have escaped the notice not only of Captain Kater, but of his own countrymen, Borda and Biot, and were unknown even to Bessel himself, who, in

1828, rediscovered the correction in question, and, for the first time, made it an integrant feature in the modern system of pendulum reductions. The light in which this correction was placed by Buat, and even in some respects by Bessel, tended not a little, in my opinion, to obscure the clear perception of its nature, by representing it as due to a certain portion of air adhering to and bodily *dragged along* by the pendulum in its motion, thus adding to its inertia without adding to its relative weight when corrected for buoyancy; and in this view, also, Mr. Baily regarded it. That this is not a complete and adequate view of the subject is easily made a matter of ocular inspection, by causing a pendulum to vibrate, or any body to move, near the flame of a candle, when it will be at once evident that the movement of the air consists in the continual transfer of a portion of air from the front to the rear of the body, by performing a circuit half round it. Its hydrodynamical investigation, therefore, is of an infinitely higher order of difficulty than the ordinary problems of resistance, which turn upon a theory of molecular impulse, simple indeed, but very far from satisfactory. It properly refers itself to the theory of sound, and has, in fact, been so investigated in an admirable memoir by Poisson.*

* If this view of the subject be correct, as I am persuaded it is, it seems not impossible that, by making a section of the pendulum coincident in form with the "wave-formed outline" of Mr. Russel's ships, the resistance correction might

But to return from this digression (which, however, will not have been without its use, if it shall tend to diffuse clear conceptions of the subject, and to disentangle from one another corrections which seem to have got unduly mixed up together in the minds of practical inquirers). No sooner were the ideas of Bessel promulgated in England than Captain Sabine, whose attention was pointedly directed to a subject which had occupied so large and active a portion of his life, resolved to ascertain the true amount of this new, or newly mentioned, correction, in the only way in which it could be effectually done, viz., by vibrating the pendulum *in vacuo*, which he accordingly effected by a series of highly interesting experiments, carried on at the Royal Observatory at Greenwich, and recorded in the "Philosophical Transactions," in a paper read March 12. 1829. His result makes the total reduction to a vacuum about one and two-thirds of that usually called "the correction for buoyancy." It should, however, be borne carefully in mind that the particular correction now in question has, in fact, nothing whatever to do with the buoyancy correction, either in its mode of production or its form of expression, and ought, therefore, to be very studiously kept apart from it

be annihilated altogether, or so nearly as to render it quite inappretiable.

I trust that, in what is said above, I shall not be supposed to undervalue M. Bessel's analytical treatment of this intricate problem, especially as it conducts to results which, regarded as a first approximation, represent sufficiently well the results of experience.

in all theoretical views, though of course they must be numerically amalgamated in the "reduction to a vacuum."

Meanwhile the attention of Mr. Baily had, about the same time, been called to the pendulum, in consequence of the contemplated expedition about to sail under the command of Captain Foster, on that memorable and most unfortunate expedition which cost him his life. It was on this occasion, and with a view to the use of this expedition, that Mr. Baily (still acting for the Astronomical Society, whose aid had been requested in suggesting useful objects of inquiry) devised that capital improvement in the system of itinerant pendulum observation, which consists in making each transferable pendulum a convertible one, by the simple addition of another knife-edge, and in doing away with extra-apparatus of tail-pieces, sliders, &c., by the initial adjustments of the instrument. And I may here incidentally remark, that the general principles of reducing, as far as possible, the number of moveable parts in every instrument intended for standard determinations of whatever kind, is one which cannot be too strongly recommended, and has been successfully acted on by the present Astronomer Royal in more than one recent construction. Two pendula, a copper and an iron one, on Mr. Baily's principle, were furnished by the Society for this expedition, an account of which may be found in the "Notices" of the Society for June 13. 1828.

The adjustment and trial of these pendula previous to the sailing of the expedition, were performed by Mr. Baily at his own house, and, thus engaged in actual experiment, he at once became led on into a minute examination of all the possible sources of practical error in the experiments, and consequent uncertainty in the important results of which they had become the basis. It was in this stage of his experience that he became acquainted with Professor Bessel's results, which determined him (as it had already done Captain Sabine) to go into the whole subject of the new correction by experiments performed *in vacuo*. But not content with assuming any fixed proportionality between it and the buoyancy correction, he resolved so to vary the form, magnitude, and materials of the vibrating masses, as to make its true nature and amount an object of inductive experimental inquiry; thus, though adopting the language of Buat and Bessel, disengaging himself in effect from any theoretical view of the *modus operandi* or mechanical process by which the effect was produced.

The result of these inquiries was a very elaborate and masterly paper read to the Royal Society, on the 31st of May, 1832, containing the results of experiments in air and *in vacuo*, on upwards of eighty pendulums of various forms and materials by which the new correction is clearly shown to depend not only on the dimensions but on the form and situation of the vibrating body. Independent of the excellence of this paper as a specimen of delicate experimental inquiry and induction, in which, to use the expression of one best capable of estimating and admiring them, his generalizing powers seem to have been held in abeyance till the right moment for their exercise arrived, it had the further merit of bringing into distinct notice a number of minute circumstances, chiefly relative to the mode of suspension (important, however, from their influence on results), which it is absolutely necessary to attend to in these delicate and difficult inquiries, if the pendulum be ever again resorted to as a means of verifying or fixing anew the standard of length.

The return of the Chanticleer in 1831, without its lamented commander, threw the whole task of arranging and digesting for publication Captain Foster's pendulum observations on Mr. Baily—a labour of love, prompted by the warmest friendship, and which he executed in the spirit of one determined to erect a monument to the fame of that truly amiable and talented officer, of the most durable and precious materials. His Report on the subject to the Admiralty was presented by the Lords Commissioners to the Council of the Astronomical Society, and printed at the expense of Government as the seventh volume of our "Memoirs." In this report the observations are given in full, and with the most scrupulous fidelity, and those at each of the numerous stations discussed with the utmost care. The final re-examination of the pendulums in London was also personally executed by Mr. Baily, and the whole series of stations combined into a general result, which gives for the ellipticity of the earth $\frac{1}{289\cdot48}$. Not content with this, he has here also collected into one synoptic view the results obtained at various stations all over the globe with the invariable pendulum, by observers of all nations, so as to place them in comparison with each other, and to deduce from them a general result. Of these, by far the most numerous and prominent, in every respect, are those of our own countrymen, Captains Foster and Sabine, and nothing can be more gratifying, in estimating our own national share in this sublime application of science, than to find these principal authorities, whose observations were made and reduced with the most absolute independence of each other, agreeing at all the stations where they admit of comparison, with a precision truly admirable. In fact, the greatest disagreement of each of their final results, from a mean of them both, amounts to a quantity less than half a vibration out of 86,400, or in a mean solar day.*

Standard of Length.—From the pendulum to the standard of length, or the fixation of the *scientific unit*, the transition is easy; and in Mr. Baily's case, was unavoidable. For, being once satisfied

* The stations of comparison are London, Maranham, Ascension, and Trinidad. Taking London for a term of departure, each station affords a ratio whose extremes (see "Report," p. 86) differ only by 0·0000103, the half of which multiplied by 86,400 gives $0^{s}\cdot44446$.

by experience of the innumerable minute circumstances on which perfect precision in these inquiries depends, and finding the parliamentary enunciation of the relation between the conventional and natural standards nullified, as it were, under his eye, he felt himself irresistibly urged to inquire how far the conventional unit itself might be depended upon, and within what limits of error it might certainly be reproduced in copies. His first step in this direction was to obtain the most perfect possible representative of this unit, and (as the Astronomical Society was now identified with almost all his undertakings) justly considering the possession of such a standard by that body as a thing in itself desirable, and the instrument itself likely, if thoroughly well executed, to become in its hands of universal scientific reference, he procured himself, to be named by the Council, a Committee for superintending its execution, and comparing it with the most authentic standards at present existing in this country. Perhaps there is no subject of inquiry more perplexing, or one whose investigation calls for more patience and perseverance, than the detection and exact estimation of those minute sources of error which influence these delicate measurements, which can only be satisfactorily performed by endless repetition and systematic variation of every circumstance by which error can possibly be introduced. Another and peculiar source of annoyance, and even vexation, consisted in the rough and careless usage to which those precious instruments, on which the conservation of our national units depends, had been subjected in too many instances; by which rude and ignorant hands had irrecoverably marred some of those refined productions of human workmanship, which ought not even to be approached but with precaution, or touched but with the utmost delicacy. Few things seem to have excited Mr. Baily's indignation more than the continual occurrence of evidence, only too palpable, of the small respect in which these standards appear to have been held by those under whose protection they had been placed, and of the violence which has been repeatedly suffered to be perpetrated on them.

I shall by no means go into any minute analysis of the admirable "Report" to the Council of this Society, which contains his account of the construction of our standard scale, its comparison with the parliamentary standard, and its most authentic existing representatives—and, with the French metre, as we have it represented in this country by two platina metres, in the possession of the Royal Society; or the means taken to secure it from loss, by the formation of carefully compared copies, two of which have been sent abroad, and two retained in England. Suffice it so say, that the delicacy of the means employed, the minuteness of the precautions used, and the multiplicity of the comparisons, surpassed every thing of the kind which had ever before been done in this country. This Report, too, is valuable in another way. Under the modest title of "A short History of the Standard Measures of this Country," it presents a summary of the subject so complete as almost to obviate the necessity of referring elsewhere for *historical* information.*

The immediate result of this useful and most laborious undertaking has been to put this Society in possession of, perhaps, the most perfect standard measure and divided scale in existence, in which every division, even to the individual inches, has been micrometrically verified, and their errors ascertained and placed on record. It would almost seem, too, as if a prophetic spirit had actuated the undertaking, and urged it to its completion without any of those delays which so often and proverbially attend the construction and optical examination of delicate instruments. For the comparison of the new scale with the imperial standard yard had hardly been completed six months, when the latter, together with the other original standard by Bird (that of 1758), as well as the imperial standard of weight, were destroyed

* Mr. Baily was assisted in the actual comparisons by several Fellows of the Society, among whom the late Lieut. Murphy was conspicuous, an observer whose temper and scientific habits peculiarly fitted him for co-operating with Mr. Baily, and whose name would probably have occurred more than once in this memoir but for his untimely death, which took place in the service of Astronomy in a distant region, and was probably the unfortunate consequence of over-exertion in its cause.

in the conflagration of the Houses of Parliament in October 1834. Thus the operation in question has been the fortunate means of preserving, to the latest posterity, that unit which has pervaded all our science, almost from the first dawn of exact knowledge.

The scientific unit is indeed preserved; but the nation remained, and remains up to this moment *, without a legal standard either of weight or measure. In the early part of 1838, however, in consequence (as I have been led to understand) of some communications on the subject between Mr. Baily, Mr. Bethune, and the Astronomer Royal, the latter was induced to draw the attention of Government to the subject, an occasion having arisen which rendered the mention in an official form unavoidable. And on the 11th of May of the same year a commission was appointed, consisting of seven † members (Mr. Baily being one), to report on the course most advisable to be pursued under these circumstances. To this duty, which involved the hearing of a vast deal of evidence and much personal attendance, Mr. Baily gave his unceasing attention; suggesting many valuable points, both practical and theoretical; and, on the Report of the Commission being agreed on, and the practical formation of new standards, in conformity with the view therein taken of the subject, being referred by Government to the same com-

* 1844.

† An eighth was subsequently added.

missioners, Mr. Baily undertook, to the general satisfaction of the whole body, and at their particular request, the delicate and important task of reconstructing the standard of length—a task which, unhappily, he did not live to complete. On whomsoever may * devolve the completion of this standard, it will be satisfactory to the members of this Society to know that, among the evidence adduced for its restoration, the scale prepared for it by Mr. Baily necessarily forms a most important and prominent feature.

Density of the Earth.—The accurate determination of one fundamental quantity naturally leads to inquiry into others. To make our globe the basis of measurement for the dimensions of the planetary system and of the visible universe, its form and magnitude must first be accurately known. To make it afford a scale by which the masses and attractive forces of the sun and planets can be expressed in terms conveying a positive meaning, its density must be ascertained, as compared with that of substances which occur on its surface, with which our experience is familiar, and from which our notions of material existence are drawn. The

* "The task was undertaken by Mr. Sheepshanks, one of Mr. Baily's most devoted friends, who gave it, during eleven years, an amount of thought and labour which will be but poorly collected even from the Report of his proceedings now preparing. The number of *recorded* micrometer observations falls but 500 short of 90,000. Mr. Sheepshanks died August 4. 1855, almost on the day on which his results received a legal sanction."—*Note of Professor De Morgan,* 1856.

fine experiment of Cavendish, confirmed as it was, in its general result, by the operations on Schehallien, had satisfactorily demonstrated the continuity of the Newtonian law of gravity, from such vast distances as astronomy is conversant with, through the intermediate steps of the diameters of the earth, and of a mountain, down to the minute intervals between the parts of a philosophical apparatus, and their agreement within as moderate limits as could have reasonably been expected, had even led to something like a probable estimate of the earth's density, which, however, could never be regarded as satisfactory, otherwise than as a first step towards more precise determinations. Mr. Baily's labours, therefore, on the pendulum were hardly brought to a conclusion when he was led to enter upon this subject, the immediate occasion of his doing so being an incidental suggestion at the council table by Mr. De Morgan, of the desirableness of repeating the experiment of Cavendish* — a suggestion immediately seconded both by the Astronomer Royal and by Mr. Baily. The experience

* *Fiat justitia, ruat cœlum.* The original design of this beautiful experiment was Michell's, who actually constructed the identical apparatus which Cavendish used, but died before he could execute the experiment. The apparatus came, after his death, into the possession of the Rev. W. H. Wollaston, D.D., *who gave it* to Cavendish, who used it, indeed, to excellent purpose, but who assuredly neither devised the experiment, nor invented, nor constructed, nor even, so far as I can perceive, materially improved the apparatus. All this is distinctly stated by Cavendish himself, who is, therefore, noway to blame for any misconception which may prevail on the subject.

of the latter had shown him how indispensably necessary, in such inquiries, are extensive repetition and variation of circumstance. The Schehallien experiment, from its very nature, admitted of neither; and, on carefully examining Cavendish's record of his own experiment, he found abundant reason to perceive how much was left to be desired, in both these respects, even in that form of the inquiry.

In resolving on a repetition of this experiment. the difficulty of the undertaking itself, and his own preparation for it, must have been, and no doubt were, very seriously considered. However confident in his own resources and perseverance, it was no holiday task in which he was now about to engage. The pendulum experiments, with all their delicacy, could hardly be regarded as more than an elementary initiation into the extreme minuteness necessary for this inquiry. There are two branches of research in physical astronomy which task to the utmost the resources of art, the delicacy of manipulation, and the perseverance of the inquirer—the parallax of the fixed stars and the density of the earth. In both, an immense object has to be seized by the smallest conceivable handle. But, of the two problems, the latter is probably that which throws the greatest burden on the inquirer, inasmush as it is not merely a series of observations to be carried on under well-ascertained circumstances and known laws, but a course of experiments to be entered on for eliminating or controlling influences which war against success in every part of the process, and where every element, nay, even the elementary powers of heat, electricity, magnetism, the molecular movements of the air, the varying elasticity of fibres, and a host of ill-understood disturbing causes, set themselves in opposing array in their most recondite and unexpected forms of interference. Nor could it have been overlooked by him that it was necessary, not merely to do over again what Cavendish had done before him, a thing in itself not easy, but to do it much more thoroughly and effectually.

Mr. Baily, however, was not to be discouraged by such considerations. He saw that there existed a blank in our list of exact data which it was necessary to fill, and he felt himself in possession of those gifts of nature and position which enabled him to fill it. Accordingly, in 1835, on the occasion above alluded to, the Astronomical Society appointed a committee to consider the subjest; and Mr. Baily having offered to perform the experiment, in 1837, the Government (at the instance of Mr. Airy) granted the liberal sum of 500*l.* to defray the cost of the experiment.

This great work was brought to a satisfactory conclusion in 1842, and a complete account, with a full detail of the experiments, printed in one volume, published in 1843, forming the fourteenth of the series of "Memoirs" of this Society. The experiments were varied with balls of different materials, and with suspensions no less various, combined so as to form no less than 62 distinct series, embodying the results of 2153 experiments; and which, formed into groups according to the nature of the combination, afford 36 distinct results, taking those only in which the balls were used, the extremes of which are 5·847 and 5·507, and the most probable mean 5·660, none of them being so low as Cavendish's mean result, 5·448. The probable error of the whole (0·0032) shows that the mean specific gravity of this our planet is, in all human probability, quite as well determined as that of an ordinary hand-specimen in a mineralogical cabinet,—a marvellous result, which should teach us to despair of nothing which lies within the compass of number, weight, and measure. I ought not to omit mentioning that, of all the five determinations of this element we possess, Mr. Baily's is the highest.*

Though it would be equally remote from my

* The five determinations alluded to are, in order of magnitude, as follows:—

Schehallien experiment from Playfair's data and calculations .	Max...4·867 Min...4·559	Mean. 4·713

Carlini, from pendulum on Mount Cenis, corrected by Giulio...4·950
Reich, repetition of Cavendish's experiment (most probable combination)...5·438
Cavendish, computation corrected by Baily....................5·448
Baily (most probable combination).................................5·660

Since this memoir was written, Mr. Airy has added another determination to this list, the result of an elaborate series of observations on the pendulum in the Harton coal-pit. (*Phil. Tr.* 1856.) The result is higher than any of the foregoing, viz.: 6·565. (*H.* 1857.)

present purpose, and superfluous in presence of such an assembly, to enter minutely into a discussion of these experiments, there is one point in their conduct which I cannot pass over in silence. The experiments had been carried on for eighteen months, a vast number of preliminary trials had been made, and upwards of 1000 registered results obtained, when it became apparent that the coincidence of Cavendish's results, one with another, was rather to be attributed to the paucity of his trials than to any especial accuracy in his observations or felicity in his mode of operating. Even in the few experiments made by Cavendish, discordances had shown themselves, of which no account could be given other than by reference to the movements of included air; but, on Mr. Baily's extensive scale of operation, the limits of disagreement obviously arising from this cause became so enormous as to render it hardly possible to draw any line for the reception and rejection of results. In fact, at one period he had almost begun to despair of bringing the matter to any positive conclusion. The happy suggestion of Mr. Forbes, *to gild* the torsion-box and leaden balls, at once dispelled all this vagueness and uncertainty, and reduced the results to a high degree of uniformity.*

* This was not, however, the *only* precaution used. Mr. Baily carried out the suggestion, by swathing the torsion-box in flannel, and applying over this defence an exterior *gilded* case. Should the experiment ever again be repeated, it should be attempted *in vacuo*.

Most experimenters would have been content to reject the discordant results. Mr. Baily unhesitatingly sacrificed the whole, and began anew, without appearing to regard with an instant's regret the time and labour lost. The gold medal of this Society was awarded to him for this important memoir.

Revision of Catalogues of the Stars.—The contributions of Mr. Baily to this branch of sidereal astronomy are so numerous and so important, as alone would suffice to rank him among the greatest benefactors to the science, since, without being himself an observer, he has conferred, by his indefatigable industry and perseverance in collating authorities, rescuing original observations from oblivion, and rectifying printed errors, a vast and unhoped-for accession of value to the works of all those on whom he has commented. In fact, this, which may be termed the archæology of practical astronomy, formed his staple and standing work, which, though from time to time interrupted by other subjects, was always resumed, always with increasing interest, and always on a larger and more effective scale, up to the very year of his death. His object appears to have been, so far as is now practicable, to destroy the gap which separates us from the elder astronomers, and to multiply, or at least to preserve from further destruction, the links which connect us with them; to ascertain *all* that *has really been* recorded of the stars, and to make that totality of knowledge the common property of astronomers — a precious and a pious labour, of which we have no examples, except in that spirit of loyal reverence which prompted Ptolemy to secure from oblivion the observations of Hipparchus, and make them the foundation of all future astronomy; and in that which animated Bessel, when on the basis of Bradley's observations he may be said to have afforded the means of reconstructing the whole fabric of the science.

The catalogues which Mr. Baily has re-edited are those of Ptolemy, Ulugh Beigh, Tycho Brahe, Halley, Hevelius, Flamsteed, Lacaille, and Mayer: a mass of commentation, expurgation, and minute inquiry before which the most stout-hearted might quail, since there is not one of them in which each individual star has not been made the subject of a most scrupulous and searching examination, and in which errors that had escaped all prior detection,—errors of reading, errors of entry, of copying, of calculation, of printing, out of number,—have not been detected and corrected. But for these labours, the catalogues of Ptolemy and Ulugh, indeed, must have remained sealed books to any but professed antiquaries; and although we can now hardly ever have occasion to appeal to these earliest authorities for any practical purpose, we cannot but look on the labour thus cheerfully bestowed in embalming and consecrating their venerable relics as the sure pledge that our own works, if really worthy, will not be suffered to perish by time and neglect.

But while we admire both the diligence and the

scrupulous exactness, of which the notes appended to these catalogues bear ample evidence, we must not omit to mention, that there are two of them, those of Mayer and Flamsteed, in respect of which Mr. Baily's researches have been pushed far beyond the mere duties of comparison and comment, having been extended to the conservation and minute examination of the original records from which the catalogues were formed. In the case of Mayer, his influence with the late Board of Longitude secured the publication of the original observations of that eminent astronomer at Göttingen, which had never before seen the light.* In the case of Flamsteed, his labours were much more extensive, and require a more particular statement, inasmuch as not only Flamsteed's greatest work, the "British Catalogue," found in him its restorer to that high rank, as an astronomical document, which it is justly entitled to hold, but the fame and character of its author their defender and rescuer from grievous misapprehension and misstatement.

In 1832 it happened, by a most singular coincidence, that Mr. Baily became aware of the existence, in the possession of his opposite neighbour in the same street, E. Giles, Esq., of the whole of Flamsteed's autograph letters to Abraham Sharp, and was permitted to peruse and copy them. Their perusal convinced him that Flamsteed's life, astronomical labours, and personal character, had

* In 1826.

never been fairly placed before the world, and induced him to examine with care the mass of his papers preserved (or rather neglected and mouldering) at Greenwich. His first care was to arrest the progress of their further decay. His next, to avail himself of the original entries of the observations, and of the manuscript records of the computations founded on them, to trace out the sources, and to rectify the numerous errors and inconsistencies, of the "British Catalogue" as it then stood before the world, and to present it to the public under quite a new aspect — as a noble monument of its author's skill and devotion, and a work worthy of the age and country which produced it. Among the papers thus examined, however, were also found an almost complete autobiography of Flamsteed, and a voluminous correspondence illustrative of those points so painfully at issue between Flamsteed, Newton, and Halley, relative to the publication of the Catalogue and observations, and to other matters of a more personal nature, which had hitherto all along been stated in an infinitely more unfavourable light towards Flamsteed than that which appears, from Mr. Baily's thorough and voluminous exposition of the whole affair, and the evidence of the almost innumerable letters which he has printed at length, truly and properly to belong to them. Indeed it seems impossible not to admit, on the evidence here produced, that great and grievous injustice was done, and hardship imposed, in these transactions, on Flamsteed, whose character stands forward, on the whole showing, as that of a most devoted and painstaking astronomer, working at extreme disadvantage, under most penurious arrangements on the part of government, making every sacrifice, both personal and pecuniary, and embroiled (as I cannot help considering, by the misrepresentations and misconduct of Halley) with the greatest man of his own or any other age, holding a position with respect to the Observatory, as Visitor, which, under mistaken impressions of the true bearings of the case, might cause severity to assume the guise of public duty.

The volume which contains this important work of Mr. Baily was commenced (as we have seen) in 1832, and published in 1835, a rapidity of execution truly astonishing, when we consider that the volume extends to nearly 800 pages quarto; that the notes to the Catalogue alone occupy no less than 144 of them closely printed, not a line of which but involves some question of identity, of nomenclature, of arithmetical inquiry, or of reference to other authorities; that the examination and selection of the letters and other biographical matter for publication was an affair of the utmost delicacy and responsibility; and that the preface, which contains Mr. Baily's own summary of Flamsteed's life, the introduction to the Catalogue, and the Supplement, in further vindication of Flamsteed's character and justification of his own views of it, — are all of them works of a very elaborate nature, and of the highest interest.

Catalogues of Lacaille and Lalande. — But Mr. Baily's views were not confined to the mere correction of existing catalogues. The labour of the commentator and collator, which has filled and satisfied so many minds, was to him only a means to an end of real practical importance. His aim was to render readily available to every astronomer all recorded observations of the sidereal heavens which could be depended on. Two great masses of observation might be said to exist buried under their own weight, and affording matter of grief and reproach to astronomy, now to be exchanged for congratulation and triumph. These were Lacaille's observations at the Cape of nearly 10,000 stars, and those of D'Agelet and Michel Lefrancais Lalande at Paris of nearly 50,000. Neither of these collections of observations had been more than partially reduced. Lacaille himself had performed this task for 1942 of his stars. A considerable number of the stars of the "Histoire Céleste" (Lalande's observations) had also been reduced and catalogued by Bode. But the great mass of both remained unreduced and unarranged, though it is true that Lacaille had accompanied each page of his observations with a table of reductions, and that in 1825, Professor Schumacher had published and dedicated to this Society a volume of assistant tables, enabling any one, with little trouble, to reduce any single observation of the "Histoire Céleste." Still they remained unreduced, and, therefore, useless, except on those rare occasions when, for special reasons, it might be necessary to search out and reduce any particular object.

Thus was a treasure of great value held in abeyance. This Mr. Baily perceived, and after some correspondence with the French Bureau des Longitudes, which, however, led to no result, he resolved to bring the subject before the British Association. That liberal and energetic body at once acceded to his views, and in 1838 appointed two committees, each with funds at their disposal, to execute the reductions and prepare the catalogues. The reduction and arrangement of Lacaille's stars was executed under the superintendence of Mr. Henderson, that of Lalande's under Mr. Baily, the arrangement of the work in both (if I mistake not) having been effected on a plan concerted and matured by the latter. Both works were reported as complete (the prefaces alone excepted) in 1843, and it only remained to provide for their printing. This also was done by the liberality of the British government, who assigned 1000*l*. for the purpose; and this work was especially placed under Mr. Baily's direction. These Catalogues, unhappily, he did not live to see published. The printing, however, of each was found advanced at his decease as far as 8320 stars*, and is now being continued under the more immediate inspection and superintendence of Mr. Stratford.

Catalogue of the British Association.—I have yet

* The total number of stars in the two Catalogues respectively will amount to 9766, and 47,490.

to speak of another and a magnificent work undertaken and brought to a successful conclusion by Mr. Baily; a work which, perhaps, deserves to be considered as the greatest boon which could have been conferred on practical astronomy in its present state, and whose influence will be felt in all its ramifications, giving to them a coherence and a unity which it could hardly gain from any other source. I allude to the General Standard Catalogue of nearly 10,000 stars, which the British Association are about to publish, at the instance of Mr. Baily. The plan of this great and useful work is an extension of that of the Astronomical Society, of which I have already spoken. The stars (selected by Mr. Baily) form a universal system of zero-points, comprehending probably every star of the sixth and higher magnitudes in the whole heavens. All the coefficients for their reduction are tabulated, and the greatest pains bestowed upon their exact identification and synonymes in other catalogues; so that this, in all human probability, will become the catalogue of universal reference. Prefixed to it is a valuable preface from the pen of Mr. Baily, —his last contribution to astronomical science.

A very important feature of this and the two catalogues last noticed is their nomenclature. The system adopted is the same in all; and *that*, a system not capriciously adopted or servilely copied, put founded on a most searching and careful revision of all existing catalogues, and of the charts of Bayer, Flamsteed, and Lacaille: rectifying the boundaries of constellations which had become strangely confused; correcting innumerable errors of naming, numbering, and lettering; and reducing, in short, to order and regularity, a subject which had become almost hopelessly entangled. The way is thus at length opened to a more rational distribution of the heavens into constellations, and that final step, which must sooner or later be taken, of introducing a systematic nomenclature into sidereal astronomy, rendered easy, whensoever astronomers shall be prepared on other grounds to take it. The trouble and difficulty attending this part of the work exceeds what any one unused to such tasks can easily imagine.

There are two papers by Mr. Baily relating to sidereal astronomy, of which mention ought to be made here; viz., one "On the Proper Motions of the Stars," which was read before the Astronomical Society on the 9th December, 1831, in which a list of about 200 stars, whose proper motion appears sufficiently sensible to merit further inquiry, is discussed. In drawing up this list, he was much aided by a series of transit observations by Dr. Robinson, observed expressly with a view to this inquiry. But as no positive conclusion of a general nature is arrived at in this memoir, and as the subject is yet hardly ripe for a complete discussion, I shall dilate no further on it. The other paper to which I allude (which was read also to this Society on the 14th November, 1834,) states the result of an examination of Dr. Halley's MSS. at the Royal Observatory. The appointment of Astronomer Royal was held by Halley twenty-two years, and though for the two first of them the Observatory was entirely deprived of instruments, and for the next four a five-feet transit only was available, it might, at least, have been expected that he should have used diligently the means he did possess, or, at all events, have recorded the observations he did make in a regular, methodical, and intelligible manner. From Mr. Baily's examination of these papers, however, this appears to have been very far indeed from the case; and that, with the exception of differences of right ascension between the moon and planets and neighbouring fixed stars, which alone he seems to have considered worthy of attention, little of interest could be expected to repay the trouble and expense of their reduction. Of these papers Mr. Baily, ever anxious for the preservation of records, and mindful of the dormant value which they so often possess, obtained from the Admiralty a transcript, which, being carefully collated with, and corrected by, the original MSS., is now deposited in our library.

The mention of the Royal Observatory induces me to notice here a change which has been lately made in the constitution of that noble institution, by a revision of the royal warrant, defining the number and mode of appointment of the Visitors, and placing this Society on a similar and equal footing with the Royal Society in the discharge of that important duty. This change was made at

Mr. Baily's suggestion, with the entire concurrence however, of the then President of the Royal Society, as to its expediency, on the occasion of the demise of the crown by the death of George IV., which rendered a new warrant necessary. The new system has been found to work admirably well, and to have secured a perfect harmony of feeling between the Visitors and the eminent individual who now fills the post of Astronomer Royal, as well as entire confidence in the recommendations and suggestions of that body on the part of government. Aware, as all are now, of the fatal and soporific influence of routine in public institutions, they have only henceforward to guard against the opposite extreme; to which end, they cannot do better than take for their guide and example that admirable combination of energy, gentleness, and judgment, which distinguished Mr. Baily, no less on every public occasion than in his conduct as a Visitor, in which capacity, under both the old and the new system of visitation, he was an invariable attendant, being never absent during a period of twenty-eight years from any meeting but the last.

About the end of June, 1841, an accident happened to him which had very nearly proved fatal. Crossing Wellington Street for the purpose of taking some MSS. to a printer, a deafness, which had for some years been increasing on him, rendered him unaware of a rider recklessly urging his horse to furious speed, who either did not see him or was unable to pull up. In consequence a collision took place, and Mr. Baily received a stunning fall, accompanied with a severe scalp-wound. So violent, indeed, was the shock, that he lay for a whole week senseless, and for an equal period after his life was considered in imminent danger. His sound and excellent constitution, however, carried him through it, and no ill consequences remained. By the end of September he was enabled to resume the observations of the Cavendish experiment, which this unfortunate occurrence had interrupted, and a few weeks' residence in the country completed the cure.

On the 8th of July, 1842, he was gratified by the observation of a phenomenon which it had from his youth upwards been one of his most ardent wishes to witness, viz. a total eclipse of the sun. To this he looked forward, indeed, with a curiosity peculiarly intense ; having, on the occasion of the annular eclipse of May 15, 1836, which he travelled to Scotland to observe, and which he succeeded in observing under very favourable circumstances at Jedburgh, noticed a very singular phenomenon attending the formation of the annulus. I mean the appearance of beads of light, alternating finally with long, straight, dark threads, cutting across the narrow line of the sun's limb, which he described in a highly interesting paper read to this Society on the 9th December, 1836. On the occasion of the total eclipse he selected Pavia for his station, that town lying in the path of the centre of the shadow. There, by especial good fortune, he obtained an excellent view of it, and there he witnessed, not only a repetition of the phenomenon of the beads, but that much more astonishing and previously unheard-of one, of the flame-like, or conical rose-coloured protuberances, seen to project, as it were, from the hidden disk of the sun beyond the border of the moon. This truly wonderful appearance (which was corroborated by several other observers at different places, among others by Mr. Airy, at Turin,) was described by him, on his return from Italy, in a paper read to this Society on the 11th Nov. 1842; and it is not a little singular that the two most remarkable solar eclipses on record should thus have furnished the subjects of his first and last astronomical memoirs.

On his return from this journey he resumed his astronomical labours on the catalogues, as we have seen, which he continued, as well as his usual unremitted attendance to the business and at the meetings of this Society, till the spring of the present year, when his health began to decline, and several weeks of serious illness, a thing utterly unknown to him at any former period of his life (except as a result of accident), gave intimation of a failing constitution. For the first time since the reorganization of the visitation of the Royal Observatory he was unable to attend the annual meeting of the Visitors in June. He, however, rallied somewhat, so as to be able to be present at the commemoration at Oxford on July 2nd, on which occasion the honorary degree of Doctor of Civil Law was conferred on him by that university, as well as on Mr. Airy and Professor Struve. On his return from Oxford his health again rapidly declined, and all efforts of medical skill proving unavailing to relieve an internal complaint, which had at length declared itself, he expired, after a protracted, but happily not painful illness, during which he was fully sensible of his approaching end, in a state of the utmost calmness and composure, at half-past nine o'clock in the evening of the 30th of August, at the age of seventy years and four months.

In passing in review, as I have attempted to do, the scientific works of Mr. Baily, and noticing, as we cannot help doing, the gradual expansion of his views, and the progressively increasing importance of the objects they embraced, we are naturally led to ask by what means he was enabled thus to live as it were two distinct lives, each so active and successful, yet so apparently incompatible with each other ? how, in what is generally regarded as the decline of life, he could not only accomplish so much with such apparent ease to himself, but go on continually opening out wider and wider plans of useful exertion in a manner which seems only to belong to the freshness of youth ? The answer to such an inquiry is, no doubt, partly to be found in his uninterrupted enjoyment of health, which was so perfect that he has been heard to declare himself a stranger to every form of bodily ailment, and even to those inequalities of state which render most men at some hours of the day or night less fit for business or thought than at others. But though this is in itself a blessing of the most precious kind, and, if properly used, a vantage ground of power and success to any one favoured enough to possess it, it must be regarded in his case as subordinate to, though, no doubt, intimately connected with, a gift of a much higher order,— that of an equable and perfectly balanced intellectual and moral nature,— the greatest of gifts, which has been regarded, and justly, as the only one really worthy to be asked of Heaven in this life, — *mens sana, in corpore sano.* Few men, indeed, have ever enjoyed a state of being so habitually serene and composed, accompanied with so much power, and disposition to exert it. A calm, the reverse of apathy, a moderation having nothing in common with indifference, a *method* diametrically opposed to routine, pervaded every part of his sentiments and conduct. And hence it arose that every step which he took was measured and consequent —one fairly secured before another was put in progress. Such is ever the march of real power to durable conquest. Hence, too, it arose that a clear natural judgment, and that very uncommon gift, a sound common sense viewing all things through a medium unclouded by passion or prejudice, gave to his decisions a certainty from which few were ever found to dissent, and to his recommendations a weight which few though it *right* to resist.

It is very difficult in speaking of Mr. Baily's character to convey a true impression through the medium of a language so exaggerative as that which men now habitually use. Its impressiveness was more felt on reflection than on the instant, for it consisted in the absence of all that was obtrusive or imposing, without the possibility of that absence being misconstrued into a deficiency,—like a sphere whose form is perfect simply because nothing is protuberant. Equal to every occasion which arose, either in public or private life, yet, when not called forth, or when others occupied the field, content to be unremarked; to speak of his conduct as unassuming would convey but a faint idea of the perfect simplicity with which he stood aside from unnecessary prominence or interference.

Hardly less inadequate would it be to say of his temper that, always equable and cheerful, it was a source of peace and happiness to himself and others. It was much more,—it was a bond of kindness and union to all around him, and infused an alacrity of spirit into every affair in which the co-operation of others was needed, which was more than a simple reflex of his own good humour. It rendered every relation between himself and others easy and natural, and brought out all the latent warmth of every disposition. One would have been ashamed to evade a duty or refuse a burden when it was seen how lightly his share was borne, how readily he stept out of his way to offer aid wherever he saw it needed, and how frankly every suggestion was received, and every aid from others accepted and acknowledged. This is the secret of all successful co-operation.

Order, method, and regularity, are the essence of business, and these qualities pervaded all proceedings in which he took a part, and, indeed, all his habits of life. In consequence, all details found their right place and due provision for their execution, in every matter in which he engaged. This was not so much the result of acquired habits, as a man of business, as the natural consequence of his practical views, and an emanation of that clear, collected spirit, of which even his ordinary handwriting was no uncertain index. Among hundreds of his letters which I possess, there is hardly an erasure or correction to be found, but everywhere, on whatever subject, or whatever the haste, the same clear, finished, copper-plate characters.

Of his choice of life I have spoken something. Fortune he regarded as a mean to an end, but that end he placed very high; and fortune, he well knew, though a mean to its attainment, was not the only or the chief mean. As a member of civilized society, to add something to civilization, to ennoble his country and improve himself, by enlarging the boundaries of knowledge, and to provide for his own dignity and happiness by a pursuit capable of conferring both,—these were the ends which he proposed and accomplished. In choosing the particular line which he did, it is impossible to estimate too highly the self-knowledge and judgment which enabled him to see and adopt those objects best adapted to his powers, and on which they could be, on the whole, most availably and usefully employed. Both in his public and private capacity he was liberal and generous in the extreme; and both his purse and his influence were ever ready, whether to befriend merit or to promote objects of public, and, especially, of scientific utility.

To term Mr. Baily a man of brilliant genius or great invention, would in effect be doing him wrong. His talents *were* great, but rather solid and sober than brilliant, and such as seized their subject rather with a tenacious grasp than with a sudden pounce. His mind, though, perhaps, not excursive, was yet always in progress, and by industry, activity, and using to advantage every ray of light as it broke in upon his path, he often accomplished what is denied to the desultory efforts of more imaginative men. Whatever he knew he knew thoroughly, and enlarged his frontier by continually stepping across the boundary and making good a new and well-marked line between the cultivation within and the wilderness without. But the frame of his mind, if not colossal, was manly in the largest sense. Far-sighted, clear-judging, and active; true, sterling, and equally unbiassed by partiality and by fear; upright, undeviating, and candid, ardently attached to truth, and deeming no sacrifice too great for its attainment;

—these are qualities which throw what is called genius, when unaccompanied, or but partially accompanied, with them, quite into the shade.

In speaking of his conduct with respect to this Society, and the infinite obligations we owe to him, we must regard him in the first place as the individual to whom, more than to any other, we owe the titles of a parent and a protector, and our early consolidation into a compact, united, and efficient body. As Secretary *pro tempore*, the draft of our Rules and the first Address explanatory of our objects, circulated at the commencement of our existence, were entirely, or in great measure, prepared by him; and, governed by these rules with hardly any change, we have continued to flourish for twenty-four years, which is the best test of their adaptation to our purposes. As I have already stated, he acted as Secretary during the first three years of our existence, during which period the business of our meetings and of our council was brought into that systematic and orderly train of which the benefit has never since ceased to be felt. On retiring from this office he was elected Vice-President, and on the next biennial demise of the chair he became our President, an office which he afterwards filled for three subsequent periods for two years, including that of his lamented death. Altogether, during eight years as President and eleven as Vice-President, he filled the highest offices of our institution, and was never off the Council, nor was there any Committee on which he did not sit as one of its most active and efficient members.

With the exception of the meeting of May 12, 1836, when he was in Scotland observing the annular eclipse, he was never absent from any Council, or from any Ordinary, General, or Committee Meeting, until finally prevented by illness. Nor during the whole period of the Society's existence was there any matter in which its interests were concerned in which he was not a mover, and, indeed, the principal mover and operator. Nor was this care of our interests and respectability confined to formal business or to matters of internal management. On every external occasion which offered he bore those interests in mind. He watched and seized the precise opportunity to procure for us from Government the commodious apartments we occupy. He obtained for us the respected and dignified position of Joint-Visitors of the Royal Observatory. He let no opportunity pass of enriching our library with attested copies of the most valuable astronomical documents, such as "Flamsteed's Letters" and "Halley's Recorded Observations." He husbanded and nursed our finances with the utmost judgment and economy, thereby rendering us rich and independent. He printed at his own cost the thirteenth volume of our "Memoirs," and procured to be defrayed by Government the expense of the seventh, and, by subscription among the members, without entrenching on the funds of the Society, that of the computation and printing of our Catalogue. He prepared all our annual Reports, and his addresses from the chair will always be read with pleasure and instruction. He also prepared all Committee's Reports, and translated for reading at our meetings numerous notices and communications in the German language: among others the memoir relating to the Berlin charts. In fine, he superintended every thing in every department. But it was the manner and delicate tact of this superintendence which gave it its value and rendered it efficient. In respect of this point, I may, perhaps, be permitted to use the expressions of a distinguished member of our body, to whom we owe many and great obligations, and who has witnessed the working of its machinery from the beginning, an advantage of which for some years I have myself been deprived by non-residence in London and absence from England. "Of his management of our Society," says Mr. Sheepshanks, "it is diffcult to speak so as to convey a correct idea. No assumption, no interference with other people, no martinet spirit (which seems almost natural to all good business men), but every thing carried on smoothly and correctly, and without bustle. He hit, better than any chairman I have ever seen, the mean between strictness and laxity, and, while he kept every thing going in its proper channel, he also kept every body in good humour. This natural tact was a great gift, but there was another quality which I never saw in any one but him, and that was his readiness to give precedence and room to every one who wished to do anything useful, and his equal readiness to supply every deficiency and do the work of every body else. He was also the person who never was asleep, and never forgot any thing, and who contrived by his good humour, hospitality, and good sense, to keep every thing in train." To much of this view, as a matter of general character, I have given my own independent expression, but I could not deny myself the satisfaction of corroborating my own judgment by that of one so well qualified, from intimate knowledge, to form opinions.

Mr. Baily, as I have already stated, was a member of the Royal, Geological, and Linnean Societies, to which I may also add the Royal Geographical Society and the Royal Irish Academy. In the Royal Society his eminence as an astronomer and a man of general science made his presence valuable, and the universal respect in which he was held gave him much influence. He filled in that body the office of Vice-President for six years, of Treasurer for three, and was fifteen times elected on the Council. I have already mentioned two of the three papers he contributed to its "Transactions." The third contains a minute account of the standard barometer of that Society, fixed up in their apartments in the year 1837, in which he enters into every particular of its construction, mode of registry, and corrections. It was read on the 16th of November, 1837. He was also a member and one of the trustees of the British Association, at whose meetings he was an occasional attendant, and acted, as we have seen, on some important committees. In 1835, the University of Dublin conferred on him the honorary title of Doctor of Civil Law, as I have already stated, was also done by Oxford in 1844. Among the Foreign Academies, which in honouring him honoured themselves, I find him to have been a correspondent of the Royal Institute of Sciences of Paris, and of the Royal Academies of Berlin, Naples, and Palermo, as well as the American Academy of Arts and Sciences at Boston.

His portrait by Phillips, presented by some Fellows of the Society, has long adorned, and, though for the present removed from its frame, will speedily again adorn, our meeting-room. May his mantle descend on our future presidents, and his spirit long continue to preside over our councils and animate our exertions in the cause he had so so much at heart!

The following Epitaph is inscribed on his tomb.

H. S. J.

FRANCISCUS BAILY

LL.D. R.S.S. L. ET ED. ET HIB. SOCIUS

SOC. REG. ASTRONOMICÆ LONDIN

PRÆSES ET COLUMEN

NATUS NEUBURIÆ APRIL. XXVIII. MDCCLXXIV.

OBIIT LONDINI AUGUST. XXX. MDCCCXLIV.

ÆQUO SEMPER ANIMO MORIBUSQUE

PURIS SIMPLICIBUS COMMODIS

IPSE BEATUS CARUS VIXIT SUIS

NEGOTIANDI OLIM CURIS FELICITER EXPEDITUM

AD SUBLIMIORES ASTRONOMIÆ CALCULOS

SUCCESSU NON MINUS FELICI

SESE CONTULISSE TESTANTUR

TERRA EXPENSA

STELLÆ EX ORDINE NUMERATÆ

VIS GRAVITATIS EMENSA

MODULUS SUMMÂ ARTE DEFINITUS

HUNC TALEM VIRUM PATRIOS PROPE CINERES

PULVERIS EXIGUI COHIBET MUNUS.

AN ADDRESS

DELIVERED AT THE ANNIVERSARY MEETING OF THE ASTRONOMICAL SOCIETY, FEB. 9. 1849, BY THE PRESIDENT (SIR J. F. W. HERSCHEL, BART.) ON PRESENTING THE HONORARY MEDAL OF THE SOCIETY TO WILLIAM LASSELL, ESQ., OF LIVERPOOL.

GENTLEMEN,

THE Report of the Council having been read, in which the astronomical discoveries of the year, and epecially that of the planet *Metis*, have been clearly and eloquently commemorated, it is now my pleasing duty to state to you the grounds on which it has been agreed by us to award the gold medal of the Society for this year to Mr. Lassell. And this duty, pleasing in itself, I execute with the greater satisfaction, because I have a sort of hereditary fellow-feeling with Mr. Lassell, seeing that he belongs to that class of observers who have created their own instrumental means, — who have felt their own wants, and supplied them in their own way. I believe that this greatly enhances the pleasure of observing, especially when accompanied by discovery, and gives a double interest in the observer's eyes, and perhaps, too, in some degree, an increased one in those of the public, to every accession to the stock of our knowledge which his instruments have been the means of revealing; upon the same principle that the fruit which a man grows in his own garden, cultivated with his own hands, is enjoyed with a far higher zest than what he purchases in the market. Nor is this feeling by any means a selfish one. It arises from the natural and healthy excitement of successful exertion, and is part of that happy system of compensation by which Providence sweetens effort, and honours well-directed labour. If this be true of the labour of a man's hands in the mere production of material and perishable objects, it is so in a far superior sense, when the faculties of the intellect are called into exercise, and works elaborated with rare skill, and wrought to an extraordinary pitch of perfection, have yet a higher ulterior, intellectual object, to which their existence is subordinate, as means to an end.

Mr. Lassell has long been advantageously known to us as an ardent lover of astronomy, and as a diligent and exact observer, in which capacity he has appeared before us, as a reference to our "Memoirs" and "Notices" will testify, on numerous other occasions besides those to which I shall more particularly call your attention presently. In the year 1840, he erected an observatory at his residence near Liverpool, bearing the appropriate name of Starfield, which has ever since been the scene of his astronomical labours. Even at its first erection, this observatory presented features of novelty and interest. In addition to a good transit, it was furnished, instead of a meridian instrument or an ordinary equatorial achromatic, with a Newtonian reflecting telescope of nine inches aperture, and rather more than nine feet in focal length, equatorially mounted, the specula of which were of his own construction, and the mode of mounting devised by himself. This was already a considerable step, and forms an epoch in the history of the astronomical use of the reflecting telescope. Those only who have had experience of the annoyance of having to keep an object long in view, especially with high magnifying powers, and in micrometrical measurements, with a reflector mounted in the usual manner, having merely an altitude and azimuth motion, can duly feel and appretiate the advantage thus gained. But the difficulties to be surmounted in the execution of such a mode of mounting were very considerable — much more so than in the case of an achromatic, — owing partly to the non-coincidence of the centre of gravity of the telescope and mirror with the middle of the length of the tube, and partly to the necessity of supporting the mirror itself within the tube in a uniform bearing free from lateral constraint, and guaranteed against flexure and disturbance of its adjustment by alteration of its bearings. These difficulties, however, Mr. Lassell overcame; the latter, which is the most formidable, by an ingenious adaptation of the balancing principle first devised, if I am not mistaken, by Fraunhofer and Reichenbach for the prevention of flexure in the tubes of telescopes — a principle which has not received half the applications of which it is susceptible, and which, by throwing the whole strain of the weight of instruments on axes which may be made of unlimited strength, may be employed to destroy the distorting force of gravity on every other part.*

The success of this experiment was such, and the instrument was found to work so well, that Mr. Lassell conceived the bold idea of constructing a reflector of two feet in aperture, and twenty feet in focal length, and mounting it upon the same principle. The circumstances of his local situation, in the centre of manufacturing industry and mechanical construction, were eminently favourable to the success of this undertaking; and in Mr. Nasmyth he was fortunate enough to find a mechanist capable of executing in the highest perfection all his conceptions; and prepared, by his own love of astronomy and practical acquaintance with astronomical observation and with the construction of specula, to give them their full effect. It was of course, however, the construction and polishing of the large reflector which constituted the chief difficulty of this enterprise. To ensure success, Mr. Lassell spared neither pains nor cost.

* As, for example, the divided limbs of circles, and the spokes connecting them with their centres; an easy and simple mechanism, which, devised some time ago, and approved by the late M. Bessel, I may, perhaps, take some future opportunity to submit to the Society.—(*Note added in the printing.*)

As a preliminary step, he informs us that he visited the Earl of Rosse, at Birr Castle, and besides being favoured with more than one opportunity of satisfying himself of the excellent performance of that nobleman's three-foot telescope, enjoyed the high privilege of examining the whole machinery for grinding and polishing the large speculum, and returned so well satisfied as to resolve on the immediate execution of his own ideas.

The mode of casting and grinding the mirror, differing in some of the details, though proceeding generally on the same principle as Lord Rosse's (*i. e.* by a chilled casting), has been described in a communication read to this Society on the 8th of December last. The polishing was performed on a machine almost precisely similar to that of his lordship. But finding after many months' trial that he could not succeed in obtaining a satisfactory figure, he was led to contrive a machine for imitating as closely as possible those evolutions of the hand by which he had been accustomed to produce perfect surfaces on smaller specula. This machine has been described (and a model of it, as well as Mr. Nasmyth's finished working drawings of it, exhibited) in the paper to which I have already referred, of which an abstract has been printed in our "Notices," and must by this time be in the hands of every fellow here present, so that it cannot be necessary for me to recapitulate its contents. Suffice it to say that I have carefully examined both the drawings and the model, and having myself had some experience in the working and polishing of reflecting specula approaching (though inferior) in magnitude to Mr. Lassell's, I am enabled to say that the machine seems to unite every requisite for obtaining a perfect command over the figure; and when executed with that finish which belongs to every work of Mr. Nasmyth, from the steam-hammer down to the most delicate product of engineering and mechanical skill, cannot fail to secure, by the oily smoothness and equability of its movements, the ultimate perfection of polish, and the most complete absence of local irregularities of surface. The only part which I do not quite like about it, or perhaps I should rather say which seems open to an *à priori* objection, refutable, and, in point of fact, refuted by the practical results of its operation, is the wooden polisher, owing to the possibility of warping, should moisture penetrate the coating of pitch with which it is (I presume) enveloped on every side. Some unhygrometric, non-metallic substance, such as, for instance, earthenware, porcelain, biscuit, or slate, would be free from this objection, though possibly open to others of more importance.

Both Mr. Lassell and Lord Rosse appear to be fully aware of the vital importance of supporting the metal, not only while in use, but also during the process of polishing, in a perfectly free and equable manner; but the former has adopted a mode of securing a free bearing on the supports, by suspending the mirror, which is a great and manifest improvement on the old practice of allowing it to rest on its lower edge, by which not only is the figure necessarily injured by direct pressure, but the metal is prevented from playing freely to and fro, and taking a fair bearing on its bed. As I have, however, on another occasion, enlarged on the necessity of making provisions against these evils, by a mechanism almost identical in principle, I need not dwell upon this point further than to recommend it to the particular attention of all who may engage in similar undertakings.

It is right that I should now say something of the performance of the nine-inch and two-feet reflectors. And first, as regards the success of the system of mounting adopted in securing the peculiar advantages of the equatorial movement. This appears to have been very complete. The measurements, both differential and micrometrical, made with them, and recorded in our "Notices," show that in this respect they may be considered on a par with refractors, and in facility of setting and handling, they appear no wise inferior. Of the optical power of the former instrument, two facts will enable the meeting to form a sufficient judgment. With this Mr. Lassell, independently and without previous knowledge of its existence, detected the sixth star of the trapezium of θ *Orionis*. And with this, under a magnifying power of 450, and in very unfavourable circumstances of altitude, both himself and Mr. Dawes became satisfied of the division of the exterior ring of *Saturn* into two distinct annuli, a perfectly clear and satisfactory view of the division being obtained.

The feats performed by the larger instrument have been much more remarkable and important. It has established the existence of at least one of the four satellites of *Uranus*, which, since its announcement by Sir W. Herschel, has been seen by no other observer, viz., the innermost of all the series, and afforded strong presumptive evidence of the reality of another, intermediate between the most conspicuous ones. The observations of M. Otto Struve, if they really refer to the same satellite, are of nearly a month later date.

To Mr. Lassell's observations with this telescope, we also owe the discovery of a satellite of *Neptune*. The first occasion on which this body was seen, was on the 10th of October, 1846; but owing to the then rapid approach of the planet to the end of its visibility for the season, it could not be satisfactorily followed until the next year, when on the 8th and 9th of July, observations decisive as to its reality as a satellite were made, and in August and September, full confirmation was obtained. This important discovery has since been verified, both in Russia and in America. I call it so, because, in fact, the mass of *Neptune* is a point of such moment, that it is difficult to overrate the value of any means of definitively settling it. Unfortunately, the exact measurement of the satellite's distance from the planet is of such extreme difficulty, that, up to the present time, astronomers are still considerably at issue as to the result.

I come now to the most remarkable of Mr. Lassell's discoveries; one of the most remarkable, indeed, as an insulated fact, which has occurred in modern astronomy; though, indeed, it can hardly be regarded as an insulated fact, when considered in all its relations. I need hardly say that I allude to the discovery of an eighth satellite of *Saturn;* a discovery the history of which is in the highest degree creditable, not only to the increased power of the instruments with which observatories are furnished in these latter days of astronomy, but also to the vigilance of observers. If I am right in the principle that discovery consists in the certain knowledge of a new fact or a new truth, a knowledge grounded on positive and tangible evidence, as distinct from bare *suspicion* or *surmise* that such a fact exists, or that such a proposition is true — if I am right in assigning as the moment of discovery, that moment when the discoverer is first enabled to say to himself, or to a bystander, "I *am sure* that such is the fact,—and I am sure of it, *for such and such reasons,*" reasons subsequently acquiesced in as valid ones when the discovery comes to be known and acknowledged — if, I say, I am right in this principle (and I really can find no better), then I think the discovery of this satellite must be considered to date from the 19th of September last, and to have been made simultaneously, putting difference of longitude out of the question, on both sides of the Atlantic. In speaking thus, I desire of course to be understood as expressing only my own private opinion, and in no way as backing that opinion by the authority of the Society whose chair I for the moment occupy. The Astronomical Society receives with equal joy the intelligence of advances made in that science from whatever quarter emanating, and accords the meed of its approbation to diligence, devotion, and talent, with equal readiness wherever it finds them — but declines entering into *nice* questions of personal or national priority, and would, I am sure, emphatically disavow the assumption of any title to lay down authoritative rules for guidance of men's judgments in such matters. The medal of this day is awarded to Mr. Lassell, not on account of this discovery alone, and as such, but as taken in conjunction with the many other striking proofs he has afforded of successful devotion to our science — both in the improvement and in the use of instruments. And among the motives which have induced your Council to place Professor Bond on the list of our Associates (I trust not long to be the only one of his countrymen by whom that honour is enjoyed), though this discovery has had its due and just weight, we have not been unheedful of his general merits, both as an observer and as a theoretical astronomer — merits of which the "Memoirs" which have recently reached us convey the most abundant evidence in both departments.

I have observed that, when taken in all its relations, the discovery of an eighth satellite of *Saturn* cannot be regarded as quite an insulated fact. Between *Iapetus* and *Titan* there existed a great gap unfilled, in which (as formerly between *Mars* and *Jupiter*) it was not in itself unlikely that some additional member of the Saturnian system might exist. The extreme minuteness of *Hyperion* forcibly recalls the analogous features of the asteroids, and it would be very far from surprising if a further application of the same instrumental powers should carry out this analogy in a plurality of such minute attendants.

Mr. Lassell, as you are all well aware, is bound to astronomy by no other tie than the enjoyment he receives in its pursuit. But in *our* estimation of his position as an Amateur Astronomer it must not be left out of consideration, that his worldly avocations are such as most men consider of an engrossing nature, and which entitle them in their moments of relaxation, as they conceive, to enjoyments of a very different kind from those which call into fresh and energetic exertion all their faculties, intellectual and corporeal. It is no slight and desultory exercise of those faculties which will enable any man to carry into effect so much thoughtful combination, and to avail himself with so much consecutiveness of their results when produced. And however we may and must acknowledge that such a course of action is really calculated to confer a very high degree of enjoyment and happiness, we ought not to feel the less gratefully towards those who, by their personal example, press forward the advent of that higher phase of civilization which some fancy they see not indistinctly dawning around them; a civilization founded on the general and practical recognition of the superiority of the pleasures of mind over those of sense; a civilization which may dispense with luxury and splendour, but not with the continual and rapid progress of knowledge in science and excellence in art.

I think I should hardly be doing full justice to my subject or to the grounds taken by the Council in the award, if I were to conclude what I have to say otherwise than in the pointed and emphatic words of a Report officially embodying the prominent features of the case. "The simple facts," says that document, "are, that Mr. Lassell cast his own mirror, polished it by machinery of his own contrivance, mounted it equatorially in his own fashion, and placed it in an observatory of his own engineering: that with this instrument he discovered the satellite of *Neptune*, the eighth satellite of *Saturn*, and re-observed the satellites of *Uranus*. A private man, of no large means, in a bad climate" (nothing, I understand, can be much worse), "and with little leisure, he has anticipated, or rivalled, by the work of his own hands, the contrivance of his own brain and the outlay of his own pocket, the magnificent refractors with which the Emperor of Russia and the citizens of Boston have endowed the observatories of Pulkowa and the Western Cambridge."

(*The President then, delivering the medal to Mr. Lassell, addressed him in the following terms*): —

And now, Mr. Lassell, all that remains for me is to place the medal in your hands, and to congratulate you on your success and on the noble prospect of future discovery which lies before you, now that, free from the preliminary labour of construction, your whole attention can be devoted to using the powerful means you have created. In the examination of the nebulæ, in the measurement of the closest double stars, and the discovery of others which have hitherto defied separation — in the physical examination of the planets and comets of our own system, there is a wide field open and the sure promise of an ample harvest; and I can only add, that we all heartily wish you health and long life to reap it.

AN ADDRESS

TO THE BRITISH ASSOCIATION FOR THE ADVANCEMENT OF SCIENCE AT THE OPENING OF THEIR MEETING AT CAMBRIDGE, JUNE 19TH, 1845.

GENTLEMEN,

THE terms of kindness in which I have been introduced to your notice by my predecessor * in the office which you have called on me to fill, have been gratifying to me in no common degree—not as contributing to the excitement of personal vanity (a feeling which the circumstances in which I stand, and the presence of so many individuals every way my superiors, must tend powerfully to chastise), but as the emanation of a friendship begun at this University when we were youths together, preparing for our examinations for degrees, and contemplating each other, perhaps, with some degree of rivalry (if that can be called rivalry from which every spark of jealous feeling is absent). That friendship has since continued, warm and unshadowed for a single instant by the slightest cloud of disunion, and among all the stirring and deep-seated remembrances which the

* The Very Reverend the Dean of Ely, President for the year 1844-5.

sight of these walls within which we are now assembled arouse, I can summon none more every way delightful and cheering than the contemplation of that mutual regard. It is, therefore, with no common feelings that I find myself now placed in this chair, as the representative of such a body as the British Association, and as the successor of such a friend and of such a man as its late President.

Gentlemen,—There are many sources of pride and satisfaction, in which *self* has no place, which crowd upon a Cambridge man in revisiting for a second time this University, as the scene of our annual labours. The developement of its material splendour which has taken place in that interval of twelve years, vast and noble as it has been, has been more than kept pace with by the triumphs of its intellect, the progress of its system of instruction, and the influence of that progress on the public mind and the state of science in England. When I look at the scene around me—when I see the way in which our Sections are officered in so many instances by Cambridge men, not out of mere compliment to the body which receives us, but for the intrinsic merit of the men, and the pre-eminence which the general voice of society accords them in their several departments—when I think of the large proportion of the muster-roll of science which is filled by Cambridge names, and when, without going into any details, and confining myself to only one branch of public instruction, I look back to the vast and extraordinary developement in the state of mathematical cultivation and power in this University, as evidenced both in its examinations and in the published works of its members, now as compared with what it was in my own time — I am left at no loss to account for those triumphs and that influence to which I have alluded. It has ever been, and I trust it ever will continue to be, the pride and boast of this University to maintain, at a conspicuously high level, that sound and thoughtful and sobering discipline of mind which mathematical studies imply. Independent of the power which such studies confer as instruments of investigation, there never was a period in the history of science in which their moral influence, if I may so term it, was more needed, as a corrective to that propensity which is beginning to prevail widely, and, I fear, balefully, over large departments of our philosophy, the propensity to crude and overhasty generalization. To all such propensities the steady concentration of thought, and its fixation on the clear and the definite which a long and stern mathematical discipline imparts, is the best, and, indeed, the only proper antagonist. That such habits of thought exist, and characterize, in a pre-eminent degree, the discipline of this University, with a marked influence on the subsequent career of those who have been thoroughly imbued with it, is a matter of too great notoriety to need proof. Yet, in illustration of this disposition, I may be allowed to mention one or two features of its scientific history, which seem to me especially worthy of notice on this occasion. The first of these is the institution of the Cambridge University Philosophical Society, that body at whose more especial invitation we are now here assembled, which has now subsisted for more than twenty years, and which has been a powerful means of cherishing and continuing those habits among resident members of the University, after the excitement of reading for academical honours is past. From this Society have emanated eight or nine volumes of Memoirs, full of variety and interest, and such as no similar collection, originating as this has done in the bosom, and, in great measure, within the walls of an academical institution, can at all compare with; the Memoirs of the École Polytechnique of Paris, perhaps, alone excepted. Without undervaluing any part of this collection, I may be allowed to particularize, as adding largely to our stock of knowledge of their respective subjects, the Hydrodynamical contributions of Professor Challis — the Optical and Photological papers of Mr. Airy — those of Mr. Murphy, on Definite Integrals — the curious speculations and intricate mathematical investigations of Mr. Hopkins on Geological Dynamics — and, more recently, the papers of Mr. De Morgan on the foundations of Algebra, which, taken in conjunction with the prior researches of the Dean of Ely and Mr. Warren on the geometrical interpretation of imaginary symbols in that science, have effectually dissipated every obscurity which heretofore prevailed on this subject. The elucidation of the metaphysical difficulties in question, by this remarkable train of speculation, has, in fact, been so complete, that henceforward they will never be named as difficulties, but only as illustrations of principle. Nor does its interest end here, since it appears to have given rise to the theory of Quaternions of Sir W. Hamilton, and to the Triple Algebra of Mr. De Morgan himself, as well as to a variety of interesting inquiries of a similar nature on the part of Mr. Graves, Mr. Cayley, and others. Conceptions of a novel and refined kind have thus been introduced into analysis—new forms of imaginary expression rendered familiar—and a vein opened which I cannot but believe will terminate in some first-rate discovery in abstract science.

Neither are inquiries into the logic of symbolic analysis, conducted as these have been, devoid of a bearing on the progress even of physical science. Every inquiry, indeed, has such a bearing which teaches us that terms which we use in a narrow sphere of experience, as if we fully understood them, may, as our knowledge of nature increases, come to have superadded to them a new set of meanings and a wider range of interpretation. It is thus that modes of action and communication, which we hardly yet feel prepared to regard as strictly of a material character, may, ere many years have passed, come to be familiarly included in our notions of

Light, Heat, Electricity, and other agents of this class; and that of the transference of physical causation from point to point in space—nay, even the generation or development of attractive, repulsive, or directive forces at their points of arrival may come to be enumerated among their properties. The late marvellous discoveries in actino-chemistry and the phenomena of muscular contraction as dependent on the will, are, perhaps, even now preparing us for the reception of ideas of this kind.

Another instance of the efficacy of the course of study in this University, in producing not merely expert algebraists, but sound and original mathematical *thinkers*—(and, perhaps, a more striking one, from the generality of its contributors being men of comparatively junior standing), is to be found in the publication of "The Cambridge Mathematical Journal," of which already four volumes, full of very original communications, are before the public. It was set on foot in 1837, by the late Mr. Gregory, Fellow of Trinity College, whose premature death has bereft Science of one who, beyond a doubt, had he lived, would have proved one of its chief ornaments, and the worthy representative of a family already so distinguished in the annals of mathematical and optical science. His papers on the "Calculus of Operations," which appeared in that collection, fully justifies this impression, while they afford an excellent illustration of my general position. Nor ought I to omit mentioning the

Chemical Society, of whom he was among the founders, as indicative of the spirit of the place, untrammelled by abstract forms, and eager to spread itself over the whole field of human inquiry.

Another great and distinguishing feature in the scientific history of this place, is the establishment of its Astronomical Observatory, and the regular publication of the observations made in it. The science of astronomy is so vast, and its objects so noble, that its practical study for its own sake is quite sufficient to insure its pursuit wherever civilization exists. But such institutions have a much wider influence than that which they exercise in forwarding their immediate object. Every astronomical observatory which publishes its observations, becomes a nucleus for the formation around it of a school of exact practice—a standing and accessible example of the manner in which theories are brought to their extreme test—a centre, from which emanate a continual demand for and suggestion of refinements and delicacies, and precautions in matter of observation and apparatus which re-act upon the whole body of science, and stimulate, while they tend to render possible, an equal refinement and precision in all its processes. It is impossible to speak too highly of the mode in which the business of this institution is carried on under its present eminent director—nor can it be forgotten in our appretiation of what it has done for science, that in it our present Astronomer-Royal first proved and familiarized himself with that admirable system of astronomical observation, registry, and computation, which he has since brought to perfection in our great national observatory, and which have rendered it, under his direction, the pride and ornament of British science and the admiration of Europe.

Gentlemen, I should never have done if I were to enlarge on, or even attempt to enumerate the many proofs which this University has afforded of its determination to render its institutions and endowments efficient for the purposes of public instruction, and available to science. But such encomiums, however merited, must not be allowed to encroach too largely on other objects which I propose to bring before your notice, and which relate to the more immediate business of the present meeting, and to the general interests of science. The first and every way the most important, is the subject of the Magnetic and Meteorological Observatories. Every member of this Association is, of course, aware of the great exertions which have been made during the last five years, on the part of the British, Russian, and several other foreign governments, and of our own East India Company, to furnish data on the most extensive and systematic scale, for elucidating the great problems of Terrestrial Magnetism and Meteorology, by the establishment of a system of observatories all over the world, in which the phenomena are registered at instants strictly simultaneous, and at intervals of two hours throughout both day and night. With the particulars of these national institutions, and of the multitude of local and and private ones of a similar nature, both in Europe, Asia, and America, working on the same concerted plan, so far as the means at their disposal enable them, I need not detain you: neither need I enter into any detailed explanation of the system of Magnetic Surveys, both by sea and land, which have been executed or are in progress, in connexion with, and based upon the observations carried on at the fixed stations. These things form the subject of Special Annual Reports, which the Committee appointed for the purpose have laid before us at our several meetings, ever since the commencement of the undertaking; and the most recent of which will be read in the Physical Section of the present meeting, in its regular course. It is sufficient for me to observe, that the result has been the accumulation of an *enormous* mass of most valuable observations, which are now and have been for some time in the course of publication, and when thoroughly digested and discussed, as they are sure to be, by the talent and industry of magnetists and meteorologists, both in this country and abroad, cannot fail to place those sciences very far indeed in advance of their actual state. For such discussion, however, time must be allowed. Even were all the returns from the several observatories before the public, (which they are not, and are very far from being,) such is the mass of matter to be grappled with, and such the multitude of ways in which the observations will necessarily have to be grouped and combined to elicit mean results and quantitative laws, that several years must elapse before the full scientific value of the work done can possibly be realized.

Meanwhile, a question of the utmost moment arises, and which *must* be resolved, so far as the British Association is concerned, before the breaking up of this meeting. The second term of three years, for which the British Government and the East India Company have granted their establishments—nine in number—will terminate with the expiration of the current year, at which period, if no provision be made for their continuance, the observations at those establishments will of course cease, and with them, beyond a doubt, those at a great many—probably the great majority—of the foreign establishments, both national and local, which have been called into existence by the example of England, and depend on that example for their continuance or abandonment. Now, under these circumstances, it becomes a very grave subject for the consideration of our Committee of Recommendations, whether to suffer this term to expire without an effort on the part of this Association to influence the Government for its continuance, or whether, on the other hand, we ought to make such an effort, and endeavour to secure either the continuance of these establishments for a further limited time, or the perpetuity of this or some equivalent system of observation in the same or different localities, according to the present and future exigencies of science. I term this a grave subject of deliberation, and one which will call for the exercise of their soundest judgment; *because*, in the first place, this system of combined observation is by far the greatest and most prolonged effort of scientific co-operation which the world has ever witnessed; *because*, moreover, the spirit in which the demands of science have been met on this occasion by our own Government, by the Company, and by the other governments who have taken part in the matter, has been, in the largest sense of the words, munificent and unstinting; and *because* the existence of such a spirit throws upon us a solemn responsibility to recommend nothing but upon the most entire conviction of very great evils consequent on the interruption, and very great benefits to accrue to Science from the continuance of the observations.

Happily we are not left without the means of forming a sound judgment on this momentous question. It is a case in which, connected as the science of Britain is with that of the other co-operating nations, we cannot and ought not to come to any conclusion without taking into our counsels the most eminent magnetists and meteorologists of other countries who have either taken a direct part in the observations, or whose reputation in those sciences is such as to give their opinions, in matters respecting them, a commanding weight. Accordingly it was resolved, at the York meeting last year, to invite the attendance of the eminent individuals I have alluded to at this meeting, with the especial objects of conference on the subject. And in the interval since elapsed, knowing the improbability of a complete personal reunion from so many distant quarters, a circular has been forwarded to each of them, proposing certain special questions for reply, and inviting, besides, the fullest and freest communication of their views on the general subject. The replies received to this circular, which are numerous and in the highest degree interesting and instructive, have been printed and forwarded to the parties replying, with a request for their reconsideration and further communication, and have also been largely distributed at home to every member of our own Council, and the Committee of Recommendations, and to each member of the Council and Physical Committee of the Royal Society, which, conjointly with ourselves, memorialized Government for the establishment of the observatories.

In addition to the valuable matter thus communicated, I am happy to add, that several of the distinguished foreigners in question have responded to our invitation, and that in consequence this meeting is honoured by the personal presence of M. Kupffer, the Director-General of the Russian System of Magnetic and Meteorological Observation; of M. Ermann, the celebrated circumnavigator and meteorologist; of Baron von Senftenberg, the founder of the Astronomical, Magnetic, and Meteorological Observatory of Senftenberg; of M. Kreil, the director of the Imperial Observatory at Prague; and of M. Boguslawski, director of the Royal Prussian Observatory of Breslau, all of whom have come over for the express purpose of affording us the benefit of their advice and experience in this discussion. To all the conferences between these eminent foreigners and our own Magnetic and Meteorological Committee, and such of our members present as have taken any direct theoretical or practical interest in the subjects, all the members of our Committee of Recommendations will have free access for the purpose of enabling them fully to acquaint themselves with the whole bearing of the case, and the arguments used respecting all the questions to be discussed, so that when the subject comes to be referred to them, as it must be if the opinion of the conference should be favourable to the continuance of the system, they may be fully prepared to make up their minds on it.

I will not say one word from this chair which can have the appearance of in any way anticipating the conclusion which the conference thus organized may come to, or the course to be adopted in consequence. But I will take this opportunity of stating my ideas generally on the position to be assumed by this Association and by other scientific bodies in making demands on the national purse for scientific purposes. And I will also state, quite irrespective of the immediate question of magnetic co-operation, and therefore of the fate of this particular measure, what I conceive to be the objects which might be accomplished, and ought to be aimed at in the establishment of PHYSICAL OBSERVATORIES, as part of the integrant institutions of each nation calling itself civilized, and as its contribution to Terrestrial Physics.

It is the pride and boast of an Englishman to pay his taxes cheerfully when he feels assured of their application to great and worthy objects. And as civilization advances, we feel constantly more and more strongly, that, after the great objects of national defence, the stability of our institutions, the due administration of justice, and the healthy maintenance of our social state, are provided for, there is no object greater and more noble — none more worthy of national effort than the furtherance of Science. Indeed, there is no surer test of the civilization of an age or nation than the degree in which this conviction is felt. Among Englishmen it has been for a long time steadily increasing, and may now be regarded as universal among educated men of all classes. No government, and least of all a British government, can be insensible to the general prevalence of a sentiment of this kind; and it is our good fortune, and has been so for several years, to have a government, no matter what its denomination as respects party, impressible with such considerations, and really desirous to aid the forward struggle of intellect, by placing at its disposal the material means of its advances.

But to do so with effect, it is necessary to be thoroughly well informed. The mere knowledge that such a disposition exists, is sufficient to surround those in power with every form of extravagant pretension. And even if this were not so, the number of competing claims, which cannot be all satisfied, can only harass and bewilder, unless there be somewhere seated a discriminating and selecting judgment, which, among many important claims, shall fix upon the most important, and urge them with the weight of well-established character. I know not where such a selecting judgment can be so confidently looked for as in the great scientific bodies of the country, each in its own department, and in this Association, constituted, in great measure, out of, and so representing them all, and numbering besides, among its members, abundance of men of excellent science and enlightened minds who belong to none of them. The constitution of such a body is the guarantee both for the general soundness of its recommendations, and for the due weighing of their comparative importance, should ever the claims of different branches of science come into competition with each other.

In performing this most important office of suggesting channels through which the fertilizing streams of national munificence can be most usefully conveyed over the immense and varied fields of scientific culture, it becomes us, in the first place, to be so fully impressed with a sense of duty to the great cause for which we are assembled, as not to hesitate for an instant in making a recommendation of whose propriety we are satisfied, on the mere ground that the aid required is of great and even of unusual magnitude. And on the other hand, keeping within certain reasonable limits of total amount, which each individual must estimate for himself, and which it would be unwise and indeed impossible to express in terms, it will be at at once felt that *economy in asking* is quite as high a "distributive virtue" as economy in *granting*, and that every pound recommended unnecessarily is so much character thrown away. I make these observations because the principles they contain cannot be too frequently impressed, and by no means because I consider them to have been overstepped in any part of our conduct hitherto. In the next place, it should be borne in mind that, in recommending to Government, not a mere grant of money, but a scientific enterprise or a national establishment, whether temporary or permanent, not only is it our duty so to place it before them that its grounds of recommendation shall be thoroughly intelligible, but that its whole proposed extent shall be seen — or at least if that cannot be, that it should be clearly stated to be the possible commencement of something more extensive — and besides, that the printing and publication of results should, in every such case, be made an express part of the recommendation. And, again, we must not forget that our interest in the matter does not cease with such publication. It becomes our duty to forward, by every encouragement in our power, the due consideration and scientific discussion of results so procured — to urge it upon the science of our own country and of Europe, and to aid from our own resources those who may be willing to charge themselves with their analysis, and to direct or execute the numerical computations or graphical projections it may involve. This is actually the predicament in which we stand, in reference to the immense mass of data already accumulated by the magnetic and meteorological observatories. Let the science of England, and especially the rising and vigorous mind which is pressing onward to distinction, gird itself to the work of grappling with this mass. Let it not be said that we are always to look abroad whenever industry and genius are required to act in union for the discussion of great masses of raw observation. Let us take example from what we see going on in Germany, where a Dove, a Kamtz, and a Mahlmann are battling with the meteorology, a Gauss, a Weber, and an Ermann with the magnetism of the world. The mind of Britain is equal to the task — its mathematical strength, developed of late years to an unprecedented extent, is competent to any theoretical analysis or technical combination. Nothing is wanting but the resolute and persevering devotion of undistracted thought to a single object, and that will not be long wanting when once the want is declared and dwelt upon, and the high prize of public estimation held forth to those who fairly and freely adventure themselves in this career. Never was there a time when the mind of the country, as well as its resources of every kind, answered so fully and readily to any call reasonable in itself and properly urged upon it. Do we call for *facts?* they are poured upon us in such profusion as for a time to overwhelm us, like the Roman maid who sank under the load of wealth she called down upon herself. Witness the piles of unreduced meteorological observations which load our shelves and archives; witness the immense and admirably arranged catalogues of stars which have been and still are pouring in from all quarters upon our astronomy so soon as the want of extensive catalogues came to be felt and declared. What we now want is *thought*, steadily directed to single objects, with a determination to eschew the besetting evil of our age — the temptation to squander and dilute it upon a thousand different lines of inquiry. The philosopher must be wedded to his subject if he would see the children and the children's children of his intellect flourishing in honour around him.

The establishment of astronomical observatories has been, in all ages and nations, the first public recognition of science as an integrant part of civilization. Astronomy, however, is only one out of many sciences, which can be advanced by a combined system of observation and calculation carried on uninterruptedly; where, in the way of experiment, man has no control, and whose only handle is the continual observation of Nature as it developes itself under our eyes, and a constant collateral endeavour to concentrate the records of that observation into empirical laws in the first instance, and to ascend from those laws to theories. Speaking in a utilitarian point of view, the globe which we inhabit is quite as important a subject of scientific inquiry as the stars. We depend for our bread of life and every comfort on its climates and seasons, on the movements of its winds and waters. We guide ourselves over the ocean, when astronomical observations fail, by our knowledge of the laws of its magnetism; we learn the sublimest lessons from the records of its geological history; and the great facts which its figure, magnitude, and attraction, offer to mathematical inquiry, form the very basis of Astronomy itself. Terrestrial Physics, therefore, form a subject every way worthy to be associated with Astronomy as a matter of universal interest and public support, and one which cannot be adequately studied except in the way in which Astronomy itself has been — by permanent establishments keeping up an unbroken series of observation: — but with this difference, that whereas the chief data of Astronomy might be supplied by the establishment of a very few well worked observatories properly disposed in the two hemispheres — the gigantic problems of meteorology, magnetism, and oceanic movements can only be resolved by a far more extensive geographical distribution of observing stations, and by a steady, persevering, systematic attack, to which every civilized nation, as it has a direct interest in the result, ought to feel bound to contribute its contingent.

I trust that the time is not far distant when such will be the case, and when no nation calling itself civilized will deem its institutions complete without the establishment of a permanent physical observatory, with at least so much provision for astronomical and magnetic observation as shall suffice to make it a local centre of reference for geographical determination and trigonometrical and magnetic surveys — which latter, if we are ever to attain to a theory of the secular changes of the earth's magnetism, will have to be repeated at intervals of twenty or thirty years for a long while to come. Rapidly progressive as our colonies are, and emulous of the civilization of the mother country, it seems not too much to hope from them, that they should take upon themselves, each according to its means, the establishment and maintenance of such institutions both for their own advantage and improvement, and as their contributions to the science of the world. A noble example has been set them in this respect, within a very few months, by our colony of British Guiana, in which a society recently constituted, in the best spirit of British co-operation, has established and endowed an observatory of this very description, furnishing it partly from their own resources and partly by the aid of Government with astronomical, magnetic, and meteorological instruments, and engaging a competent observer at a handsome salary to work the establishment — an example which deserves to be followed wherever British enterprise has struck root and flourished.

The perfectly unbroken and normal registry of all the meteorological and magnetic elements — and of tidal fluctuations where the locality admits — would form the staple business of every such observatory, and, according to its means of observation, periodical phenomena of every description would claim attention, for which the list supplied by M. Quetelet, which extends not merely to the phases of inanimate life, but to their effects on the animal and vegetable creation, will leave us at no loss beyond the difficulty of selection. The division of phenomena which magnetic observation has suggested, into periodical, secular, and occasional, will apply *mutatis mutandis* to every department. Under the head of occasional phenomena, storms, magnetic disturbances, auroras, extraordinary tides, earthquake movements, meteors, &c., would supply an ample field of observation — while among the secular changes, indications of the varying level of land and sea would necessitate the establishment of permanent marks, and the reference to them of the actual mean sea level which would emerge from a series of tidal observations, carried round a complete period of the moon's nodes with a certainty capable of detecting the smallest changes.

The abridgment of the merely mechanical work

of such observatories by self-registering apparatus is a subject which cannot be too strongly insisted on. Neither has the invention of instruments for superseding the necessity of much arithmetical calculation by the direct registry of *total* effects received anything like the attention it deserves. Considering the perfection to which mechanism has arrived in all its departments, these contrivances promise to become of immense utility. The more the merely mechanical part of the observer's duty can be alleviated, the more will he be enabled to apply himself to the theory of his subject, and to perform what I conceive ought to be regarded as the most important of all his duties, and which in time will come to be universally so considered — I mean the systematic deduction from the registered observations of the mean values and local coefficients of diurnal, menstrual, and annual change. These deductions, in the case of permanent institutions, ought not, if possible, to be thrown upon the public, and their effective execution would be the best and most honourable test of the zeal and ability of their directors.

Nothing damps the ardour of an observer like the absence of an object appretiable and attainable by himself. One of my predecessors in this chair has well remarked, that a man may as well keep a register of his dreams as of the weather, or any other set of daily phenomena, if the spirit of grouping, combining, and eliciting results be absent. It can hardly be expected indeed, that, observers of facts of this nature should themselves reason from them up to the highest theories. For that their position unfits them, as they see but locally and partially. But no other class of persons stands in anything like so favourable a position for working out the first elementary laws of phenomena, and referring them to their immediate points of dependence. Those who witness their daily progress, with that interest which a direct object in view inspires, have in this respect an infinite advantage over those who have to go over the same ground in the form of a mass of dry figures. A thousand suggestions arise, a thousand improvements occur — a spirit of interchange of ideas is generated, the surrounding district is laid under contribution for the elucidation of innumerable points, where a chain of corresponding observation is desirable; and what would otherwise be a scene of irksome routine, becomes a school of physical science. It is needless to say how much such a spirit must be excited by the institution of provincial and colonial scientific societies, like that which I have just had occasion to mention. Sea as well as land observations are, however, equally required for the effectual working out of these great physical problems. A ship is an itinerant observatory; and, in spite of its instability, one which enjoys several eminent advantages — in the uniform level and nature of the surface, which eliminate a multitude of causes of disturbance and uncertainty, to which land observations are liable.

The exceeding precision with which magnetic observations can be made at sea, has been abundantly proved in the Antarctic Voyage of Sir James Ross, by which an invaluable mass of data has been thus secured to science. That voyage has also conferred another and most important accession to our knowledge in the striking discovery of a permanently low barometric pressure in high south latitudes over the whole Antarctic Ocean — a pressure actually inferior, by considerably more than an inch of mercury, to what is found between the Tropics. A fact so novel and remarkable will of course give rise to a variety of speculations as to its cause; and I anticipate one of the most interesting discussions which have ever taken place in our Physical Section, should that great circumnavigator favour us, as I hope he will, with a *vivâ voce* account of it. The voyage now happily commenced under the most favourable auspices for the further prosecution of our Arctic discoveries under Sir John Franklin, will bring to the test of direct experiment a mode of accounting for this extraordinary phenomenon thrown out by Colonel Sabine, which, if realized, will necessitate a complete revision of our whole system of barometric obser vation in high latitudes, and a total reconstruction of all our knowledge of the laws of pressure in regions where excessive cold prevails. This, with the magnetic survey of the Arctic seas, and the not improbable solution of the great geographical problem which forms the chief object of the expedition, will furnish a sufficient answer to those, if any there be, who regard such voyages as useless. Let us hope and pray, that it may please Providence to shield him and his brave companions from the many dangers of their enterprise, and restore them in health and honour to their country.

I cannot quit this subject without reverting to and deploring the great loss which science has recently sustained in the death of the late Prof. Daniell, one of its most eminent and successful cultivators in this country. His work on Meteorology is, if I mistake not, the first in which the distinction between the aqueous and gaseous atmospheres, and their mutual independence, was clearly and strongly insisted on as a highly influential element in meteorological theory. Every succeeding investigation has placed this in a clearer light. In the hands of M. Dove, and more recently of Colonel Sabine, it has proved the means of accounting for some of the most striking features in the diurnal variations of the barometer. The continual generation of the aqueous atmosphere, at the Equator, and its destruction in high latitudes, furnishes a *motive power* in meteorology, whose mode of action, and the mechanism through which it acts, have yet to be inquired into. Mr. Daniell's claims to scientific distinction were, however, not confined to this branch. In his hands, the voltaic pile became an infinitely more powerful and manageable instrument than had ever before been thought possible; and his improvements in its construction (the effect not of accident, but of patient and persevering experimental inquiry,) have in effect changed the face of Electro-Chemistry. Nor did he confine himself to these improvements. He applied them: and among the last and most interesting inquiries of his life, are a series of electro-chemical researches which may rank with the best things yet produced in that line.

The immediate importance of these subjects to one material part of our business at this meeting, has caused me to dwell more at length than perhaps I otherwise should on them. I would gladly use what time may remain, without exciting your impatience, in taking a view of some features in the present state and future prospects of that branch of science to which my own attention has been chiefly directed, as well as to some points in the philosophy of science generally, in which it appears to me that a disposition is becoming prevalent towards lines of speculation, calculated rather to bewilder than enlighten, and, at all events, to deprive the pursuit of science of that which, to a rightly constituted mind, must ever be one of its highest and most attractive sources of interest, by reducing it to a mere assemblage of marrowless and meaningless facts and laws.

The last year must ever be considered an epoch in Astronomy, from its having witnessed the successful completion of the Earl of Rosse's six-feet reflector—an achievement of such magnitude, both in itself as a means of discovery, and in respect of the difficulties to be surmounted in its construction (difficulties which perhaps few persons here present are better able from experience to appretiate than myself), that I want words to express my admiration of it. I have not myself been so fortunate as to have witnessed its performance, but from what its noble constructor has himself informed me of its effects on one particular nebula, with whose appearance in powerful telescopes I am familiar, I am prepared for any statement which may be made of its optical capacity. What may be the effect of so enormous a power in adding to our knowledge of our own immediate neighbours in the universe, it is of course impossible to conjecture; but for my own part I cannot help contemplating, as one of the grand fields open for discovery with such an instrument, those marvellous and mysterious bodies, or systems of bodies, the Nebulæ. By far the major part, probably at least nine tenths, of the nebulous contents of the heavens consist of nebulæ of spherical or elliptical forms presenting every variety of elongation and central condensation. Of these a great number have been resolved into distinct stars, and a vast multitude more have been found to present that mottled appearance which renders it almost a matter of certainty that an increase of optical power would show them to be similarly composed. A not unnatural or unfair induction would therefore seem to be, that those which resist such resolution do so only in consequence of the smallness and closeness of the stars of which they consist; that, in short, they are only optically and not physically nebulous. There is, however, one circumstance which deserves especial remark; and which, now that my own observation has extended to the nebulæ of both hemispheres, I feel able to announce with confidence as a general law, viz., that the character of easy resolvability into separate and distinct stars, is almost entirely confined to nebulæ deviating but little from the spherical form; while, on the other hand, very elliptic nebulæ, even large and bright ones, offer much greater difficulty in this respect. The cause of this difference must, of course, be conjectural; but, I believe, it is not possible for any one to review *seriatim* the nebulous contents of the heavens without being satisfied of its reality as a physical character. Possibly the limits of the conditions of dynamical stability in a spherical cluster may be compatible with less numerous and comparatively larger individual constituents than in an elliptic one. Be that as it may, though there is no doubt a great number of elliptic nebulæ in which stars have *not* yet been noticed, yet there are so many in which they *have*, and the gradation is so insensible from the most perfectly spherical to the most elongated elliptic form, that the force of the general induction is hardly weakened by this peculiarity; and for my own part I should have little hesitation in admitting all nebulæ of this class to be, in fact, congeries of stars. And this seems to have been my father's opinion of their constitution, with the exception of certain very peculiar looking objects, respecting whose nature all opinion must for the present be suspended. Now, among all the wonders which the heavens present to our contemplation, there is none more astonishing than such close compacted families or communities of stars, forming systems either insulated from all others, or in binary connexion, as double clusters whose confines intermix, and consisting of individual stars nearly equal in apparent magnitude, and crowded together in such multitudes as to defy all attempts to count or even to estimate their numbers. What *are* these mysterious families? Under what dynamical conditions do they subsist? Is it conceivable that they can exist at all, and endure under the Newtonian law of gravitation without perpetual collisions? And, if so, what a problem of unimaginable complexity is presented by such a system if we should attempt to dive into its perturbations and its conditions of stability by the feeble aid of our analysis. The existence of a luminous matter, not congregated into massive bodies in the nature of stars, but disseminated through vast regions of space in a vaporous or cloud-like state, undergoing, or awaiting the slow process of aggregation into masses by the power of gravitation, was originally suggested to the late Sir W. Herschel in his reviews of the nebulæ, by those extraordinary objects which his researches disclosed, which exhibit no regularity of outline, no systematic gradation of brightness, but of which the wisps and curls of a cirrus cloud afford a not inapt description. The wildest imagination can conceive nothing more capricious than their forms, which in many instances seem totally devoid of plan, as much so as real clouds, — in others offer traces of a regularity hardly less uncouth and characteristic, and which in some cases seems to indicate a cellular, in others a sheeted structure, complicated in folds as if agitated by internal winds.

Should the powers of an instrument such as Lord Rosse's succeed in resolving these also into stars, and, moreover, in demonstrating the starry nature of the regular elliptic nebulæ, which have hitherto resisted such decomposition, the idea of a *nebulous matter*, in the nature of a shining fluid, or condensible gas, must, of course, cease to rest on any support derived from actual observation in the sidereal heavens, whatever countenance it may still receive in the minds of cosmogonists from the tails and atmospheres of comets, and the zodiacal light in our own system. But though all idea of its being ever given to mortal eye to view aught that can be regarded as an outstanding portion of primæval chaos, be dissipated, it will by no means have been even then demonstrated that among those stars, so confusedly scattered, no aggregating powers are in action, tending to draw them into groups and insulate them from neighbouring groups; and, speaking from my own impressions, I should say that, in the structure of the Magellanic Clouds, it is really difficult not to believe we see distinct evidences of the exercise of such a power. This part of my father's general views of the construction of the heavens, therefore, being entirely distinct from what has of late been called "the nebulous hypothesis," will still subsist as a matter of rational and philosophical speculation — and perhaps all the better for being separated from the other.

Much has been said of late of the Nebulous Hypothesis, as a mode of representing the origin of our own planetary system. An idea of Laplace, of which it is impossible to deny the ingenuity, of the successive abandonment of planetary rings, collecting themselves into planets by a revolving mass gradually shrinking in dimension by the loss of heat, and finally concentrating itself into a sun, has been insisted on with some pertinacity, and supposed to receive almost demonstrative support from considerations to which I shall presently refer. I am by no means disposed to quarrel with the nebulous hypothesis even in this form, as a matter of pure speculation, and without any reference to final causes; but if it is to be regarded as a demonstrated truth, or as receiving the smallest support from any observed numerical relations which actually hold good among the elements of the planetary orbits, I beg leave to demur. Assuredly, it receives no support from observation of the effects of sidereal aggregation, as exemplified in the formation of globular and elliptic clusters, supposing *them* to have resulted from such aggregation. For we see this cause, working itself out in thousands of instances, to have resulted, *not* in the formation of a single large central body, surrounded by a few much smaller attendants, disposed in one plane around it,—but in systems of infinitely greater complexity, consisting of multitudes of nearly equal luminaries, grouped together in a solid elliptic or globular form. So far, then, as any conclusion from our observations of nebulæ can go, the result of agglomerative tendencies *may*, indeed, be the formation of families of stars of a general and very striking character; but we see nothing to lead us to presume its further result to be the surrounding of those stars with planetary attendants. If, therefore, we go on to push its application to that extent, we clearly theorize in advance of all inductive observation.

But if we go still farther, as has been done in a philosophical work of much mathematical pretension, which has lately come into a good deal of notice in this country, and attempt "to give a mathematical consistency" to such a cosmogony by the "*indispensable criterion*" of "a numerical verification,"—and to exhibit, as "necessary consequences of such a mode of formation," a series of numbers which observation has established independent of any such hypothesis, as primordial elements of our system—if, in pursuit of this idea, we find the author first computing the time of rotation the sun must have had about its axis so that a planet situate on its surface and forming a part of it should not press on that surface, and should therefore be in a state of indifference as to its adhesion or detachment—if we find him, in this computation, throwing overboard as troublesome all those essential considerations of the law of cooling, the change of spheroidical form, the internal distribution of density, the probable non-circulation of the internal and external shells in the same periodic time, on which alone it is possible to execute such a calculation correctly; and avowedly, as a short cut to a result, using as the basis of his calculation "the elementary Huyghenian theorems for the evaluation of centrifugal forces in combination with the law of gravitation;" —a combination which, I need not explain to those who have read the first book of Newton, leads direct to Kepler's law;—and if we find him then gravely turning round upon us, and adducing the coincidence of the resulting periods compared with the distances of the planets with this law of Kepler, as *being* the numerical verification in question,—where, I would ask, is there a student to be found who has graduated as a Senior Optime in this University, who will not at once lay his finger on the fallacy of such an argument *, and declare it a

* M. Comte ("Philosophie Positive," ii. 376, &c.), the author of the reasoning alluded to, assures us that his calculations lead to results agreeing only approximately with the exact periods, a difference to the amount of 1/45, the part more or less exist-

vicious circle? I really should consider some apology needed for even mentioning an argument of the kind to such a meeting, were it not that this very reasoning, so ostentatiously put forward, and so utterly baseless, has been eagerly received among us * as the revelation of a profound analysis. When such is the case, it is surely time to throw in a word of warning, and to reiterate our recommendation of an early initiation into mathematics,

ing in all. As he gives neither the steps nor the data of his calculations, it is impossible to trace the origin of this difference,—which, however, *must* arise from error *somewhere*, if his fundamental principles be really from what he states. For the Huyghenian measures of centrifugal force $\left(F \times \frac{V^2}{R}\right)$ "combined" with the "law of gravitation" $\left(F \times \frac{M+m}{R^2}\right)$ replacing V by its equivalent $\frac{R}{P}$ can result in no other relation between P and R than what is expressed in the Keplerian law, and is incompatible with the smallest deviation from it.

Whether the sun threw off the planets or not, Kepler's law *must* be obeyed by them when once fairly detached. How, then. can their actual observance of this law be adduced in proof of their origin, one way or the other? How is it proved that the sun must have thrown off planets *at those distances, and at no others*, where we find them,—no matter in what times revolving? *That*, indeed, would be a powerful presumptive argument; but what geometer will venture on such a *tour d'analyse?* And, lastly, how can it be adduced as *a numerical coincidence of an hypothesis with observed fact*, to say that, at an unknown epoch, the sun's rotation (*not observed*) *must have been* so and so, *if* the hypothesis were a true one?

* Mill, "Logic," ii. 28; also "Vestiges of the Creation," p. 17.

and the cherishing a mathematical habit of thought, as the safeguard of all philosophy.

A very great obstacle to the improvement of telescopes in this country has been happily removed within the past year by the repeal of the duty on glass. Hitherto, owing to the enormous expense of experiments to private individuals not manufacturers, and to the heavy excise duties imposed on the manufacture, which has operated to repress all attempts on the part of practical men to produce glass adapted to the construction of large acromatics, our opticians have been compelled to resort abroad for their materials—purchasing them at enormous prices, and never being able to procure the largest sizes. The skill, enterprise and capital of the British manufacturer have now free scope, and it is our own fault if we do not speedily rival, and perhaps outdo the far-famed works of Munich and Paris. Indeed, it is hardly possible to over-estimate the effect of this fiscal change on a variety of other sciences to which the costliness of glass apparatus has been hitherto an exceeding drawback, not only from the actual expense of apparatus already in common use, but as repressing the invention and construction of new applications of this useful material.

A great deal of attention has been lately, and I think very wisely, drawn to the philosophy of science, and to the principles of logic, as founded not on arbitrary and pedantic forms, but on a careful inductive inquiry into the grounds of human belief, and the nature and extent of man's intellectual faculties. If we are ever to hope that science will extend its range into the domain of social conduct, and model the course of human actions on that thoughtful and effective adaptation of means to their end, which is its fundamental principle in all its applications (the *means* being here the total devotion of our moral and intellectual powers — the *end*, our own happiness and that of all around us) — if such be the far hopes and long protracted aspirations of science, its philosophy and its logic assume a paramount importance in proportion to the practical danger of erroneous conceptions in the one, and fallacious tests of the validity of reasoning in the other.

On both these subjects works of first-rate importance have of late illustrated the scientific literature of this country. On the philosophy of science, we have witnessed the production, by the pen of a most distinguished member of this University, of a work so comprehensive in its views, so vivid in its illustrations, and so right-minded in its leading directions, that it seems to me impossible for any man of science, be his particular department of inquiry what it may, to rise from its perusal without feeling himself strengthened and invigorated for his own especial pursuit, and placed in a more favourable position for discovery in it than before, as well as more competent to estimate the true philosophical value and import of any new views which may open to him in its prosecution. From the peculiar and *à priori* point of view in which the distinguished author of the work in question has thought proper to place himself before his subject, many may dissent; and I own myself to be of the number; — but from this point of view it is perfectly possible to depart without losing sight of the massive reality of that subject itself: on the contrary, that reality will be all the better seen and understood, and its magnitude felt, when viewed from opposite sides, and under the influence of every accident of light and shadow which peculiar habits of thought may throw over it.

Accordingly, in the other work to which I have made allusion, and which, under the title of a "System of Logic," has for its object to give "*a connected view of the principles of evidence and the methods of scientific investigation*," its acute, and in many respects profound author, — taking up an almost diametrically opposite station, and looking to experience as the ultimate foundation of all knowledge — at least, of all scientific knowledge — in its simplest axioms as well as in its most remote results — has presented us with a view of the inductive philosophy, very different indeed in its general aspect, but in which, when carefully examined, most essential features may be recognized as identical, while some are brought out with a salience and effect which could not be attained from the contrary point of sight. It cannot be expected that I should enter into any analysis or comparison of these remarkable works; but it seemed to me impossible to avoid pointedly mentioning them on this occasion, because they certainly, taken together, leave the philosophy of science, and indeed *the principles of all general reasoning*, in a very different state from that in which they found them. Their influence, indeed, and that of some other works of prior date, in which the same general subjects have been more lightly touched upon, has already begun to be felt and responded to from a quarter where, perhaps, any sympathy in this respect might hardly have been looked for. The philosophical mind of Germany has begun, at length, effectually to awaken from the dreamy trance in which it had been held for the last half-century, and in which the jargon of the Absolutists and Ontologists had been received as oracular. An "anti-speculative philosophy" has arisen and found supporters — rejected, indeed, by the Ontologists, but yearly gaining ground in the general mind. It is something so new for an English and a German philosopher to agree in their estimate either of the proper objects of speculation or of the proper mode of pursuing them, that we greet, not without some degree of astonishment, the appearance of works like the Logic and the New Psychology of Beneke, in which this false and delusive philosophy is entirely thrown aside, and appeal at once made to the nature of things as we find them, and to the laws of our intellectual and moral nature, as our own consciousness and the history of mankind reveal them to us.*

Meanwhile, the fact is every year becoming more broadly manifest, by the successful application of scientific principles to subjects which had hitherto been only empirically treated (of which agriculture may be taken as perhaps the most conspicuous instance), that the great work of Bacon was not the completion, but, as he himself foresaw and foretold, only the commencement of his own philosophy ; and that we are even yet only at the threshold of that palace of Truth which succeeding generations will range over as their own — a world of scientific inquiry, in which not matter only and its properties, but the far more rich and complex relations of life and thought, of passion and motive, interest and actions, will come to be regarded as its legitimate objects. Nor let us fear that in so regarding them we run the smallest danger of collision with any of those great principles which we regard, and rightly regard, as sacred from question. A faithful and undoubting spirit carried into the inquiry, will secure us from such dangers, and guide us, like an instinct, in our paths through that vast and entangled region which intervenes between those ultimate principles and their extreme practical applications. It is only by working our way *upwards towards* those principles as well as

* *Vide* Beneke, "Neue Psychologie," s. 300 *et seq.* for an admirable view of the state of metaphysical and logical philosophy in England.

downwards from them, that we can ever hope to penetrate such intricacies, and thread their maze; and it would be worse than folly — it would be treason against all our highest feelings — to doubt that to those who spread themselves over these opposite lines, each moving in his own direction, a thousand points of meeting and mutual and joyful recognition will occur.

But if science be really destined to expand its scope, and embrace objects beyond the range of merely material relation, it must not altogether and obstinately refuse, even within the limits of such relations, to admit conceptions which at first sight may seem to trench upon the immaterial, such as we have been accustomed to regard it. The time seems to be approaching when a merely mechanical view of nature will become impossible —when the notion of accounting for *all* the phenomena of nature, and even of mere physics, by simple attractions and repulsions fixedly and unchangeably inherent in material centres (granting any conceivable system of Boscovichian alternations), will be deemed untenable. Already we have introduced the idea of *heat atmospheres* about particles to vary their repulsive forces according to definite laws. But surely this can only be regarded as one of those provisional and temporary conceptions which, though it may be useful as helping us to laws, and as suggesting experiments, we must be prepared to resign if ever such ideas, for instance, as *radiant stimulus* or *conducted influence* should lose their present vagueness, and come to receive some distinct scientific interpretation. It is one thing, however, to suggest that our present language and conceptions should be held as provisional—another to recommend a general unsettling of all received ideas. Whatever innovations of this kind may arise, they can only be introduced slowly, and on a full sense of their necessity; for the limited faculties of our nature will bear but little of this sort at a time without a kind of intoxication, which precludes all rectilinear progress—or, rather, all progress whatever, except in a direction which terminates in the wildest vagaries of mysticism and clairvoyance.

But, without going into any subtleties, I may be allowed to suggest that it is at least high time that philosophers, both physical and others, should come to some nearer agreement than appears to prevail as to the meaning they intend to convey in speaking of causes and causation. On the one hand we are told that the grand object of physical inquiry is to explain the phenomena of nature, by referring them to their causes: on the other, that the inquiry into causes is altogether vain and futile, and that science has no concern, but with the discovery of *laws*. Which of these is the truth? Or are both views of the matter true on a different interpretation of the terms? Whichever view we may take or whichever interpretation adopt, there is one thing certain,—the extreme inconvenience of such a state of language. This can only be reformed by a careful analysis of this widest of all human generalizations, disentangling from one another the innumerable shades of meaning which have got confounded together in its progress, and establishing among them a rational classification and nomenclature. Until this is done we cannot be sure, that by the relation of cause and effect one and the same kind of relation is understood. Indeed, using the words as we do, we are quite sure that the contrary is often the case; and so long as uncertainty in this respect is suffered to prevail, so long will this unseemly contradiction subsist, and not only prejudice the cause of science in the eyes of mankind, but create disunion of feeling, and even give rise to accusations and recriminations on the score of principle among its cultivators.

The evil I complain of becomes yet more grievous when the idea of *law* is brought so prominently forward as not merely to throw into the background that of *cause*, but almost to thrust it out of view altogether; and if not to assume something approaching to the character of direct agency, at least to place itself in the position of a substitute for what mankind in general understand by *explanation:* as when we are told, for example, that the successive appearance of races of organized beings on earth, and their disappearance, to give place to others, which Geology teaches us, is a result of some certain law of developement, in virtue of which an unbroken chain of gradually exalted organization from the crystal to the globule, and thence, through the successive stages of the polypus, the mollusk, the insect, the fish, the reptile, the bird, and the beast, up to the monkey and the man (nay, for ought we know, even to the angel), has been (or remains to be) evolved. Surely, when we hear such a theory, the natural, human craving after *causes*, capable in some conceivable way of giving rise to such changes and transformations of organ and intellect,—*causes why* the developement at different parts of its progress should divaricate into different lines,—*causes*, at all events, intermediate between the steps of the developement—becomes importunate. And when nothing is offered to satisfy this craving, but loose and vague reference to *favourable circumstances* of climate, food, and general situation, which no experience has ever shown to convert one species into another; who is there who does not at once perceive that such a theory is in no respect more *explanatory*, than that would be which simply asserted a miraculous intervention, at every successive step of that unknown series of events, by which the earth has been alternately peopled and dispeopled of its denizens?

A *law* may be a *rule* of action, but it is not *action*. The Great First Agent may lay down a rule of action for himself, and that rule may become known to man by observation of its uniformity: but constituted as our minds are, and having that conscious knowledge of causation, which is forced upon us by the reality of the distinction between *intending* a thing, and *doing* it, we can never substitute the *Rule* for the *Act*. Either directly, or through delegated agency, whatever takes place is not merely *willed*, but *done*, and what is done we then only declare to be explained, when we can trace a process, and show that it consists of steps analogous to those we observe in occurrences which have passed often enough before our own eyes to have become familiar, and to be termed *natural*. So long as no such process can be traced and analyzed out in this manner, so long the phenomenon is unexplained, and remains equally so whatever be the number of unexplained steps inserted between its beginning and its end. The transition from an inanimate crystal to a globule capable of such endless organic and intellectual developement, is *as* great a step — *as* unexplained a one — *as* unintelligible to us — and in any human sense of the word as *miraculous* as the immediate creation and introduction upon earth of every species and every individual would be. Take these amazing facts of geology which way we will, we must resort elsewhere than to a mere speculative law of developement for their explanation.

Visiting as we do once more this scene of one of our earliest and most agreeable receptions — as travellers on the journey of life brought back by the course of events to scenes associated with exciting recollections and the memory of past kindness—we naturally pause and look back on the interval with that interest which always arises on such occasions, "How has it fared with you meanwhile?" we fancy ourselves asked.—"How have you prospered?"—"Has this long interval been well or ill spent?"—"How is it with the cause in which you have embarked?"—"Has it flourished or receded, and to what extent have you been able to advance it?" To all these questions we may, I believe, conscientiously, and with some self-gratulation, answer—"Well!" The young and then but partially fledged institution has become established and matured. Its principles have been brought to the test of a long and various experience, and been found to work according to the expectations of its founders. Its practice has been brought to uniformity and consistency, on rules which, on the whole, have been found productive of no inconvenience to any of the parties concerned. Our calls for reports on the actual state and deficiencies of important branches of science, and on the most promising lines of research in them, have been answered by most valuable and important essays from men of the first eminence in their respective departments, not only condensing what is known, but adding largely to it, and in a multitude of cases entering very extensively indeed into original inquiries and investigations—of which Mr. Scott Russell's "Report on Waves," and Mr. Carpenter's on the "Structure of Shells," and several others in the most recently published volume of our Reports, that for the York meeting last summer, may be specified as conspicuous instances.

Independent of these Reports, the original communications read or verbally made to our several Sections, have been in the highest degree interesting and copious; not only as illustrating and extending almost every branch of science, but as having given rise to discussions and interchanges of idea and information between the members present, of which it is perfectly impossible to estimate sufficiently the influence and value. Ideas thus communicated fructify in a wonderful manner on subsequent reflection, and become, I am persuaded, in innumerable cases, the germs of theories, and the connecting links between distant regions of thought, which might have otherwise continued indefinitely dissociated.

How far this Association has hitherto been instrumental in fulfilling the ends for which it was called into existence, can, however, be only imperfectly estimated from these considerations. Science, as it stands at present, is not merely advanced by speculation and thought; it stands in need of material appliances and means; its pursuit is costly, and to those who pursue it for its own sake, utterly unremunerative, however largely the community may benefit by its applications, and however successfully practical men may turn their own or others' discoveries to account. Hence arises a wide field for scientific utility in the application of pecuniary resources in aid of private research, and one in which assuredly this Association has not held back its hand. I have had the curiosity to cast up the sums which have been actually paid, or are now in immediate course of payment, on account of grants for scientific purposes by this Association since its last meeting at this place, and I find them to amount to not less than 11,167*l*. And when it is recollected that in no case is any portion of these grants applied to cover any personal expense, it will easily be seen how very large an amount of scientific activity has been brought into play by its exertions in this respect, to say nothing of the now very numerous occasions in which the attention and aid of Government has been effectually drawn to specific objects at our instance.

As regards the general progress of science within the interval I have alluded to, it is far too wide a field for me now to enter upon, and it would be needless to do so in this assembly, scarcely a man of which has not been actively employed in urging on the triumphant march of its chariot wheels, and felt in his own person the high excitement of success joined with that noble glow which is the result of companionship in honourable effort. May such ever be the prevalent feeling among us! True science, like true religion, is wide-embracing in its extent and aim. Let interests divide the worldly and jealousies torment the envious! We breathe, or long to breathe, a purer empyrean. The common pursuit of Truth is of itself a brotherhood. In these our annual meetings, to which every corner of Britain — almost every nation in Europe, sends forth as its representative some distinguished cultivator of some separate branch of knowledge; where, I would ask, in so vast a variety of pursuits which seem to have hardly anything in common, are we to look for that acknowledged source of delight which draws us together and inspires us with a sense of unity? That astronomers should congregate to talk of stars and planets — chemists of atoms—geologists of strata—is natural enough; but what is there of *equal* mutual interest, *equally* connected with and *equally* pervading all they are engaged upon, which causes their hearts to burn within them for mutual communication and unbosoming? Surely, were each of us to give utterance to all he feels, we would hear the chemist, the astronomer, the physiologist, the electrician, the botanist, the geologist, all with one accord, and each in the language of his own science, declaring not only the wonderful works of God disclosed by it, but the delight which their disclosure affords him, and the privilege he feels it to be to have aided in it. This is indeed a magnificent induction —a consilience there is no refusing. It leads us to look onward, through the long vista of time, with chastened but confident assurance that science has still other and nobler work to do than any she has yet attempted; work which, before she is prepared to attempt, the minds of men must be prepared to *receive* the attempt,—prepared, I mean, by an entire conviction of the wisdom of her views, the purity of her objects, and the faithfulness of her disciples.

THE WALK.

[TRANSLATED FROM THE GERMAN OF SCHILLER.]

DER SPAZIERGANG.

SEY mir gegrüsst mein Berg mit dem röthlich strahlenden Gipfel,
Sey mir, Sonne, gegrüsst, die ihn so lieblich bescheint,
Dich auch grüss ich belebte Flur, euch säuselnde Linden,
Und den fröhlichen Chor, der auf den Aesten sich wiegt,
Ruhige Bläue dich auch, die unermesslich sich ausgiesst
Um das braune Gebirg, über den grünenden Wald
Auch um mich, der endlich entflohn des Zimmers Gefängniss
Und dem engen Gespräch freudig sich rettet zu dir,
Deiner Lüfte balsamicher Strom durchrinnt mich erquickend,
Und den durstigen Blick labt das energische Licht,
Kräftig auf blühender Au erglänzen die wechselnden Farben,
Aber der reizende Streit löset in Anmuth sich auf.
Frey empfängt mich die Wiese mit weithin verbreitetem Teppich
Durch ihr freundliches Grün schlingt sich der ländliche Pfad
Um mich summt die geschaftige Biene, mit zweifelndem Flügel.
Wiegt der Schmetterling sich über dem röthlichen Klee.
Glühend trifft mich der Sonne Pfeil, still liegen die Weste
Nur der Lerche Gesang wirbelt in heiterer Luft.
Doch jetzt braust's aus dem nahen Gebüsch, tief neigen der Erlen
Kronen sich, und im Wind wogt das versilberte Gras,
Mich umfängt ambrosische Nacht; in duftende Kühlung
Nimmt ein prächtiges Dach schattender Buchen mich ein,
In des Waldes Geheimniss entflieht mir auf einmal die Landschaft
Und ein mystischer Pfad leitet mich steigend empor.

THE WALK.

HAIL to thee, Mountain Mine! with thy crest all purple and glowing.
  Hail to thy beams, O Sun, falling so sweet on its slope.
Life-teeming fields, all hail! and ye gently whispering Lime-trees —
  Peopled with many a bird rocking aloft in your boughs.
Hail! thou blue and tranquil expanse, whose fathomless concave
  Folds round the dark brown hill—sinks o'er the shadowy wood—
Me too receive! Escaped from my chamber's narrow confinement—
  Gladly to thee I fly from the world's wearisome themes.
Rich are thy breezes of balm my inmost bosom reviving!
  Strong is thy lively light poured on my rapturous glance.
Where the wide-carpeted mead with friendly welcome receives me
  Free the green path I trace, rurally winding along;
Bright on the blooming plain the changeful colours are playing,
  Now contrasting, and now melting and blending in grace.
Hark! to the bees' busy hum all around. The butterfly flitting
  O'er the red clover skims, fickle, in objectless dance.
Now the Sun darts his glow, and the west wind hushed into stillness
  Mars not the lark's clear strain cheerfully warbled on high:
Now from the copse, and aloft in the crowns of the deep-nodding
  Rustles the coming breeze, curling in silver the grass. [alders,
Deep in ambrosial night I plunge, where freshness and odours
  Breathe 'neath the beechen roof broad over-arching in shade.
Lost is the landscape at once in the dark wood's secret recesses
  Where a mysterious path leads up the winding ascent.

Nur verstohlen durchdringt der Zweige laubigtes Gitter
Sparsames Licht, un est blickt lachend das Blaue herein.
Aber plötzlich zerreisst der Flor. Der geöffnete Wald giebt
Überraschend des Tags blendendem Glanz mich zurück.
Unabsehbar ergiesst sich vor meinen Blicken die Ferne,
Und ein blaues Gebirg endigt im Dufte die Welt.
Tief an des Berges Fuss, der gählings unter mir abstürzt,
Wallet des grünlichten Stroms fliessender Spiegel vorbei.
Endlos unter mir seh' ich den Aether, über mir endlos,
Blicke mit Schwindeln hinauf, blicke mit Schaudern hinab,
Aber zwischen der ewigen Höh' und der ewigen Tiefe
Trägt ein geländerter Steig sicher den Wandrer dahin.
Lachend fliehen an mir die reichen Ufer vorüber,
Und den fröhlichen Fleiss rühmet das prangende Thal.
Jene Linien, sieh! die des Landmanns Eigenthum scheiden,
In den Teppich der Flur hat sie Demeter gewirkt.
Freundliche Schrift des Gesetzes, des Menschenerhaltenden Gottes,
Seit aus der ehernen Welt fliehend die Liebe verschwand,
Aber in freieren Schlangen durchkreuzt die geregelten Felder,
Jetzt verschlungen vom Wald, jetzt an den Bergen hinauf
Klimmend, ein schimmernder Streif die Länder verknüpfende Strasse,
Auf dem ebenen Strom gleiten die Flösse dahin,
Vielfach ertönt der Heerden Geläut im belebten Gefilde,
Und den Wiederhall weckt einsam des Hirten Gesang.
Muntre Dörfer bekränzen den Strom, in Gebüschen verschwinden
Andre, vom Rücken des Bergs stürzen sie gäh dort herab.
Nachbarlich wohnet der Mensch noch mit dem Acker zusammem,
Seine Felder umruhn friedlich sein ländliches Dach,
Traulich rankt sich die Reb' empor an dem niedrigen Fenster,
Einen umarmenden Zweig schlingt um die Hütte der Baum,
Glückliches Volk der Gefilde! noch nicht zur Freiheit erwachet,
Theilst du mit deiner Flur fröhlich das enge Gesetz.
Deine Wünsche beschränkt der Aernten ruhiger Kreislauf,
Wie dein Tagewerk, gleich, windet dein Leben sich ab!

There through crossing boughs the noonday dimly admitting,
Smiling with furtive glance scarce the blue heaven looks in.
Suddenly rent is the veil — All startled I view with amazement
Through the wood's opening glade, blazing in splendour, the day.
Heavens! what a prospect extends, till the sight bewildered and failing
Rests on the world's last hill, shimmering in distance and mist —
Deep at my feet, where sheer to its base the precipice plunges,
Lo! where the glassy stream glides through its margin of green —
Boundless, above and around and below me, the Æther is rolling,
Giddy aloft I gaze, shuddering recoil from beneath.
Yet 'twixt the yawning gulph, and the cliff in horror impending,
Led by a rock-built path, safely the wanderer descends:
Safely and swift, while the laughing shores fly past in their richness,
And the luxuriant vale industry's triumph proclaims.
Hedgerows there, with tracery neat, on its velvety carpet
Broidered by Ceres' hand, limit each rural domain.
Legible lines of Justice and Law, whose firm interdiction
(Love from the world being fled) curbs the encroachments of man —
There with a freer sweep, far-stretching o'er field and o'er meadow,
Commerce her high-way leads, land interlinking with land;
Now in dark woods ingulfed, now crowning the crest of the mountain
While the raft-laden stream glides in its easy descent.
Wide o'er the peopled mead the lowing herds are resounding,
And the rude herdsman's song wakes the lone echoes afar,
Bordering villages deck the gay banks, or in sheltering woodlands
Shrink — or shelf over shelf climb the projecting ascent.
Man on the land which he tills, in peace contentedly dwelling,
Sees the loved fields of his youth stretched round his rustic abode;
Where the confiding vine up the lowly window is climbing,
Where the old friendly tree wraps its protecting embrace:
Blest, thrice blest is his lot! Not yet to false freedom awakened,
Pleased he respects the law, sovereign o'er him and his field,
Bounded in thought and in wish by the peaceful round of his har-
Calm as his daily toil glides his existence away. [vests,

Aber wer raubt mir auf einmal den lieblichen Anblick? Ein fremder
 Geist verbreitet sich schnell über die fremdere Flur!
Spröde sondert sich ab was kaum noch liebend sich mischte,
 Und das Gleiche nur ist's, was an das Gleiche sich reyht.
Stände seh ich gebildet. Der Pappeln stolze Geschlechter
 Ziehn in geordnetem Pomp vornehm und prächtig daher,
Regel wird alles und alles wird Wahl und alles Bedeutung,
 Dieses Dienergefolg meldet den Herrscher mir an.
Prangend verkündigen ihn von fern die beleuchteten Kuppeln
 Aus dem felsigten Kern, hebt sich die thürmende *Stadt.*
In die Wildniss hinaus sind des Waldes Faunen verstossen,
 Aber die Andacht leyht höheres Leben dem Stein.
Näher gerückt ist der Mensch an den Menschen, Enger wird um ihn
 Reger erwacht, es umwälzt rascher sich in ihm die Welt.
Sieh, da entbrennen in feurigem Kampf die eifernden Kräfte
 Grosses wirket ihr Streit, grösseres wirket ihr Bund.
Tausend Hände belebt Ein Geist, hoch schlaget in tausend
 Brüsten, von einem Gefühl glühend, ein einziges Herz,
Schlägt für das Vaterland und glüht für der Ahnen Gesetze,
 Hier auf dem theuren Grund ruht ihr verehrtes Gebein.
Nieder steigen vom Himmel die seligen Götter, und nehmen
 In dem geweihten Bezirk festliche Wohnungen ein,
Herrliche Gaben bescheerend erscheinen sie; Ceres vor allen
 Bringet des Pfluges Geschenk, Hermes den Anker herbei,
Bacchus die Traube, Minerva des Oehlbaums grünende Reiser
 Auch das kriegrische Ross führet Poseidon heran,
Mutter Cybele spannt an des Wagens Deichsel die Löwen,
 In das gastliche Thor zieht sie als Bürgerinn ein.
Heilige Steine! Aus euch ergossen sich Pflanzer der Menschheit,
 Fernen Inseln des Meers sandtet ihr Sitten und Kunst,
Weise sprachen das Recht an diesen geselligen Thoren,
 Helden stürzten zum Kampf für die Penaten heraus.

Fare ye well, sweet scenes! A stranger spirit is breathing
  O'er the transformèd plains, snatching your charms from my view:
Harshly springing asunder from forced and unequal alliance,
  What shall coërce the strong, when at the lovely it spurns?
Classes behold, and ranks. In long and stately perspective
  Lo! where the poplar's pomp sweeps in aspiring array—
All is Rule and Arrangement and Choice. Each feature has meaning;
  Such an impressive train tells of THE RULER at hand:
Brightly yon gleaming domes his presence announce, where the city
  High o'er its nest of rocks soars in its towery pride.
Far from their ancient haunts the Fauns complaining are driven;
  (What though piety lend holier life to the stone;)
Man pressed closer to Man, finds his being concentred, his feelings
  Broader awake. His world rolls in a swifter career.
There in contention fierce blaze forth antagonist powers,
  Great, opposed in their strife—greater in union linked.
Linking a thousand hands in a single effort; a thousand
  Hearts in a single pulse; thoughts in a single resolve;
Burning with patriot love, and with long ancestral devotion,
  There on the hallowed spot where the loved ashes repose;
Where the immortal Gods their glorious temples have chosen,
  Drawn by established rites down from their Heavenly abodes.
Fraught with blessings they come. First, Ceres, Mother of harvests,
  Brings the productive plough—Hermes the anchor affords—
Bacchus the grape—Minerva the genial fruit and the graceful
  Frond of the olive bough—Neptune the warrior steed—
Cybele borne through the welcoming gate on her lion-yoked chariot
  Enters, an honoured guest,—dwells, a protectress and friend.
Sacred Walls! from whose bosom the seeds of humanity, wafted
  Ev'n to the farthest isles, morals and arts have conveyed.
Sages in these thronged gates in justice and judgment have spoken;
  Heroes to battle have rushed hence for their altars and homes:

Auf den Mauren erschienen, den Saügling im Arme, die Mütter
 Blickten dem Heerzug nach, bis ihn die Ferne verschlang.
Betend stürzten sie dann vor der Götter Altären sich nieder,
 Flehten um Ruhm und Sieg, flehten um Rückkehr für euch.
Ehre ward euch und Sieg, doch der Ruhm nur kehrte zurücke,
 Eurer Thaten Verdienst meldet der rührende Stein:
" Wanderer, kommst du nach Sparta, verkündige dorten, du habest
 " Uns hier liegen gesehn, wie das Gesetz es befahl."
Ruhet sanft ihr Geliebten! Von eurem Blute begossen
 Grünet der Oelbaum, es keimt lustig die köstliche Saat.
Munter entbrennt, des Eigenthums froh, das freye Gewerbe,
 Aus dem Schilfe des Stroms winket der blaülichte Gott.
Zischend fliegt in den Baum die Axt, es erseufzt die Dryade,
 Hoch von des Berges Haupt stürzt sich die donnernde Last.
Aus dem Felsbruch' wiegt sich der Stein, vom Hebel beflügelt,
 In der Gebirge Schlucht taucht sich der Bergmann hinab.
Mulcibers Ambos tönt von dem Takt geschwungener Hämmer
 Unter der nervigten Faust sprützen die Funken des Stahls,
Glänzend umwindet der goldne Lein die tanzende Spindel,
 Durch die Saiten des Garns sauset das webende Schiff,
Fern auf der Rhede ruft der Pilot, es warten die Flotten,
 Die in der Fremdlinge Land tragen den heimischen Fleiss,
Andre ziehn frohlockend dort ein, mit den Gaben der Ferne,
 Hoch von dem ragenden Mast wehet der festliche Kranz.
Siehe da wimmeln die Märkte, der Krahn von fröhlichem Leben,
 Seltsamer Sprachen Gewirr braust in das wundernde Ohr.
Auf den Stapel schüttet die Aernten der Erde der Kaufmann,
 Was dem glühenden Strahl Afrikas Boden gebiert,
Was Arabien kocht, was die aüsserste Thule bereitet,
 Hoch mit erfreuendem Gut füllt Amalthea das Horn.
Da gebieret das Glück dem Talente die göttlichen Kinder,
 Von der Freiheit gesäugt wachsen die Künste der Lust

Mothers the while (their infants in arms) from the battlements gazing
  Follow with tears the host, till in the distance it fades;
Then to the temples crowding, and prostrate flung, at the altars
  Pray for their triumph and fame—pray for their joyful return.
Triumph and fame are theirs, but in vain their welcome expects them:
  Read how the exciting stone tells of their glorious deserts.
"Traveller! when to Sparta thou comest, declare thou hast seen us,
  "Each man slain at his post,—even as the law hath ordained."
Soft be your honoured rest! with your precious life-blood besprinkled;
  Freshens the olive bough—Sparkles with harvests the plain.
Commerce awakes, by freedom inspired, by security nurtured;
  Beckons the azure God, pleased, from the reeds of his stream.
Gashing, the broad axe flies—while the Dryad shrieks—and in ruin
  Down from the mountain's brow, crashes the thundering tree.
Winged by the lever's force, the stone nods forth from the quarry,
  Deep in its innermost gorge plunges the miner beneath.
Hark to the rude Vulcanian music from anvil and hammer,
  Where at each nervous blow flashes the bickering steel:
Hark to the whirling reel, with its flaxen burden surrounded,
  And the swift shuttle's play, brushing the weft as it flies:
Hark to the Pilot's hail in the distant road, where a navy
  Waits to transport abroad industry's costly results.
Others arrive, deep laden, from far, and jovially cheering,
  Garland and streamer on high float from the towering mast;
Rises o'er all the mart's busy din—the bustle of commerce;
  Barbarous tongues uncouth strike on the wondering ear.
Hither the harvests of Earth are consigned. Here heapeth the mer-
  All that Africa's soil yields to the ripening sun; [chant
All that Arabia distils—all that uttermost Thule can proffer;
  Fair Amalthea's horn brims with exuberant wealth:
Wealth which, to Genius wedded, a godlike offspring produces—
  Arts which strengthen and grow nurtured by freedom and taste,

Mit nachahmendem Leben erfreuet der Bildner die Augen,
Und vom Meissel beseelt redet der fühlende Stein,
Künstliche Himmel ruhn auf schlanken Ionischen Säulen
Und den ganzen Olymp schliesset ein Pantheon ein,
Leicht wie der Iris Sprung durch die Luft, wie der Pfeil von der
Hüpfet der Brücke Joch über den brausenden Strom. [Senne
Aber im stillen Gemach entwirft bedeutende Zirkel
Sinnend der Weise, beschleicht forschend den schaffenden Geist,
Prüft der Stoffe Gewalt, der Magnete Hassen und Lieben, [Strahl;
Folgt durch die Lüfte dem Klang, folgt durch den Aether dem
Sucht das vertraute Gesetz in des Zufalls grausenden Wundern,
Sucht den ruhenden Pol in der Erscheinungen Flucht.
Körper und Stimme leyht die Schrift dem stummen Gedanken,
Durch der Jahrhunderte Strom trägt ihn das redende Blatt.
Da zerrinnt vor dem wundernden Blick der Nebel des Wahnes,
Und die Gebilde der Nacht weichen dem tagenden Licht.
Seine Fesseln zerbricht der Mensch. Der Beglückte! Zerriss er
Mit den Fesseln der Furcht nur nicht den Zügel der Schaam!
Freiheit ruft die Vernuft, Freiheit die wilde Begierde,
Von der heil'gen Natur ringen sie lüstern sich los.
Ach, da reissen im Sturm die Anker, die an dem Ufer
Warnend ihn hielten, ihn fasst mächtig der flutende Strom;
Ins Unendliche reisst er ihn hin, die Küste verschwindet,
Hoch auf der Fluten Gebirg wiegt sich entmastet der Kahn;
Hinter Wolken erlöschen des Wagens beharrliche Sterne,
Bleibend ist nichts mehr, es irrt selbst in dem Busen der Gott.
Aus dem Gespräche verschwindet die Wahrheit, Glauben und Treue
Aus dem Leben, es lügt selbst auf der Lippe der Schwur.
In der Herzen vertraulichsten Bund, in der Liebe Geheimniss
Drängt sich der Sykophant, reisst von dem Freunde den Freund.
Auf die Unschuld schielt der Verrath mit verschlingendem Blicke
Mit vergiftendem Biss tödtet des Lasterers Zahn.

Charming the sight with emulous life spreads the painter his canvas,
  And by the sculptor* inspired feels the cold marble and speaks.
Sky-like vaults scarce press on the slender Ionian column;
  And a Pantheon's dome swells,—an Olympus on Earth!
Light as the rainbow's leap—as the vaulting flight of the arrow,
  Bounds the self-balanced bridge yoking the torrent beneath;
Science, the while, deep musing in cell over circle and figure,
  Knows and adores the Power which through creation it tracks,
Measures the forces of matter—the hates and loves of the magnets—
  Sound through its wafting breeze, Light through its Æther pursues;
Seeks in the marvels of chance the law which pervades and controls
  Seeks the reposing pole fixed in the whirl of events. [it—
Speechless thought takes body and voice from the craft of the pen-
  Down the long stream of time borne on the eloquent page. [man,
Fast from the wondering sight the mists of error are clearing:
  Chased by the dawning beam fly the dark spectres of night.
Burst are the chains which fettered mankind. O happy! if only
  Bursting the chains of fear, kept they the bridle of shame. [echoed,
Freedom the watchword — by Reason proclaimed — by Passion re-
  Rending each natural bond madly they tear themselves loose.
Cast is each anchor aside (all warning neglected) which held them
  Safe to the shore. The flood sweeps them in tumult away.
Far from the vanishing coast, on a swelling and limitless ocean
  Tossed on the mountain-wave labours dismasted their bark.
Quenched is each lode-star in cloud—no mark—no principle constant,
  Even their own bosom-god† swerves in his doubtful response.
Truth from their language, faith from their life, and confidence, vanish;
  Even on their glozing lips lies in its utterance the oath.
Into the heart's most sacred recess, love's holiest secret, [friend—
  Creeps the vile sycophant's art, severing the friend from the
Treachery scowls with withering glance on its innocent victim,
  And with envenomed death darts the fell slanderer's tooth:

* Literally *the Chisel*. † Conscience.

Feil ist in der geschändeten Brust der Gedanke, die Liebe
Wirft des freyen Gefühls göttlichen Adel hinweg,
Deiner heiligen Zeichen, o Wahrheit, hat der Betrug sich
Angemasst, der Natur köstlichste Stimmen entweiht,
Die das bedürftige Herz in der Freude Drang sich erfindet,
Kaum giebt wahres Gefühl noch durch Verstummen sich kund.
Auf der Tribune prahlet das Recht, in der Hütte die Eintracht,
Des Gesetzes Gespenst steht an der Könige Thron,
Jahre lang mag, Jahrhunderte lang die Mumie dauern,
Mag das trügende Bilde lebender Fülle bestehn,
Bis die Natur erwacht, und mit schweren ehernen Händen
An das hohle Gebäu rühret die Noth und die Zeit,
Einer Tygerinn gleich, die das eiserne Gitter durchbrochen
Und des Numidischen Wald's plötzlich und schrecklich gedenkt,
Aufsteht mit des Verbrechens Wuth und des Elends, die Menschheit
Und in der Asche der Stadt sucht die verlorne Natur.
O so öffnet euch Mauren, und gebt den Gefangenen ledig,
Zu der verlassenen Flur kehr' er gerettet zurück!
Aber wo bin ich? Es birgt sich der Pfad. Abschüssige Gründe
Hemmen mit gähnender Kluft hinter mir, vor mir, den Schritt.
Hinter mir blieb der Gärten, der Hecken vertraute Begleitung
Hinter mir jegliche Spur menschlicher Hände zurück
Nur die Stoffe seh ich gethürmt, aus welchen das Leben
Keimet, der rohe Basalt hofft auf die bildende Hand
Brausend stürzt der Giessbach herab durch die Rinne des Felsen
Unter den Wurzeln des Baums bricht er entrüstet sich Bahn.
Wild ist es hier und schauerlich öd'. Im einsamen Luftraum
Hängt nur der Adler und knüpft an das Gewolke die Welt
Hoch herauf bis zu mir trägt keines Windes Gefieder
Den verlornen Schall menschlicher Mühen und Lust.
Bin ich wirklich allein? In deinen Armen, an deinem
Herzen wieder, Natur, ach! und es war nur ein Traum

In the degraded bosom the thought is venal—the feeling
Ev'n of Love's godlike fire dies in ignoble constraint.
Where are thy characters, Truth? By artifice seized and perverted,
Every one precious sign Nature has marked for her own;
Even what the yearning heart gasps forth in the stress of emotion,
Till but by silence expressed genuine feeling is known.
Loud is the vaunt of right in the tribune—peace in the cottage;
And by the Sovereign's throne stands the vain phantom of law.
Years—aye, centuries long may the bloodless and impotent mummy,
Fixed in deceptive guise, carry the semblance of life,
Until nature awakes—and with hand of iron unsparing,
Heavy with time and fate, shatters the hollow device.
Then, like the tiger at large, when burst are the bars of his prison,
And his Numidian wild rushes in blood on his thoughts,
Trampled humanity rises, in crime' and in misery's madness;
And through the ashes of states, back to rude nature reverts.
Open ye walls! in mercy,—ye gates! fly wide to the captive;
Back to his long-lost plains forth let him rush in his rage.
Where am I wandering? the path is lost! Before and behind me,
Rifted and yawning ravines narrow the dangerous way!
Gardens and hedges withdraw their friendly and sociable guidance!
Trace of man's hand is none, save in the distance behind.
Pile upon pile, rude masses arise chaotic!—a Chaos
Pregnant! The formless basalt longs for the sculpturing hand.
Headlong now, from the cleft rock's brow, the torrent is rushing!
Now, 'neath the wreathèd root bursting indignant its way.
Savage and shudd'ringly lonely the spot! the companionless eagle
Hangs in mid-air aloft—linking the sky with the world.
Hush'd is each slumbering breeze! No Zephyr balmily stealing
Bears on its panting plume sound of man's toil or his joy.
Am I then truly alone? Kind Nature! Once more on thy bosom,
In thy protecting arm, dare I look back on the dream

Der mich schaudernd ergriff, mit des Lebens furchtbarem Bilde
  Mit dem stürzendem Thal stürzte der finstre hinab.
Reiner nehm' ich mein Leben von deinem reinen Altare,
  Nehme den fröhlichen Muth hoffender Jugend zurück!
Ewig wechselt der Wille den Zweck und die Regel, in ewig
  Wiederholter Gestalt wälzen die Thaten sich um.
Aber jugendlich immer, in immer veränderter Schöne
  Ehrst du, fromme Natur, züchtig das alte Gesetz,
Immer dieselbe, bewahrs du in treuen Händen dem Manne,
  Was dir das gaukelnde Kind, was dir der Jungling vertraut,
Nährest an gleicher Brust die vielfach wechselnden Alter;
  Unter demselben Blau, über dem nehmlichen Grün
Wandeln die nahen und wandeln vereint die fernen Geschlechter,
  Und die Sonne Homers, siehe! sie lächelt auch uns.

Which with the deepening gloom of the steep-down valley conspiring,
  Forced on my harrowed soul all the dire horrors of life.—
Pure, from thy altar pure, I drink the new breath of my being,
  And the rich glow of my youth joyous and hopeful returns.
Wild is the will of man, and changeful its course and its object;
  And in Ephemeral round action to action succeeds.
Thou, in enduring bloom, and in beauty's exhaustless succession,
  True to thine ancient law, holdst thine appointed career;
All that the sportive child, the confiding youth, hath entrusted
  Into thy faithful hands—back to the man is repaid.
Changeless in all! Each age on thine equal bosom is nurtured.
  Under the same blue vault—on the same tapestried green
Race upon race succeeding, through countless ages have wandered;
  Suns that on Homer smiled smile as benignant on us!*

* This translation has already appeared in a collection of "English Hexameter Translations, from Schiller, Goethe, Homer, Callinus, and Meleager," emanating from the Cambridge University Press, and edited by an eminent and accomplished member of that University.

NADOWESSISCHE TODTENKLAGE.

[SCHILLER.]

I.

Seht! da sitzt er auf der Matte,
Aufrecht sitzt er da
Mit dem Anstand den er hatte
Als er's Licht noch sah.

II.

Doch wo ist die Kraft der Fäuste,
Wo des Athems Hauch,
Der noch jüngst zum grossen Geiste,
Blies der Pfeife Rauch?

III.

Wo die Augen, Falkenhelle,
Die des Rennthiers Spur
Zählten auf des Grases Welle
Auf dem Thau der Flur?

IV.

Diese Schenkel, die behender
Flohen durch den Schnee
Als der Hirsch, der Zwanzigender,
Als des Berges Reh.

FUNERAL DIRGE OF A NADOWESSIE.

I.

See there upon the mat he sits
  Erect before his door,
With just the same majestic air
  As once in life he wore.

II.

But where is now his strength of limb,
  The whirlwind of his breath,
To the Great Spirit when he puffed
  The peace-pipe's mounting wreath?

III.

Where are those falcon eyes, which late
  Along the plain could trace,
Along the grass's dewy wave
  The reindeer's printed pace?

IV.

These legs which erst with matchless speed,
  Flew through the drifted snow,
Surpassed the stag's unwearied course,
  Outran the mountain roe.

V.

Diese Arme, die den Bogen,
Spannten streng und straff!
Seht, das Leben ist entflogen,
Seht, sie hängen schlaff!

VI.

Wohl ihm! Er ist hingegangen,
Wo kein Schnee mehr ist,
Wo mit Mays die Felder prangen
Der von selber spriesst.

VII.

Wo mit Vögeln alle Sträuche,
Wo der Wald mit Wild,
Wo mit Fischen alle Teiche
Lustig sind gefüllt.

VIII.

Mit den Geistern speisst er droben,
Liess uns hier allein,
Das wir seine Thaten loben,
Und ihn scharren ein.

IX.

Bringet her die letzten Gaben,
Stimmt die Todtenklag!
Alles sei mit ihm begraben
Was ihn freuen mag.

V.

These arms, once used with might and main
  The stubborn bow to twang,
See, see! their life is fled at last,
  All motionless they hang.

VI.

'Tis well with him, for he is gone
  Where snow no more is found:
Where self-sown maize, with bounteous crops,
  Gladdens the fields around.

VII.

Where wild birds sing from every spray,
  Where deer come sweeping by
Where fish from every lake afford
  A plentiful supply.

VIII.

With spirits now he feasts above,
  And leaves us here alone,
To celebrate his valiant deeds,
  And round his grave to moan!

IX.

Sound the death song! Bring the gifts,
  The last gifts of the dead.
Let all which yet may yield him joy
  Within his grave be laid.

X.

Legt ihm unters Haupt die Beile,
Die er tapfer schwang,
Auch des Bären fette Keule,
Denn der Weg ist lang.

XI.

Auch das Messer scharf geschliffen,
Das vom Feindeskopf,
Rasch mit drey geschickten Griffen,
Schälte Haut und Schopf.

XII.

Farben auch den Leib zu mahlen,
Steckt ihm in die Hand,
Dass er röthlich möge strahlen,
In der Seelen Land.

X.

The hatchet place beneath his head,
  Still red with hostile blood,
And add, because the way is long,
  The bear's fat ham for food.

XI.

Add the knife, which whetted keen,
  From heads struck down in fray,
With three dexterous slashes clean
  Tore their scalps away.

XII.

Paints, too, place within his hand
  Of bright vermillion dye,
That in the land of souls his form
  May beam triumphantly.*

* Reprinted with slight alterations from Grahame's "History of the United States of North America," Philadelphia Edition, 1845, vol. iii. p. 426.

DITHYRAMBE.

[SCHILLER.]

I.

Nimmer, das glaubt mir,
Erscheinen die Götter,
Nimmer allein.
Kaum dass ich Bacchus den Lustigen habe,
Kommt auch schon Amor, der lächelnde Knabe.
Phöbus der Herrliche findet sich ein,
        Sie nahen, sie kommen,
        Die Himmlischen alle,
        Mit Göttern erfüllt sich,
        Die irdische Halle.

II.

Sagt wie bewirth ich
Der Erdegeborne
Himmlischen Chor?
Schencket mir euer unsterbliches Leben,
Götter! Was kann euch der Sterbliche geben?
Hebet zu eurem Olymp mich empor.
        Die Freude, sie wohnt nur,
        In Jupiters Saale.
        O füllet mit Nektar,
        O reicht mir die Schale!

DITHYRAMBICS.

I.

The Gods descend from high,
But not alone they leave their blissful seat,
Hand in hand they quit the sky,
To join their votary's still retreat,
When jovial Bacchus crowns the bowl.
Then love, with laughing eyes, invades my soul,
And Phœbus makes the hallowed train complete.
They come, they come! the heavenly band,
In earthly bowers, they take their stand;
And bright with all their freshest rays,
Flash upon the Poet's gaze.

II.

The glorious guests, the heavenly quire,
Say, how shall earthborn man receive;
Untempered in celestial fire,
Their dazzling forms behold, and live?
Fill me, ye Gods, and full and high,
Your choicest draught of immortality!
To powers like yours what can a mortal give?
Fill with nectar, crown the cup!
I'll snatch the pledge and drink it up.
Then in the starry halls above
For ever dwell with bliss and Jove.

III.

Reich ihm die Schale!
Schencke dem Dichter
Hebe nur ein.
Netz' ihm die Augen mit himmlischen Thaue,
Dass er den Styx, den Verhassten, nicht schaue,
Einer der Unsern sich dünke zu seyn.
Sie rauschet, sie perlet,
Die himmlische Quelle,
Der Busen wird ruhig,
Das Auge wird helle.

III.

Fill the cup, and fill it high!
And Hebe, kiss the golden brim,
And let the poet taste of joy,
And feel that heaven was made for him.
Bathe his eyes in holy dew,
Lest Styx, detested power, should blast his view,
And let the Godhead glow through every limb!
Hark! the sacred stream descends,
Around the golden rim it bends!
I feel my sight grow clean from earthly shades,
While tranquil bliss my thrilling breast pervades.

SPRUCH DES CONFUCIUS.

[SCHILLER.]

DREYFACH ist der Schritt der Zeit,
Zögernd kommt die Zukunft hergezogen;
Pfeilschnell ist das Jetzt entflogen;
Ewig still steht die Vergangenheit.

Keine Ungeduld beflügelt
Ihren Schritt wenn sie verweilt,
Keine Furcht, kein Zweifel zügelt
Ihren Lauf, wenn sie enteilt.
Keine Reu, kein Zaubersegen
Kann die Stehende bewegen.

Möchtest du beglückt und weise
Endigen des Lebens Reise;
Nimm die Zögende zum Rath,
Nicht zum Werkzeug deiner That.
Wähle nicht die Fliehende zum Freund;
Nicht die Bleibende zum Feind.

SAYING OF CONFUCIUS.

TIME.

THREEFOLD is the march of Time,
The Future, lame and lingering, totters on;
Swift as a dart the Present hurries by;
The Past stands fixed in mute Eternity.

To urge his slow advancing pace
Impatience nought avails,
Nor fear, nor doubt, can check his race,
As fleetly past he sails.
No spell, no deep remorseful throes
Can move him from his stern repose.

Mortal! they bid thee read this rule sublime:
Take for thy councillor the lingering one;
Make not the flying visitor thy friend,
Nor choose thy foe in him that standeth without end.

SPRUCH DES CONFUCIUS.

[SCHILLER.]

DREYFACH ist des Raumes Maass.
Rastlos fort, ohn Unterlass
Strebt die Länge, fort ins Weite,
Endlos giesset sich die Breite,
Grundlos senkt die Tiefe sich.
Dir ein Bild sind sie gegeben,
Rastlos vorwärts musst du streben,
Nie ermüdet stille stehn
Willst du die Vollendung sehn.
Musst ins Breite dich entfalten
Soll sich dir die Welt gestalten,
In die Tiefe musst du steigen
Soll sich dir das Wesen zeigen.

Nur Beharrung führt zum Ziel,
Nur die Fülle führt zur Klarheit,
Und im Abgrund wohnt die Wahrheit.

SAYING OF CONFUCIUS.

SPACE.

In the triple form of space,
Threefold wisdom shalt thou trace.

Onward, onward, wearing never,
Length its line is urging ever,
Beyond where sky and ocean meet
Breadth expands its widening sheet,
Beneath where shaft or sea-lead goes,
Plunging Depth no bottom knows.
Print their lesson on thy soul!
Wouldst thou reach Perfection's goal,
Stay not! rest not! Forward strain!
Hold not hand and draw not rein!
Hedged within no narrow road,
Chart-like be thy plans, and broad.
Build thou deep as well as strong,
Or thy building lasts not long.

Perseverance strikes the mark
Expansion clears whate'er is dark,
Truth in the abyss doth dwell.
My say is said. — Now fare thee well.

LENORE.

[BÜRGER.]

I.

LENORE fuhr um's Morgenroth
Empor aus schweren Träumen;
"Bist untreu, Wilhelm, oder todt?
Wie lange willst Du säumen?"
Er war mit König Friedrich's Macht
Gezogen in die Prager Schlacht,
Und hatte nicht geschrieben
Ob er gesund geblieben.

II.

Der König und die Kaiserinn,
Des langen Haders müde,
Erweichten ihren harten Sinn,
Und machten endlich Friede;
Und jedes Heer, mit Sing und Sang,
Mit Paukenschlag und Kling und Klang,
Geschmückt mit grünen Reisern,
Zog heim zu seinen Häusern.

III.

Und überall all überall,
Auf Wegen und auf Stegen,
Zog Alt und Jung dem Jubelschall
Der Kommenden entgegen.
"Gottlob!" rief Kind und Gattinn laut,
"Willkommen!" manche frohe Braut;
Ach! aber für Lenore'n
War Gruss und Kuss verloren.

LEONORA.

I.

LEONORA starts, at morn's first red,
  From frightful dreams away.
"Art faithless, Wilhelm, or art dead?
  How long wilt thou delay?"
In royal Frederic's leaguering host,
He fought, when Prague was won and lost,
But since, by letter or by word,
No news of Wilhelm had been heard.

II.

The King and Empress now consent,
  To end their contest rude;
Their stubborn hearts, to friendship bent
  A peace at length conclude.
And either host rejoicing comes,
With trumpets' sound and rattling drums,
Crowned with green boughs, a jovial train,
To seek its native home again.

III.

By hedgerow path and trysting tree,
  Lane, road, and crowded street,
Thronged young and old with jubilee
  The coming host to meet.
"Thank Heaven!" each child and consort cried,
"Oh welcome!" many a happy bride;
But ah! to sad Leonora's heart,
Greeting nor kiss their joys impart.

IV.

Sie frug den Zug wohl auf und ab,
  Und frug nach aller Nahmen ;
Doch keiner war, der Kundschaft gab
  Von allen, so da kamen.
Als nun das Heer vorüber war,
Zerraufte sie ihr Rabenhaar,
  Und warf sich hin zur Erde
  Mit wüthiger Geberde.

V.

Die Mutter lief wohl hin zu ihr: —
  " Ach ! dass dich Gott erbarme !
Du trautes Kind, was ist mit Dir ? "
  Und schloss sie in die Arme.—
" O Mutter, Mutter ! Hin ist hin !
Nun fahre Welt und Alles hin !
  Bei Gott ist kein Erbarmen.
  O weh, O weh mir Armen ! "

VI.

" Hilf Gott, hilf ! Sieh uns gnädig auf !
  Kind, beth' ein Vaterunser !
Was Gott thut, das ist wohl gethan,
  Gott, Gott erbarmt sich unser ! "
" O Mutter, Mutter ! Eitler Wahn,
Gott hat an mir nicht wohl gethan !
  Was half, was half mein Bethen ?
  Nun ist's nicht mehr vonnöthen."

VII.

" Hilf, Gott, hilf ! Wer den Vater kennt,
  Der weiss, er hilft den Kindern.
Das hochgelobte Sakrament
  Wird deinen Jammer lindern."

IV.

She searched the host both up and down,
  She called on every name;
But not to one was Wilhelm known
  Who there rejoicing came.
When now no more the train was there,
She tore in vain her raven hair;
And on the ground, all frantic thrown,
She poured forth many an anguished groan.

V.

Her mother hastened to her aid,
  And caught her in her arms,
"God comfort thee! my own dear maid,
  What mean these mad alarms?"
"Oh, mother, mother! Gone is gone!
Vain world, and all it holds, begone!
God hath no comfort left for me,
Woe, woe is mine, and misery!"

VI.

"Help, Lord, help! Look with pity down!
  A Paternoster pray.
What God does, that is justly done,
  His grace endures for aye."
"Oh, mother! Empty mockery,
God hath *not* justly dealt by me.
Have I not begged and prayed in vain?
What boots it now to pray again?"

VII.

"Help, Lord, help! Father as He is
  Will He not help His child?
The Sacramental cup of bliss
  Will soothe thine anguish wild."

"O Mutter, Mutter! Was mich brennt
Das lindert mir kein Sakrament!
Kein Sakrament mag Leben
Den Todten wiedergeben."

VIII.

"Hör', Kind! wie, wenn der falsche Mann
Im fernen Ungerlande,
Sich seines Glaubens abgethan
Zum neuen Ehebande?
Lass fahren, Kind, sein Herz dahin!
Er hat es mimmermehr Gewinn!
Wann Seel' und Leib sich trennen
Wird ihn sein Meineid brennen."

IX.

"O Mutter, Mutter! Hin ist hin!
Verloren ist verloren!
Der Tod, der Tod ist mein Gewinn,
O wär' ich nie gebohren!
Lösch aus, mein Licht, auf ewig aus!
Stirb hin, stirb hin in Nacht und Graus!
Bei Gott ist kein Erbarmen,
O weh, O weh mir Armen!"

X.

"Hilf, Gott, hilf! Geh' nicht in's Gericht
Mit deinem armen Kinde!
Sie weiss nicht was die Zunge spricht,
Behalt' ihr nicht die Sünde!
Ach, Kind, vergiss dein irdisch Leid
Und denk an Gott und Seligkeit,
So wird doch deiner Seelen
Der Bräutigam nicht fehlen."

"Oh! mother, grief's consuming rage,
No Sacrament can e'er assuage.
No Sacrament can give again
Life to my Wilhelm, battle-slain."

VIII.

"Nay, hear! If faithless to his vows,
False to his God and thee,
Thy love have wed another spouse
In distant Hungary.
Be brave! Despise a heart so vain,
His is the loss and thine the gain.
When his false soul and body part
This perjury shall wring his heart."

IX.

"Oh, mother, mother! Gone is gone!
Once lost is lost for aye!
Comfort I crave from death alone,
Oh! had I ne'er seen day!
Out, lamp of life, for ever go!
Die, die away in night and woe,
God hath no comfort left for me.
Nought, nought but woe and misery."

X.

"Help, Lord! forgive and judge her not,
Her broken strength upraise.
Whate'er her sin, impute it not,
She knows not what she says.
Forget, my child, thine earthly love,
And think on God and bliss above;
The Bridegroom of thy soul shall come
And lead thee to His heavenly home!"

XI.

"O Mutter! was ist Seligkeit?
O Mutter! was ist Hölle?
Bei ihm, bei ihm ist Seligkeit
Und ohne Wilhelm Hölle!
Lösch aus, mein Licht, auf ewig aus!
Stirb hin, stirb hin, in Nacht und Graus!
Ohn' ihn mag ich auf Erden,
Mag dort nicht selig werden."

XII.

So wüthete Verzweifelung
Ihr in Gehirn und Adern.
Sie fuhr mit Gottes Vorsehung
Vermessen fort zu hadern;
Zerschlug den Busen und zerrang
Die Hand, bis Sonnenuntergang:
Bis auf am Himmelsbogen
Die goldnen Sterne zogen.

XIII.

Und aussen, horch! ging's trap, trap, trap
Als wie von Rosseshufen;
Und klirrend stieg ein Reiter ab,
An des Geländers Stufen;
Und horch! und horch! den Pfortenring
Ganz lose, leise, kling, kling, kling!
Dan kamen durch die Pforte
Vernehmlich diese Worte:

XIV.

"Holla, Holla! Thu' auf, mein Kind!
Schläfst, Liebchen, oder wachst Du?
Wie bist noch gegen mich gesinnt?
Und weinest oder lachst Du?"

XI.

"Oh! mother, mother, what is bliss?
Oh! mother, what is hell?
With him, with him, is happiness;
Without my Wilhelm, hell.
Out, lamp of life, for ever go!
Die, die away in night and woe!
On earth, in heaven, of him bereft,
No hope of happiness is left."

XII.

Thus wild despair and erring sense,
Racked heart, and maddening brain,
With God's eternal providence
An impious strife sustain.
Her hands she wrung, her bosom beat,
In anguish, till the sun was set,
And golden stars, with solemn beam,
In heaven's high arch began to gleam.

XIII.

What sounds without? Tramp, tramp, it rings
Like hoofs in fierce career.
Down from his horse a rider springs
And rattling mounts the stair.
And hark! the bell begins to ring,
All softly, gently, kling, kling, kling,
And through the unopened wicket gate,
These words her ear distinctly meet:

XIV.

"Halloh! halloh! These bolts remove!
Art waking, love, or sleeping?
As when we parted dost thou love?
And art thou glad or weeping?"

"Ach, Wilhelm, Du? . . . So spät bei Nacht? . . .
Geweinet hab' ich, und gewacht;
  Ach! grosses Leid erlitten!
  Wo kommst Du her geritten?"

XV.

"Wir satteln nur um Mitternacht.
  Weit ritt ich her von Böhmen.
Ich habe spät mich aufgemacht,
  Und will Dich mit mir nehmen."
"Ach, Wilhelm, erst herein geschwind!
Den Hagedorn durchsaust der Wind.
  Herein in meinen Armen
  Herzliebster zu erwarmen!"

XVI.

"Lass sausen durch den Hagedorn,
  Lass sausen, Kind, lass sausen!
Der Rappe scharrt, es klirrt der Sporn,
  Ich darf allhier nicht hausen.
Komm, schürze, spring' und schwinge Dich
Auf meinen Rappen hinter mich!
  Muss heut' noch hundert Meilen
  Mit Dir in's Brautbett eilen."

XVII.

"Ach! Wolltest hundert Meilen doch
  Mich heut' in's Brautbett tragen?
Und horch! Es brummt die Glocke noch
  Die eilf schon angeschlagen."
"Sieh hin, sieh her! der Mond scheint hell,
Wir und die Todten reiten schnell.
  Ich bringe Dich, zur Wette,
  Noch heut' in's Hochzeitbette."

"Hah! Wilhelm, thou? So late returned?
Right drearily I've watched and mourned
Full many a sorrowing hour have passed,
But say, whence comest thou at last?"

XV.

"We ride but near the midnight chime.
 On far Bohemian lea
I saddled late — have barely time;
 And thou must ride with me."
"Come in! The night is closing fast,
The hawthorn whistles in the blast.
Come, Wilhelm — let this loving arm
Fold thee awhile and keep thee warm."

XVI.

"Let the wind whistle in its speed
 Through bush and hawthorn spray;
The jingling spur, the pawing steed
 Tell me, I must away.
Come, busk, and spring, and mount behind,
My steed flies swifter than the wind.
A hundred leagues to-day we ride
To where I'll bed my bonny bride."

XVII.

"What? Ride a hundred leagues to-day
 Our bridal bed to find?
And hark! th' eleventh chime away
 Dies booming in the wind."
"Look round, look round — the moon shines bright,
We and the dead ride swift by night.
Be sure we'll reach our bridal bed,
Ere yet the midnight stroke hath sped."

XVIII.

"Sag' an, wo ist dein Kammerlein?
Wo? Wie dein Hochzeitbettchen?"
"Weit, weit von hier! . . . Still, kühl, und klein!
Sechs Bretter und zwei Brettchen!"
"Hatt's Raum für mich?" "Für Dich und mich!
Komm, schürze, spring' und schwinge dich!
Die Hochzeitgäste hoffen,
Die Kammer steht uns offen."

XIX.

Schön Liebchen schürzte, sprang und schwang
Sich auf das Ross behende,
Wohl um den trauten Reiter schlang
Sie ihre Lilienhände;
Und hurre, hurre, hop, hop, hop!
Ging's fort in sausendem Galopp,
Dass Ross und Reiter schnoben,
Und Kies und Funken stoben.

XX.

Zur rechten und zur linken Hand,
Vorbei vor ihren Blicken,
Wie flogen Anger, Heid' und Land!
Wie donnerten die Brücken!
"Graut Liebchen auch? . . . Der Mond scheint hell!
Hurrah! die Todten reiten schnell!
Graut Liebchen auch vor Todten?"
"Ach nein! . . . Doch lass die Todten."

XXI.

Was klang dort für Gesang und Klang?
Was flatterten die Raben?
Horch, Glockenklang! — horch, Todtensang!
"Lasst uns den Leib begraben!"

XVIII.

"But tell me then — where dost thou dwell?
What sort of bed is thine?"
"Far hence, a cool, still, narrow cell,
Eight boards its space define."
"And is there room?" "For thee and me!
Come, busk, and mount, and ride with me,
The nuptial chamber open stands.
The guests extend their welcoming hands."

XIX.

She busked her boune, and on the beast
Sprang lightly from the ground,
And round his rider's stalwart waist
Her lily arms she wound;
And gallop, gallop, — stop nor stay,
With breathless speed they whisk away,
Till horse and rider snort and blow,
And sparkling flints beneath them glow.

XX.

On either hand, to left, to right,
Heath, pasture, stream, and lake
Glanced dazzling by, too swift for sight —
The thundering bridges quake.
"Dost fear, my love? The moon shines bright,
Hurrah! The dead ride swift by night.
And art thou of the dead afraid?"
"Oh! no — but name them not — the dead."

XXI.

What sound is that of moan and knell?
Why doth the raven flit?
Hark! tolls the bell. Hark! dirges swell —
"To earth the dead commit."

Und näher zog ein Leichenzug
Der Sarg und Todtenbahre trug.
  Das Lied war zu vergleichen
  Dem Unkenruf in Teichen.

XXII.

"Nach Mitternach begrabt den Leib
  Mit Klang und Sang und Klage!
Jetzt führ' ich heim mein junges Weib
  Mit, mit, zum Brautgelage!
Komm Küster hier, komm mit dem Chor
Und gurgle mir das Brautlied vor!
  Komm, Pfaff, und sprich den Segen
  Eh' wir zu Bett uns legen."

XXIII.

Still Klang und Sang . . . Die Bahre schwand . .
  Gehorsam seinen Rufen,
Kam's hurre, hurre! nachgerannt
  Hart hinter's Rappen Hufen.
Und immer weiter, hop, hop, hop!
Ging's fort in sausendem Galopp,
  Dass Ross und Reiter schnoben,
  Und Kies und Funken stoben.

XXIV.

Wie flogen rechts, wie flogen links,
  Gebirge, Baüm', und Hecken!
Wie flogen links, und rechts, und links,
  Die Dörfer, Städt', und Flecken!
"Graut Liebchen auch? . . Der Mond scheint hell,
Hurrah! die Todten reiten schnell!
  Graut Liebchen auch vor Todten?"
  "Ach! Lass sie ruhn, die Todten."

With coffin, hearse, and pall, and plume,
Behold a dark procession come,
With sounds not those of earthly men,
But croak of toads in marsh and fen.

XXII.

"When midnight's past, your corpse inter
  With dirge, and moan, and wail.
Now home my youthful bride I bear,
  Come, join our festival!
Come clerk, and bring your choir along,
And croak us out a bridal song.
Come, priest, and ere we seek our bed,
Give us thy blessing, duly said."

XXIII.

Hushed was the dirge, the hearse was fled,
  Obedient to his call;
Close, close behind the flying steed
  They hurry, one and all.
And gallop, gallop, stop nor stay,
Still onward, on, they dart away,
Till horse and rider snort and blow,
And flashing flints beneath them glow.

XXIV.

How flew to left, how flew to right,
  Hedge, forest, moor and down!
How glimmered by, too swift for sight,
  Cot, hamlet, tower and town!
"Fears then my love? the moon shines bright,
Hurrah! the dead ride swift to-night.
Say, of the dead art thou afraid?"
"Ah! no. But let them rest, the dead."

XXV.

Sieh da! sieh da! — am Hochgericht
Tanzt um des Rades Spindel
Halb sichtbarlich bei Mondenlicht
Ein luftiges Gesindel.
"Sasa! Gesindel, hier! Komm hier!
Gesindel, komm und folge mir!
Tanz uns den Hochzeitreigen
Wann wir zu Bette steigen."

XXVI.

Und das Gesindel, husch, husch, husch!
Kam hinten nachgeprasselt,
Wie Wirbelwind am Haselbusch
Durch dürre Blätter rasselt.
Und weiter, weiter, hop, hop, hop!
Ging's fort in sausendem Galopp!
Dass Ross und Reiter schnoben
Und Kies und Funken stoben.

XXVII.

Wie flog, was rund der Mond beschien,
Wie flog es in die Ferne!
Wie flogen oben über hin
Der Himmel und die Sterne!
"Graut Liebchen auch? . . Der Mond scheint hell,
Hurrah! die Todten reiten schnell!
Graut Liebchen auch vor Todten?"
"O weh! Lass ruhn die Todten!"

XXVIII.

"Rapp'! Rapp'! Mich dünkt der Hahn schon ruft. . .
Bald wird der Sand verrinnen . . .
Rapp'! Rapp'! ich wettre Morgenluft
Rapp'! Tummle dich von hinnen!

XXV.

See there! around the accursed wheel,
In the pale moonbeam's glance,
All dimly seen, in spectral reel,
A ghastly rabble dance.
"Come, rabble rout — come, follow fair,
All pale and ghastly as ye are!
We want you all a jolly dance
Around our nuptial couch to prance."

XXVI.

And with a swirl, and with a rush,
The rout falls in behind,
Like dry leaves scattering from the bush
Before the wintry wind.
And gallop, gallop, fierce and strong,
In breathless race they dash along,
Till horse and rider snort and blow,
And fiery flakes beneath them glow.

XXVII.

How swift, where'er the moonbeam spreads,
Each object seemed to fly!
How glanced along above their heads
The star-bespangled sky!
"Is Love afraid? — the moon shines bright.
Hurrah! The dead ride well to-night!
Fear'st thou the dead?" "Good Heaven!" she said,
"Why talk thus wildly of the dead?"

XXVIII.

"Brave steed! 'Tis sure the cock I hear,
Soon will our sand be run;
Methinks I scent the morning air!
Push on! brave stee — push on!

Vollbracht, vollbracht ist unser Lauf!
Das Hochzeitbette thut sich auf!
Die Todten reiten schnelle!
Wir sind, wir sind zur Stelle."

XXIX.

Rasch auf ein eisern Gitterthor
Ging's mit verhängtem Zügel,
Mit schwanker Gert' ein Schlag davor
Zersprengte Schloss und Riegel.
Die Flügel flogen klirrend auf,
Und über Gräber ging der Lauf.
Es blinkten Leichensteine,
Rund um im Mondenscheine.

XXX.

Ha sieh! ha sieh! im Augenblick
Huhu! ein grässlich Wunder!
Des Reiters Koller, Stück für Stück,
Fiel ab, wie mürber Zunder.
Zum Schädel, ohne Zopf und Schopf,
Zum nackten Schädel ward sein Kopf;
Sein Körper zum Gerippe
Mit Stundenglas und Hippe.

XXXI

Hoch bäumte sich, wild schnob der Rapp',
Und sprühte Feuerfunken;
Und, hui! war's unter ihr hinab
Verschwunden und versunken.
Geheul! Geheul aus hoher Luft
Gewinsel kam aus tiefer Gruft.
Lenore's Herz, mit Beben
Rang zwischen Tod und Leben.

Our course is run, our race is o'er,
Wide yawns our nuptial chamber-door!
The dead — the dead ride fiery fast,
The destined goal is reached at last!"

XXIX.

Full on an iron gate, at length
  They drive, with loosened rein.
A furious charge, of headlong strength,
  Dashed bolt and bar in twain.
The jarring doors flew open wide,
And over new-made graves they ride,
Where, ranged beneath the moon's cold beam,
On either hand the tombstones gleam.

XXX.

Look there! look there! a grisly sight
  That horseman's form reveals!
From every limb his armour bright
  Like mouldering tinder peels!
A skull, all ghastly, cold, and dead,
A naked skull became his head.
With hourglass, scythe, and arm of bone,
Behold! a threatening skeleton!

XXXI.

High reared the raging horse — his breath
  Streamed forth in bickering flame —
And sought — down, down in earth beneath,
  Th' abyss from whence he came.
Loud howlings from the upper air,
Groans from the nether deeps were there.
With fluttering heart and stifled breath,
Leonora hung, 'twixt life and death.

XXXII.

Nun tanzten wohl bei Mondenglanz
  Rund um herum im Kreise
Die Geister einen Kettentanz,
  Und heulten diese Weise :
" Geduld ! Geduld ! Wenn's Herz auch bricht !
Mit Gott im Himmel hadre nicht !
  Des Leibes bist Du ledig,
  Gott sey der Seele gnädig ! "

XXXII.

Now in the moonlight's mystic glance,
In many a ghostly round,
Linked hand in hand the spectres dance,
And yell with dismal sound —
"Have patience! Though thy heart be riven,
Arraign not the decrees of Heaven!
Disburthened of thy Body's load,
Seek mercy for thy Soul with God!"

NOTE. — A trifling liberty is taken in Stanza xxix. The German makes the gates fly open at a touch of the spectre's whip. The version adopted seems better in character with the wild, fierce, and reckless proceedings of the Goblin horse and i der throughout.

PROSE AND VERSE.

I.

To thee, fair Science, long and early loved,
  Hath been of old my open homage paid;
Nor false, nor recreant have I ever proved,
  Nor grudged the gift upon thy altar laid.
  And if from thy clear path my foot have strayed,
Truant awhile,—'twas but to turn, with warm
  And cheerful haste; while thou didst not upbraid,
Nor change thy guise, nor veil thy beauteous form,
But welcomedst back my heart with every wonted charm.

II.

High truths, and prospect clear, and ample store
  Of lofty thoughts are thine! Yet love I well
That loftier far, but more mysterious lore,
  More dark of import, and yet not less real,
  Which Poetry reveals; what time with spell
High-wrought, the Muse, soft-plumed, and whisperingly
  Nightly descends, and beckoning leads to cell
Or haunted grove; where all inspiringly
She breathes her dirge of woe, or swells my heart with glee.

III.

Oh! rosy fetters of sweet-linked Rhyme,
  Which charm while ye detain, and hold me drowned
In rich o'erpowering rapture! Space and Time
  Forgot, I linger in the mazy round
  Of loveliest combination. Thought and Sound,
And tender images, and forms of grace,
  Flit by, and on my brow the laurel bound
Sits lightly!—who would not worse chains embrace,
So he might meet that loved inspirer, face to face?

TO THE LARK.

Mount! child of morning, mount and sing,
And gaily beat thy fluttering wing,
  And sound thy shrill alarms.
Bathed in the fountains of the dew,
Thy sense is keen, thy joys are new,
The wide world opens to thy view,
  And spreads its earliest charms.

Far showered around, the hill, the plain
Catch the glad impulse of thy strain
  And fling their veil aside;
While warm with hope and rapturous joy
Thy thrilling lay rings cheerily:
Love swells its notes, and liberty
  And youth's exulting pride.

Thy little bosom knows no ill,
No gloomy thought, no wayward will:
  'Tis sunshine all, and ease.
Like thy own plumes along the sky,
Thy tranquil days glide smoothly by,
No track behind them, as they fly,
  Proclaims departed peace.

'Twas thus my earliest hopes aspired,
'Twas thus with youthful ardour fired,
I vainly thought to soar —
To snatch from Fate the dazzling prize,
Beyond the beam of vulgar eyes:
Alas! the unbidden sigh will rise.
Those days shall dawn no more!

How glorious rose Life's morning-star!
In bright procession round his car,
How danced the heavenly train;
Truth beckoned from her radiant throne
And Fame held high her starry crown,
While Hope and Love looked smiling down,
Nor bade my toils be vain.

Too soon the bright illusion passed.
Too gay — too bright — too pure to last,
It melted from my gaze.
And, narrowing with each coming year
Life's onward path grew dark and drear,
While pride forbade the starting tear
Would fall o'er happier days.

Still o'er my soul, though changed and dead,
One lingering doubtful beam is shed,
One ray not yet withdrawn.
And still that twilight, soft and dear,
That tells of friends and former cheer,
Half makes me fain to linger here,
Half hope a second dawn.

Sing on, sing on! what heart so cold,
When such a tale of joy is told,
  But needs must sympathise?
As from some cherub in the sky
I hail thy morning melody!
Oh! could I mount with thee on high,
  And share thy ecstasies.*

* Reprinted from a collection edited by the late Mrs. Joanna Baillie.

MAN THE INTERPRETER OF NATURE.*

Say! when the world was new and fresh from the hand of its Maker,
Ere the first modelled frame thrilled with the tremors of life,
Glowed not primeval suns as bright in yon canopied azure,
Day succeeding to day in the same rhythmical march;
Roseate morn, and the fervid noon, and the purple of evening —
Night with her starry robe solemnly sweeping the sky?
Heaved not ocean, as now, to the moon's mysterious impulse?
Lashed by the tempest's scourge, rose not its billows in wrath?
Sighed not the breeze through balmy groves, or o'er carpeted verdure
Gorgeous with myriad flowers, lingered and paused in its flight?
Yet what availed, alas! these glorious forms of Creation,
Forms of transcendent might — Beauty with Majesty joined,
None to behold, and none to enjoy, and none to interpret?
Say! was the Work wrought out! Say, was the Glory complete?
What could reflect, though dimly and faint, the Ineffable Purpose
Which from chaotic powers, Order and Harmony drew?
What but the reasoning spirit, the thought and the faith and the feeling?
What, but the grateful sense, conscious of love and design?
Man sprang forth at the final behest. His intelligent worship
Filled up the void that was left. Nature at length had a soul.

* "Homo, naturæ minister et interpres." — *Bacon.*

A SCENE IN ELY CATHEDRAL.

SUNDAY, JULY 29th, 1845.

THE organ's swell was hushed,—but soft and low
  An echo more than music rang,—where he,
The doubly gifted, poured forth whisperingly,
  Highwrought and rich, his heart's exuberant flow,
Beneath that vast and vaulted canopy.
Plunging anon into the fathomless sea
  Of thought, he dived where rarer treasures grow,
  Gems of an unsunned warmth, and deeper glow.

Oh! born for either sphere, whose soul can thrill
  With all that Poësy has soft or bright,
Or wield the sceptre of the sage at will,
  (That mighty mace * which bursts its way to light)
Soar as thou wilt, or plunge — thy ardent mind
Darts on — but cannot leave our love behind.

* The symbolic analysis, of which the eminent and excellent individual supposed to be addressed, has proved himself a most consummate master.

MIRA.

'T IS not her beauty's charm (though that be dear),
'T is not her self-possessed and graceful air
(Though that be winning) — nor the gift, more rare,
Of words which flow as from a fountain clear,
Of bright, transparent thought, and on the ear
Drop harmony: but that beyond compare,
Of all her sex that breathe, her brow doth bear,
That seal of truth which owns nor guile, nor fear!

Hers the thrice royal robe of ermined proof
Whence stain glides smirchless, shame ashamed flies.
Hers the clear eye which fixed on Heaven aloof,
Glows with deep wealth from inly treasured stores,
Greatness that bows, and Lowliness that soars,
Faith that enquires, and Love that purifies.

THE PARTING DOVE.

IMPATIENT of constraint, around my Ark,
  In short and lowly flight my strength I tried,
But toil-worn, back to that o'ercrowded bark
  (No home abroad achieved), I sadly hied:
  There pruned my flagging wing for fresh essay,
And launched anew to seek, in purer air,
  A wider prospect, by a loftier way:
And caught one glimpse, and snatched one trophy rare,
  And bore it home, and mused for many a day
On sunny realms, where grew that bough so fresh and fair.

  Now fare thee well, thou dim and wave-tossed speck,
No more for me fit prison or fit lair!
No more for me fit cause of dull delay!
  Though sore 'twould grieve me yet to know thy wreck.

ON BURNING A PARCEL OF OLD MSS.

WRECKS of forgotten thought, or disapproved,
  Farewell! and as your smouldering flames ascend,
Read me a parting lesson. As the friend
  Familiar once, but since less fondly loved,
  (Dire spite of earthly chance), and wide removed
With earthquake of the heart! has ceased to blend
Warmth with my warmth, and sympathies extend,
  Where mine are linked and locked! Had I but proved
Earlier your weakness! Yet not all in vain
  Do I receive your warning. On I hie,
All unrepressed, though cautious; nor complain
  Of faint essays in tottering infancy.
Enough, if cleansed at last from earthly stain,
  My homeward march be firm, and pure my evening sky.

A DREAM WHICH WAS NOT ALL A DREAM.*

THROW thyself on thy God! nor mock Him with feeble denial,
Sure of His love, and oh! sure of His mercy at last,
Bitter and deep though the draught, yet shun not the cup of thy trial,
But in its healing effect, smile at its bitterness past.

Pray for that holier cup where sweet with bitter lies blending,
Tears in the cheerful eye, smiles on the sorrowing cheek,
Death expiring in life, when the long-drawn struggle is ending,
Triumph and joy to the strong, strength to the weary and weak.

Drink; and ere the bowl be quaffed,
Ask not why, nor what, the draught,
'T will assuage thy darkest grief,
Give thy direst pang relief.

Quaff it freely, without measure,
'T will not yield thee frantic pleasure,
'T will not wreck thy sense — nor gain
Present bliss with future pain.

* The first stanza was composed in sleep, and written down immediately on waking, Nov. 28. 1841.

Drink! The cup was filled by Love,
At the brightest fount above;
Meaner thirst shall never burn
Him who drains that sacred urn!

Close thine eye,—and opening see
All the Heaven that waits for thee!
Close thine ear. — Within shall ring
This celestial welcoming: —

Chorus of Angels.

The trial is over, the battle is past,
Joy! joy to the soldier who fought to the last,
And praise to the Rescuer who helped at his need,
And crowned his last effort with victory's meed.

Behold! how in mist and in distance recedes,
Yon speck of existence which witnessed his deeds.
How sink the low barriers which baffled his wing,
Ere he darted aloft like a shaft from the string.

Well done! thou good servant; thy service is o'er,
Now prepare with thy Master to rule evermore;
For faithful the promise, and true is the word,
Which welcomes thee home to the joy of thy Lord.

APPENDIX.*

IN compiling vocabularies from the mouths of natives, whether of written or unwritten languages, but especially of the latter, and of languages which, though reduced to writing, are so in characters (like the Chinese, &c.) illegible to Europeans, it is of the utmost importance to secure the possibility of a reasonably faithful reproduction of the sounds from the writing, when read by a third party having no personal communication with either the speaker or writer. This can only, of course, be accomplished by the adoption of a system of writing very different indeed from our ordinary English practice of spelling (which is utterly inapplicable to the purpose), fixing upon a set of letters, each of which shall express a distinct, recognized, and as nearly as possible invariable sound, and regulating their combination by simple and fixed rules.

Pending the introduction of a phonetic character free from objection, and bearing in mind that, after all, it is only a very imperfect representation of the native pronunciation which can be so conveyed (although amply sufficient if due care be taken to render the speech of a foreigner intelligible among them), the voyager or traveller will find in the "Ethnical Alphabet" of Mr. Ellis, a stock of characters prepared to his hand, capable

* From the "Admiralty Manual of Scientific Enquiry," edited by the Author in 1849.

of accomplishing, to a considerable extent, the object proposed; or he may adopt the following as a conventional system, which can therefore be at once transferred from manuscript into print at any ordinary printing-office. In the examples annexed, the letters printed in italics are those whose sounds are intended to be exemplified.

VOWELS.

1. *u* long (*uu*) as in Engl. b*oo*t; Germ. Br*u*der; Ital. verd*u*ra; Fr. *ou*vrir; — short (*u*) as in Engl. f*oo*t; Germ. r*u*nd; Ital. br*u*tte; — very short or coalescent, as in Engl. *w*ig, Germ. q*u*er; Ital. q*u*ale.
2. *o* long (*oo*) as in Engl. gh*o*st; Germ. Sch*oo*s; Ital. c*o*sa; Fr. ap*ô*tre; — short (*o*) as in Engl. res*o*lute; Germ. h*o*ld; Ital. d*o*lente; Fr. Napolé*o*n.
3. ʋ long (ʋ ʋ) as in Engl. p*u*rse; Fr. l*eu*r; Gael. l*u*gh; — short, and very short (*u*), or in Mr. Ellis's nomenclature "stopped," as in Engl. p*e*rt, c*u*t; Germ. V*e*rsuch.
4. ʋ̇ as in Germ. G*ü*te; Fr. Aug*u*ste.
5. ö as in Germ. L*ö*we, Fr. l*eu*r.
6. *ô* long (*ôô*) as in Engl. l*aw*; short, as in Engl. h*o*t; Germ. G*o*tt, k*o*mmen.
7. a long (*aa*) as in Engl. h*a*rd; Germ. H*aa*r; Ital. and*a*re; Fr. ch*a*r; — short (a) as in Engl. *A*meric*a*; Germ. Burgsch*a*ft; Ital. *a*ndare; Fr. charl*a*t*a*n.
8. *a* long (*aa*) as in Engl. w*a*ft, l*au*gh; — short (*a*) as in Engl. h*a*ve, qu*a*ff.
9. *α* as in Engl. b*a*nk, h*a*g; Fr. pr*i*nce, *ai*nsi, v*i*n.
10. *ἀ* long (*ἀἀ*) as in Engl. h*ai*l; Germ. S*ee*, St*ä*dtchen; Ital. l*ie*to; Fr. vér*i*t*é*; — short (*ἀ*) as in Engl. accur*a*te.
11. e long (*ee*) as in Engl. h*ei*r, h*a*re, h*ai*r, w*e*re; Germ. B*e*rg, St*ä*rke; Ital. v*e*ro; Fr. lumi*è*re; — short (*e*) as in

Engl. m*e*n, l*e*mon, *e*very; Germ. b*e*sser, *e*mpor; Ital. cast*e*llo; Fr. dang*e*reux, *e*ffort, *é*loigner.

12. *i* as in Engl. b*i*t, h*i*ll; Germ. G*i*ft, G*i*tter; Ital. c*i*nque.

13. ï as in Engl. p*ee*l, l*ea*ve, bel*ie*ve; Germ. L*ie*be; Ital. v*i*no; Fr. qu*i*.

14. ˌ as in Engl, peopl*e*; Germ. lieb*e*n (pïp.l, lïb.n).

DIPHTHONGS.

15. *j* as in Engl. b*i*te; Germ. b*ei*ssen.

16. *ʊ* as in Engl. br*ow*n, b*ou*nd; Germ. br*au*n; Fr. s*aou*l.

CONSONANTS, ETC.

17. *s* as in Engl. *s*oft; Germ. *s*anft; Ital. *s*olo; Fr. *s*alle.

18. *z* as in Engl. *z*inc; Germ. Ro*s*e; Ital. Ro*s*a; Fr. a*z*ur.

19. *sh* as in Engl. *sh*arp; Germ. *sch*arf; Ital. la*sc*iare; Fr. *ch*ien.

20. *zh* as in Engl. plea*s*ure; Fr. *j*ardin.

21. *th* as in Engl. *th*ing; Span. *z*apato, na*c*ion.

22. *dh* as in Engl. *th*at.

23—33. *k*, *g*, *t*, *d*, *p*, *b*, *f*, *m*, *n*, *l*, as the English, German, Italian, and French, and *v* as in the English, Italian, and French.

34. *rr* as in Engl. p*r*ay; Germ. *R*abe; Ital. *r*osa; Fr. e*rr*eur.

35. *r* as in Engl. smalle*r*.

36. *ϱ* or *rh* as in Engl. *rh*atany, *rh*ubarb.

37. *ν* as in Engl. ha*n*g; Germ. kli*n*gen; Ital. li*n*gua franca.

38. as in Fr. ai*n*si, rie*n*.

39. *ṽ* the nasal sound in Æt*n*a, D*n*eiper.

40. *h* as in Engl. *h*alt; Germ. *H*exe; Fr. *h*alte!

41. *χ* as in Germ. la*ch*en; Span. *X*imenes, relo*j*; Gael. Crua*ch*an.

42. *y* as in Germ. mor*g*en; Gael. lu*gh*.

Any supplemental letters may be used, if exactly exemplified and identified, for sounds peculiar to certain languages, as the Caffer and Hottentot *clicks*, &c.

RULES TO BE OBSERVED.

1. Do not use a running hand in writing from pronunciation, but form each letter separately; take care not to confound a and *a*.

2. For capitals, use the small characters enlarged.

3. A vowel sound is understood to be prolonged by repeating its character, according to the analogy of the German and Dutch. If the sound be really repeated, as in Oolite, insert a hyphen, O-o, thus, or an apostrophe, O'o. If a vowel be simply once written, it expresses the shortest sound conveying the *full vowel sound.* If intended to be very short, or to have that abruptness which has been called the stopped sound before a consonant, *double the consonant,* especially if the "stopped sound" be really perceptibly different as a *true vowel sound* from the "open," which in the English is *sometimes* the case.

4. Two different vowels coming together, when the first is intended to be shortened to the utmost possible degree consistent with the distinct audibility of its vowel character, it is to be prefixed singly to the other; as in the so-called English diphthongs, oi, eu,

(ôi, iuu,) or, as in such words as *w*et, *y*e, q*u*aff (uett, iï, ku*a*ff). But, if the vowels are intended to be separately and distinctly pronounced, as in the Italian pa*u*ra, an apostrophe must be interposed, as pa'uura, or if still more completely separated, a hyphen.

5. *h* means always a true aspiration, except in the combinations, *sh*, *zh*, *th*, *dh*, — for which, if any one should prefer to write ʃ, ʒ, θ, ϑ, respectively, he may do so with advantage, and with our entire approbation. The insertion of h in its true place among other consonants, is a matter of much nicety, and requires an exact and discriminating ear.

6. The "obscure vowel," No. 14., represented by a large unmistakeable full point, occurs only in such words as people, lieben, (Germ.) &c. Its nearest representative as a prolonged sound is ʋ (in the above nomenclature); but it is a great fault to use this character, or an equivalent one, in cases where a real, distinguishable, and particularly an essential etymological vowel is slurred over and obliterated by negligent and vulgar usage, as for instance, if we were to write the words *A*meric*a* (Engl.), Stuf*e* (Germ.), ventur*a* (Ital.), j*e* (Fr.), or the Indian name B*e*nares, all indiscriminately with the character (ʋ) appropriated to the vowel sound in the English word c*u*t. If, therefore, the necessity of imitating a *well-educated* usage require us to indicate (as no doubt it often does) a certain approach to this obscure ʋ, it should be done by subscribing the point beneath the appropriate representation of the true vowel, thus: Amerikạ, Stuufẹ, ventuurạ, zhẹ, Bẹnaarẹs.

7. Compound consonants, as in *ch*urch, *j*ournal, may be resolved into their elements (tsh and dzh).

8. Particular attention should be paid to the accentuation by a single mark (′) of that syllable in each word

where the prominent stress is laid in pronunciation, nor should the intonation of the voice be altogether neglected, though very difficult to reduce to any regular system of rules or signs, and rather a matter of description or musical notation than of alphabetic registry.

THE END.

LONDON:
Printed by SPOTTISWOODE & Co.
New-street Square.

Printed in the United States
By Bookmasters